MyDNA

Molly Fitzgerald-Hayes

**Department of Biochemistry and Molecular Biology
University of Massachusetts, Amherst
Amherst, MA**

Cover image © Frieda Reichsman

www.kendallhunt.com
Send all inquiries to:
4050 Westmark Drive
Dubuque, IA 52004-1840

Copyright © 2011 by Kendall Hunt Publishing Company

ISBN 978-0-7575-8925-6

All rights reserved. No part of this publication may be reproduced, stored in a retrieval system, or transmitted, in any form or by any means, electronic, mechanical, photocopying, recording, or otherwise, without the prior written permission of the copyright owner.

Printed in the United States of America
10 9 8 7 6 5 4 3 2 1

TABLE OF CONTENTS

UNIT 1: Genes are Written in DNA Language — 1
- Why is MyDNA Special To Me? — 1
- Cells And Genes Make Life Possible — 6
- Gene Expression Directs Protein Production in Cells — 10
- Proteins Do The Work in The Cells — 11
- Close-Up Look Inside a Eukaryotic Cell — 12
- Human Chromosomes have Small DNA Differences — 14
- Discovering The Structure of The DNA Molecule — 15
- Franklin's X-ray Diffraction Image Helped to Solve The DNA Puzzle — 17
- Researchers Determine The Composition and Shapes of Biomolecules — 18
- Molecules Contain Atoms Held Together by Chemical Bonds — 18
- Gene Expression Starts When DNA is Copied Into RNA — 24
- DNA "Language" Codes for Genes in The DNA Helix — 25
- Single-Stranded RNAs Form Secondary Structures in Cells — 26
- Gel Electrophoresis is a Key Technique to Separate DNA Molecules — 30
- Interpreting The Meaning of DNA Bands in a Gel — 32
- Restriction Enzymes Cut DNA at Specific Base Sequences — 34
- Restriction Enzymes can Produce "Sticky Ends" on DNA — 35
- How to Analyze DNA by Restriction Enzyme Digestion — 36
- Recombinant DNA Cloning and DNA Sequencing Strategies — 39

UNIT 2: DNA Replication and Human Genome Sequencing — 45
- Introducing Human Chromosomes — 45
- A DNA Helix Extends from End to End in a Eukaryotic Chromosome — 46
- DNA Replication and DNA Sequencing Reactions are very Similar — 47
- The Action at The DNA Replication Forks — 50
- DNA Sequencing Reactions Determine the Order of the DNA Bases — 53
- Sanger Chain Termination DNA Sequencing Reactions — 54

Automated DNA Sequencing Machines	59
The Polymerase Chain Reaction Amplifies DNA	61
RT-PCR is a Way to Amplify mRNA	63
The Human Genome Project	65
J. Craig Venter Led the Private Sector HGP	66
Lessons Learned from the Human Genome DNA Sequence	67
Most Human Protein-coding Genes Contain Introns and Exons	69
Humans Usually Inherit two Copies of each Gene	70
Gene Families Encode Proteins with Similar Structure and Function	73
Comparative Genomics	73
Comparing Genes: Are You a Mouse or a Man?	74
Most of the Human Genome is Non-coding DNA	75
Small Differences have a Big Impact: Single Nucleotide Polymorphism	76
Noncoding, Ultraconserved DNA Sequences in Vertebrate Genomes	77
Repeated DNA Sequences	77
Satellite DNA Repeats	78
Minisatellite DNA Repeats	79
Jumping Genes move around Human Chromosomes	80
Retrotransposons Populate the Human Genome	81
Each Human Genome has 1.5 million Alu I DNA Repeats	81
Long Interspersed Nuclear Elements	83

UNIT 3: Gene Expression Makes RNA and Proteins — 87

The Flow of Genetic Information: DNA to RNA to Protein	87
Expression of a Typical Human Gene	88
RNA Polymerase II Enzyme Binds to Promoter DNA	90
Gene Expression is often Controlled at the Start of Transcription	92
Processing and Splicing RNA Polymerase II Transcripts	94
Poly(A) "Tail" is used to Isolate mRNAs	95
Specific mRNAs are Analyzed by Northern Blot Transfer	96
Protein-Coding Genes Contain Intron and Exon Sequences in the Genome	98
Alternative RNA Splicing Increases Human Genome Flexibility	99

Table of Contents

Translating the mRNA Genetic Code into Protein	102
The AUG Codon in mRNA Initiates Protein Synthesis	105
Protein Folding is Essential for Protein Function	107
Water-hating and Water-loving Amino acids affect Protein Folding	110
Membrane Proteins have Essential Functions	112
Enzyme Proteins Drive Essential Biochemical Reactions in the Cell	115
RNA Ribozymes	117
Motor Proteins Transport Components Inside the Cell	118
DNA Binding Proteins	122
Histone Proteins Bind Nonspecifically to the Negatively charged DNA Helix	125

UNIT 4: Genetic Testing to Diagnose Genetic Diseases — 131

DNA Mutations Change Gene Sequences	131
DNA Mutations Alter Control Regions and Protein Coding Regions of Genes	134
The Importance of Inheriting two Copies of each Gene (Alleles)	135
Some People are Carriers of a Genetic Disease	137
Human Genetic Diseases and the Involvement of Gene Mutations	138
Inheritance Patterns of Genetic Diseases	139
Inheritance of Complex Multigene Genetic Diseases	141
Genetic Testing to Identify Mutant Genes and Genetic Diseases	142
Genetic Testing Applications include the Following:	142
DNA Testing with Molecular DNA Probes	143
Tests for Genetic Diseases Involving whole Chromosomes	144
One way to Light up Chromosomes is Fluorescence *in Situ* Hybridization	146
Chromosome Rearrangements and Translocations	148
Newborn and Prenatal Genetic Screening	150
Preimplantation Genetic Diagnosis (PGD)	152
Genome Testing on a DNA Chip	154
Genetic Testing Cautions	156
DNA Tests Reveal Genetic Carriers, Drug Reactions, and Genetic Risk	158
Why Genetic Risk is Important Information	158
Genetic Testing Traces Ancient DNA by Molecular Genealogy	159

Maternal Lineage Test Results	160
Drugstore DNA Testing and Access to Genetic Counselors	160
Sickle Cell Anemia Disease	162
Mutant Hemoglobin Complexes Stick together and Distort Cell Shape	163
Cystic Fibrosis is a Common Genetic Disease	164
Alzheimer's Disease Involves many Genes	167
Inherited Colon Cancer	169
Huntington's Disease	170
Breast Cancer Gene Mutations	171
Heart Disease is a Major Killer of People in the United States	172
Prions are a Bad Influence on 'Good' Proteins	173

UNIT 5: Embryonic Stem Cells and Animal Nuclear Cloning — 179

Human Development and Embryology	179
Embryonic Stem Cells are at the Top of the Human Stem Cell Hierarchy	180
The Origin of Human Embryonic Stem Cells	181
Stem Cell Renewal	183
Adult Progenitor Stem Cells (ASCs) are Multipotent	186
Specialized Cells in the Human Body	188
The Beginning of Human Embryonic Stem Cell (hESC) Research	189
The Promise of Human Embryonic Stem Cell Research	190
Federal Funding for Human Stem Cell Research Banned in 2001	193
Alternatives to Human Embryonic Stem Cells	194
Human ESCs are Found in Amniotic Fluid and in the Umbilical Cord	194
Bone Marrow Stem Cell Transplant	195
Ethical "Extra Embryos" and Extra ESCs?	195
Induced Pluripotent Stem Cells	197
Pluripotent iPS Cells can Generate Specialized Cells	198
Scientists Engineer Safer iPS Cells	199
hESCs are a Promising Potential Treatment for Parkinson's Disease	199
Stem Cell Transplant can Successfully Treat Spinal Cord Paralysis in Rats	202
The First Human Stem Cell Clinical Trials Begin in the United States	204
Clinical Trials to Treat Blindness in Humans with hESCs	205

Table of Contents

Human Cloning and Somatic Cell Nuclear Transfer (SCNT)	206
Animal Cloning Differs from Recombinant DNA Cloning	207
Therapeutic and Reproductive Cloning Diverge at the Blastocyst	209
The Goal of Human Therapeutic Cloning	210
Cloning Live Animals, Hello Dolly, The Cloned Sheep!	210
Cloned Pigs Make Organs for Use in Human Transplantation	213
Animal Cloning to Save Endangered Species	214
Ethical Questions, Concerns, and Risks of Animal Cloning	215
Epigenetic Genome and Nuclear Reprogramming	216

UNIT 6: DNA Forensics and Epigenetic Reprogramming — 221

Skin Fingerprints and DNA Fingerprints: What is the Difference?	221
Do Genetically Identical Twins Have Identical Skin Fingerprints?	222
DNA Fingerprinting Technology Relies on DNA Genome Differences	224
Detecting DNA Differences by DNA Fingerprint Analysis	225
DNA Fingerprinting Rides a White Horse and Wears a White Hat	227
DNA Fingerprinting Depends on Highly Specific Base Pairing	228
The Role of Southern Blot Transfer in DNA Fingerprinting	229
DNA Profiles in the FBI COmbined DNA Index System (CODIS)	235
PCR Short Tandem Repeat Analysis	235
FBI DNA Fingerprinting of Y Chromosome STRs	237
RFLP DNA Fingerprinting can Prove Paternity	237
RFLP can Distinguish Between Chromosome Copies	238
Mitochondrial DNA is a Genetic Link to Mom	241
DNA Testing Holds Up in U.S. Courts	243
The Innocence Project Lawyers and DNA Testing Save Lives	245
"But I Saw it with My Own Eyes"	246
Fraud and Misconduct in Forensic Science	247
Novel Applications of DNA Fingerprinting and DNA Testing	248
Super Bowl Footballs and Souvenirs from the Summer Olympics	248
DNA Testing to Identify United States Armed Forces Personnel	249
Identification of People Killed in the World Trade Center and Pentagon Terrorist Attacks	250
Identifying Missing and Murdered Children in Argentina	250

Identifying the Son of Louis XVI and Marie Antoinette	252
Scientists Use Mummy mtDNA to Trace very Old Family Trees	252
DNA Testing, Dog Genes and Man's Best Friend	253
DNA Testing to Help Endangered Animals	254
DNA Barcode Technology	257
Epigenetic Modifications can Alter Gene Expression for Many Generations	258
Transgenerational epigenetic inheritance	263
Household Toxins can Make Epigenetic Changes in Our Children's Children's DNA	265
BPA Exposure Alters Gene Expression in Lab Mice	266
MyDNA Book has Methylated Bookmarks!	267

MyPreface

MyAcknowledgements:

Writing the MyDNA book was very rewarding and exciting, although it was also time consuming! I could not have written *MyDNA* without the help of family, friends and colleagues.

MyThankyou's To:

- My husband Al, who is the best person I know; I will love him forever
- My son, who fills my heart every day with his smile
- My parents, who believe in me, and gave me love of science, art, teaching and the DNA helix molecule!
- My wonderful extended family and marvelous sister who cares for Mom with extra special love and attention
- My friend and colleague, Carol, who is absolutely amazing with our students; I really admire her combination of people skills and organizational abilities
- My friend and collaborator Dr. Frieda Reichsman, whose extensive computer experience, knowledge of science and excellent teaching abilities were critically important contributions to creating a computer-friendly MyDNA course
- My colleagues in the Biochemistry and Molecular Biology department at UMass who supported the MyDNA course, especially Professors Frank Cannon and David Gross, both excellent MyDNA teachers.
- My Kendall Hunt editor and friend Samantha. Not many people say, "Let me do that, it's my job", as often as Samantha does, and she means it!
- My excellent Kendall Hunt project leader, Ashley, and her top notch staff
- MyDNA began with support from the UMass Provost's Office and with a grant from the Camille and Henry Dreyfus Foundation to MF-H (SG-01-001).

MyDNA Cover Image: Dr. Frieda Reichsman created the wonderful cover image for the MyDNA book showing a leucine zipper protein with two multi-colored alpha helical proteins bound to a purple DNA helix. Thank you Frieda!

MyDNA Story

When I arrived at UMass, Amherst as an Assistant Professor in 1982 (!), I started my own research lab to study chromosome movement when cells divide. My research team discovered special DNA sequences and proteins that function at the centromeres of eukaryotic chromosomes to guide the chromosomes along the spindle fibers when the cells divide. I was very fortunate to be working as a scientist during the recombinant DNA revolution, when researchers had new DNA tools to analyze the molecular structures and functions of genes and proteins in cells.

In 1990 a collaboration of international scientists began the seemingly impossible job of determining the DNA sequence of the entire human genome, all 23 chromosomes. It was only a dozen years later that the scientists released the sequence of the 3.2 billion base pairs of DNA in the human genome, revealing answers to many questions about human genes and proteins. Another amazing scientific accomplishment occurred in 1997 with the birth of Dolly the sheep, the first animal ever cloned from an adult cell nucleus. After Dolly, more than 20 different cloned animals were made by transferring an adult cell nucleus into an empty egg cell, for implantation into a surrogate mother.

Researchers successfully grew human embryonic stem cells (hESCs) in the lab for the first time in 1998. This achievement is extremely exciting because in human embryo development, only the embryonic stem cells have the unique ability to generate all the different specialized cells needed in the human body, offering a goldmine of differentiated tissues for regenerative medicine involving novel transplant therapies. The promising future of human embryonic stem cell research in the U.S. was threatened by public controversy when President George W. Bush instated a ban on federal funding for human stem cell research in 2001. Now, more than a decade later, the first human clinical trials using stem cell therapies have been approved in the United States and both trials were sponsored by private companies that do not depend on federal funding to do their human stem cell research.

The controversy about using human embryos to make human ESCs encouraged some scientists to search for alternative sources of hESCs. In 2007 researchers generated induced pluripotent stem (iPS) cells in the lab that look and act just like hESCs but are made without using human embryos! Most important, the iPS cells can be converted into many different types of highly specialized human cells in the lab, making iPS cells an ideal source of cells to develop specialized tissues for human transplantation therapies.

MyDNA General Education Science Course

As these exciting events in scientific history took place during my first two decades at UMass, I realized that most students did not have the opportunity to learn about these awesome achievements in the life sciences, so I decided to create the MyDNA course at UMass. Because most UMass students do not major in the sciences, I created the MyDNA general education science course, which is open to everyone. Currently the MyDNA general education science course serves over 600 students each year, and attracts undergraduates from many different UMass majors.

The MyDNA students gain a fundamental understanding of the different structures and functions of genes in the human genome and proteins and cells in the human body. The MyDNA course also explores many amazing scientific accomplishments inspired by DNA science and human genetics. This information is relevant to the MyDNA students, after all my genome DNA sequence is 99.9% identical to your genome DNA sequence!

MyDNA Students

College students typically have a high school introduction to DNA, RNA, and cells but few actually have a working knowledge of human genes or understand how gene expression dictates human traits. Teaching people who have very little science background poses challenges, but I have had considerable experience teaching STEM DNA K-12 teacher workshops, giving presentations to community organizations, and teaching Elderhostel retirement groups. Outreach students are always interested in genetic diseases, the human genome, and embryonic stem cells and they ask how to trace their ancestors by DNA genealogy.

The MyDNA course also attracts many science majors who are interested in new applications of modern DNA technology and want to learn more about the career and employment opportunities available in the rapidly growing genome sciences especially in businesses supporting the associated biotechnology industries. College graduates often find success in businesses serving the niches evolving between DNA science and fields such as law, business, journalism, criminal forensics, health services, genetic testing, genetic counseling, psychology, journalism, to name just a few.

The MyDNA Book

Lectures in the MyDNA course are always accompanied by powerpoint slides with images and text that are closely integrated with the lecture material. These pictures are vital to help explain the scientific information presented in the MyDNA lectures. The visual perspective offered by the figures is essential to help students see how cellular components like DNA and proteins physically interact with each other to perform biochemical functions. In the MyDNA class we always use pictures to show the positions of the atoms in each molecular structure, which indicate the structures and functions of key biological molecules that perform special functions in cells.

The MyDNA students were also given online reading assignments and activities to supplement the lecture material. Many MyDNA students printed out the figures in the powerpoint slides and put them into 3 ring notebooks, making the pictures accessible to take notes in class. Some students asked me to recommend an appropriate book to help them to review the science presented in the MyDNA course. In part because of this need and the encouragement from my students, I decided to write the MyDNA book to accompany the MyDNA course. The MyDNA lectures and pictures were organized into six Units of information representing the central topics featured in the MyDNA book. The 25-30 figures in each Unit are closely integrated with the information presented in the book and in the lectures.

MyDNA Book

In Unit 1 the MyDNA book introduces genes, cells and the basic scientific facts about the DNA molecule. The students learn about different organisms, tissues and cells, and they see how human cells and bacterial cells differ. Early in the MyDNA course, the students become familiar with the structure of the DNA helix and understand how DNA stores and transmits genetic information in cells. They also learn how other vital biological molecules function in the cell so that they can better understand the process of generating specialized cells from embryonic stem cells and the genome reprogramming that occurs after the nucleus is transferred into an empty egg cell during cloning.

In Unit 2 students find out that the DNA synthesis reactions used to duplicate the chromosomes are very similar to the biochemical reactions that scientists use in the lab to determine the DNA sequence, the linear order of DNA bases along a DNA strand. When the DNA sequence of the entire human genome was released, it had a huge impact on our understanding of human genes. MyDNA students learn about the different structures and expression of human genes in Unit 3. They also explore the selected gene expression in different cell types that direct the synthesis of specific proteins to perform specialized functions, and this unit illustrating the flow of gene expression from DNA to RNA to proteins in eukaryotic cells. The possibility of inheriting a mutant human gene that codes for a defective mutant protein, potentially causing a genetic disease, is the focus of Unit 4. The future of human embryonic stem cell research is discussed in Unit 5, which shows how embryonic stem cells generate the specialized cells required in the embryo to produce tissues and organs in the human adult. Now alternative sources of embryonic stem cells have been identified, that do not use embryos of any kind, including iPS cells, which have re-ignited the excitement over new sources of specialized stem cells to treat a wide range of devastating diseases.

The *MyDNA* book concludes by investigating the novel applications of DNA testing. For example, DNA fingerprinting, which the public knows as a powerful forensics tool to catch criminals, is also used by the Innocence Project as a forensic tool to exonerate people who were falsely convicted and sent to jail. DNA barcoding is a new mitochondrial DNA test, which has become a valuable forensic tool for tracking and prosecuting illegal poachers. Scientists found that individual trees have unique DNA profiles that can be used to identify specific trees in a rainforest and can be detected in finished wood products many years after harvest. Similar DNA tests are also used to accurately identify the original sources of allegedly illegal seafood products confiscated from food processing centers and restaurants, providing evidence that is used to legally prosecute suspected poachers.

Changes in the DNA sequence of a gene can produce defective, mutant proteins that do not function correctly. In contrast, epigenetic control mechanisms alter gene expression by adding or removing the methyl groups associated with the DNA helix, without changing the DNA sequence of the affected gene. Epigenetic reprogramming in the nucleus transferred during animal cloning is essential for the genome to adapt to the new environment in the empty egg cell. Epigenetic mechanisms are also involved in programming the epigenome, the new DNA genome in the zygote cell resulting from fertilization.

Recent experiments with mouse embryos developing in the womb (*in utero*) have demonstrated that gene expression is altered in the embryos exposed to the household environmental toxin, BPA, which is banned in other countries but not the U.S. This research demonstrates that even though the epigenetic changes in gene expression did not change the DNA sequence, the changes in the mouse embryos were transmitted to offspring in future generations. Recent human studies indicate that *in utero* exposure to BPA and similar toxins can in fact change the expression of human genes, and these changes in gene function can be transmitted to future generations, to our children's children.

I hope that MyDNA will inspire more students to explore their own genetic future, and help to protect the genetic future of our planet!

UNIT 1

Genes are Written in DNA Language

Understanding the genes and cells in people are a major focus of the MyDNA book (Figure 1-1). People are often interested in their genes and in the outcome of gene expression, and they raise many questions. Why are her eyes blue, while her brother's eyes are green? Why is her skin color brown? Why do some people have curly hair and some have straight hair? Why do I have lots of hair but my Dad is bald? Why do I have a genetic disease? Will I give my genetic disease to my child? My father died of a heart attack at age 40, how can I avoid having a heart attack? What do my genes mean to me now and in the future?

Why is MyDNA Special To Me?

- Your DNA encodes your genes, which determine the molecular "instructions" to create an individual human, you!
- Your DNA genome contains all the genes that make you human.

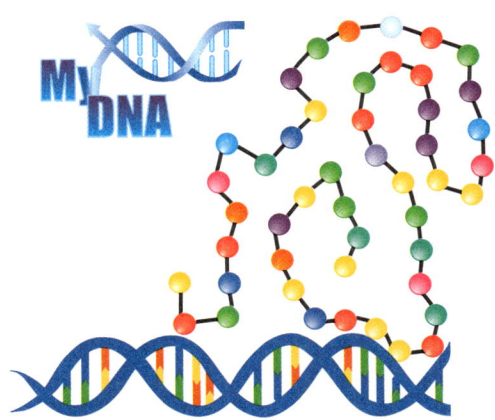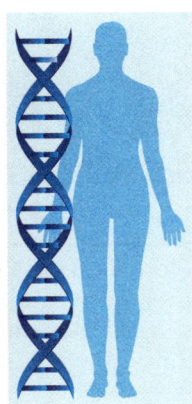

FIGURE 1-1 Welcome to MyDNA. MyDNA was written to help people learn more about their genes, their genome, and the human body.

- Your DNA is unique to you and no other human on Earth has your exact DNA genome sequence (unless you have an identical twin!)
- Your genes are influenced by your life experiences and the environment.

A gene is a fundamental unit of genetic inheritance that is represented in cells by long molecules of DNA that are carried in chromosomes. Each gene contains the DNA information necessary to produce one specific functional protein product (with interesting exceptions as we will see) (Figure 1-2). The genome in an organism contains all of the DNA sequences and genes needed to create, build, and maintain that living creature. The human genome contains all the genes that make us human.

> **KEY CONCEPT**
> Genes are written in the DNA language and contain the genetic information to make specific proteins that perform all the different functions in human cells (Unit 3).

Each human chromosome contains a linear double-stranded DNA molecule extending unbroken from one end of the chromosome to the other end of the same chromosome (Figure 1-2). The DNA sequence of the entire human genome was determined by the Human Genome Project scientists, who figured

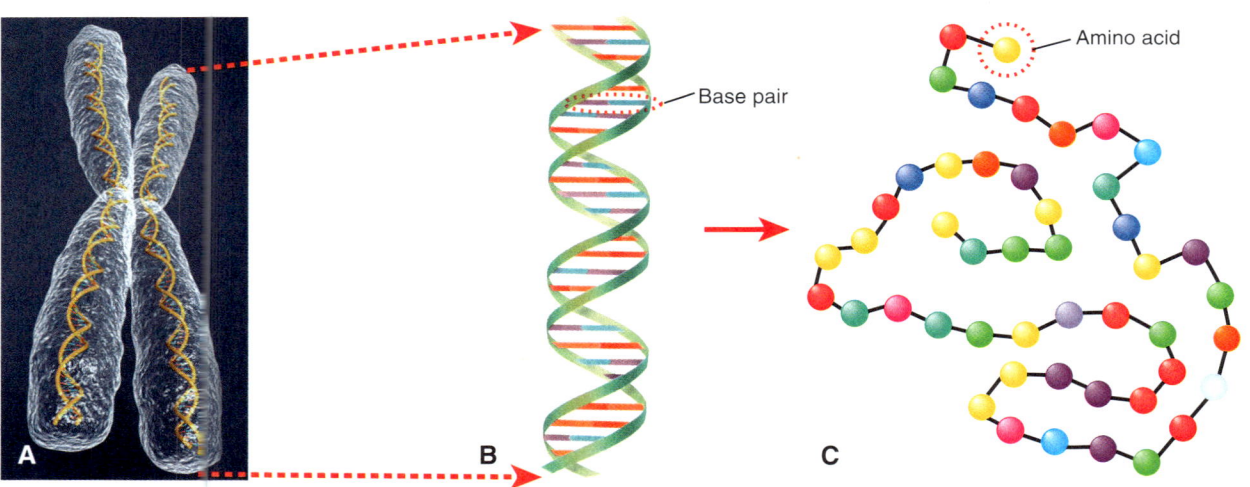

FIGURE 1-2 Chromosome DNA codes for genes and proteins. (A) Each eukaryotic chromosome contains one double-stranded DNA helix that extends from one end of the chromosome to the other end of the same chromosome. The chromosome shown has duplicated. (B) This picture represents a close-up of about 25 base pairs of a DNA helix with one base pair circled by a dotted red line. This is not drawn to scale as there are millions of base pairs in a typical eukaryotic chromosome. (C) Chromosome DNA codes for many genes that carry the genetic information to make proteins in the cell. Proteins are composed of amino acids linked together in a linear chain with an amino acid indicated by a dotted red circle. This is not drawn to scale as proteins in the cell are encoded by genes that are significantly longer than the 25 base pairs depicted in the figure.

out the exact order of DNA letters (A, G, C, T) in all 23 human chromosomes. Most human body cells are called somatic cells and are diploid cells, which contain 46 chromosomes each. Human egg and sperm cells are haploid cells that contain 23 human chromosomes each.

> **KEY CONCEPT**
> The number of chromosomes contained in a cell is an important characteristic of each type of cell.

The DNA sequences of all the human genes were determined and the location of each gene was mapped along the length of the DNA helix molecule in each human chromosome. With some exceptions, each gene codes for a unique protein product made up of a unique sequence of amino acids (Figure 1-3). Some proteins function alone, as monomers, while other proteins must bind to each other to form a functional multi-protein complex (Figure 1-3). Protein complexes are like tiny machines composed of even smaller molecular parts. The molecular parts are proteins that function together once they have been assembled into a molecular machine.

Protein and DNA molecules are very different from each other in structure and function but both represent biological macro-molecules that perform

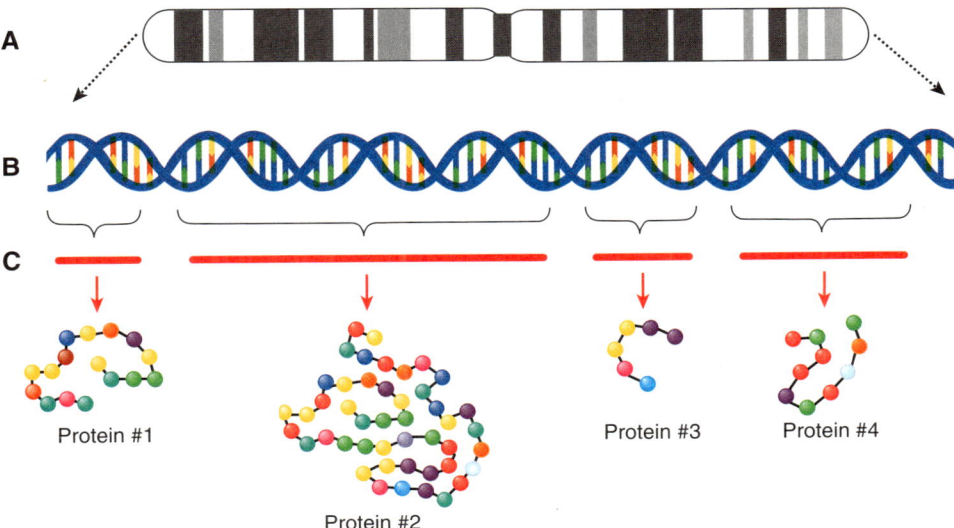

FIGURE 1-3 Each DNA gene codes for a different protein (with special exceptions as you will see) (A) Chromosome DNA encodes genes and large amounts of non-coding DNA. The banding pattern on this chromosome represents the general banding patterns that result from staining condensed human chromosomes and is not an actual chromosome pattern. (B) Genes are located along the length of a DNA molecule and can be encoded on either of the two DNA strands in the helix. (C) Each gene is copied into an RNA strand that codes for one protein (with special exceptions as you will see). Each protein is a single molecule made up of a linear chain of amino acids.

Continued

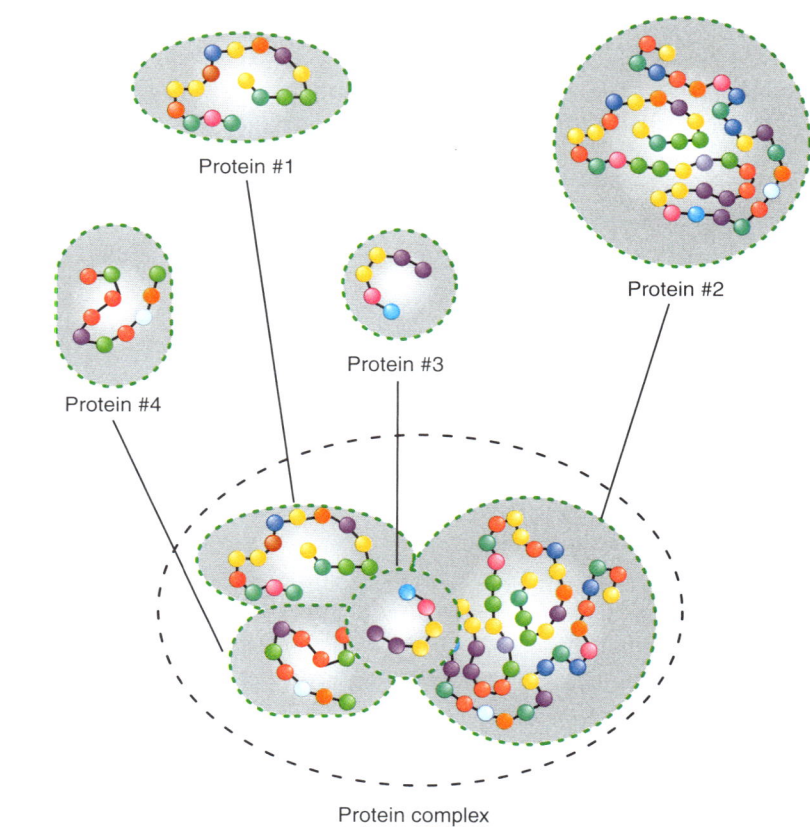

D

FIGURE 1-3 Each DNA gene codes for a different protein (with special exceptions as you will see) (con't). (D) A protein can function alone, as a monomer. However, many proteins can only function when bound to other proteins to form a multi-protein complex. In the example shown, each protein is a separate molecule that is not connected to the other proteins in the complex by permanent chemical bonds.

essential functions in living cells. The functions of these biological molecules are a result of the chemical properties of the amino acids in proteins and the nucleotide bases in DNA. The specific order of the amino acids in each protein can be determined indirectly by reading the order of the DNA letters (base pairs) in the gene encoding the protein (Figure 1-2). The order of the amino acids in the protein influences how each amino acid chain folds into a specific three-dimensional (3D) shape that is essential for the protein to perform a given function. Genetic diseases are often caused by mutations that prevent proteins from folding properly, rendering the protein nonfunctional. Different proteins encoded by separate genes can come together in the cell and assemble into protein complexes that perform specialized functions requiring the cooperation of the subunit proteins in the complex.

Some protein complexes contain several copies of the same protein subunit, while other complexes contain more than one type of protein subunit (Figure 1-4). Each subunit is made up of one protein represented by one strand of amino acids. A protein complex that our bodies need every day is hemoglobin, which makes our blood red and carries oxygen to cells throughout the human body (Figure 1-5).

UNIT 1 Genes are Written in DNA Language

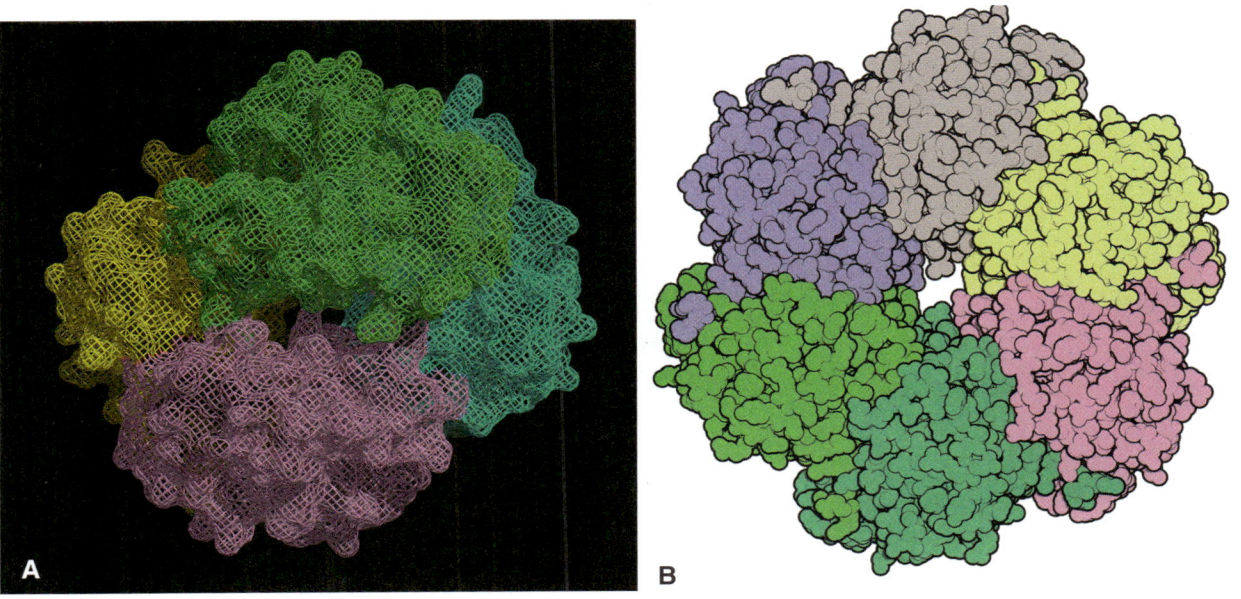

FIGURE 1-4 Folded proteins can have unique molecular surface topology (A) The hemoglobin protein complex has 4 protein subunits, two identical alpha globin proteins and two identical beta globin proteins (shown in four different colors). When a protein folds into functional 3D shape in the cell, the surface of the folded protein (or the protein complex) adopts a molecular topology that is unique to that particular molecule. The outside surfaces of these cellular components are very important for specific protein-protein interactions and binding. (B) This shows the molecular topology of the outer surface of a complex containing six protein subunits represented here in six different colors.

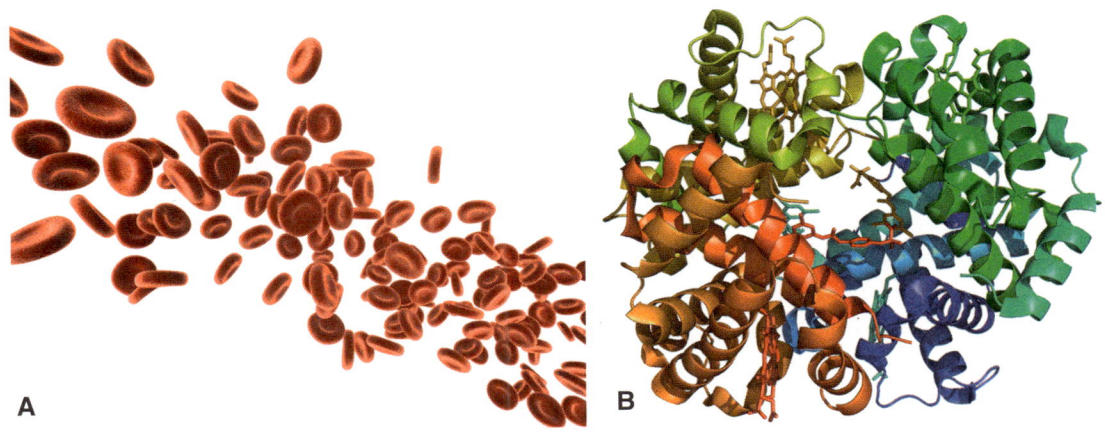

FIGURE 1-5 The hemoglobin protein complex carries oxygen in red blood cells Red blood cells (RBCs) are disc-shaped cells that circulate in the bloodstream and carry oxygen to the cells in the human body. (B) The four protein subunits in a hemoglobin complex are shown in different colors in this high resolution ribbon model of the proteins.

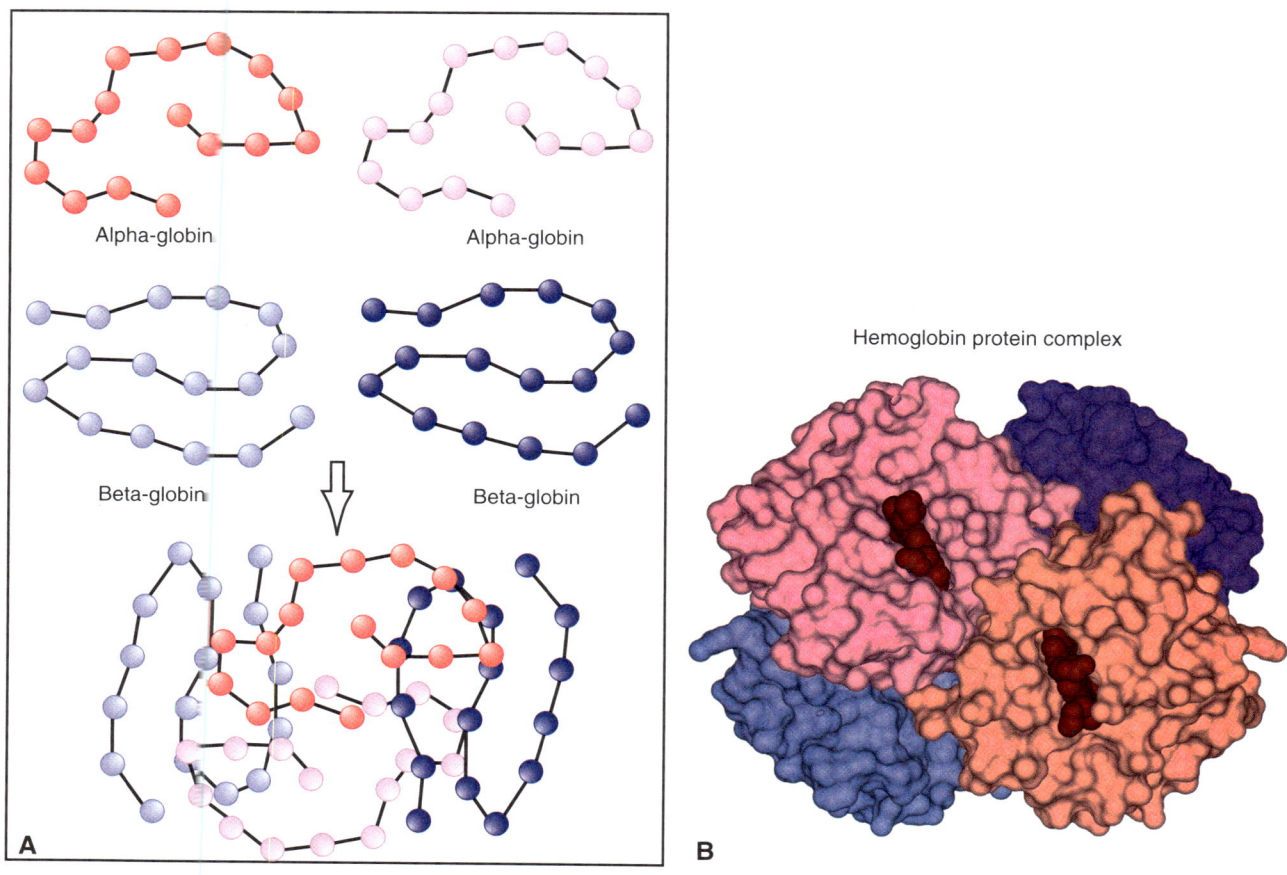

FIGURE 1-6 Each subunit in hemoglobin is a separate protein (A) A hemoglobin complex contains two alpha-globin proteins and two beta-globin proteins, depicted here as short amino acid chains (globin proteins are actually much longer). Both alpha-globin proteins are encoded by the same alpha-globin gene, and both beta-globin proteins are encoded by the same beta-globin gene. (B) This hemoglobin complex shows the locations of two of the four heme prosthetic groups (dark red) that allow hemoglobin to carry oxygen in the RBCs. Each hemoglobin subunit contains one heme group, an iron atom held in the center of a cyclic ring structure by chemical bonds. The heme groups bind to oxygen for transport in the RBCs and the iron atoms give blood its characteristic red color.

Hemoglobin is not a single protein molecule but is actually composed of four protein molecules, two alpha-globin proteins, and two beta-globin proteins, which are encoded by separate genes in the genome (Figure 1-6). The globin genes and proteins have been very well studied not only because hemoglobin is vital to human life but also because the mutations affecting globin genes cause a genetic disease called sickle cell anemia (Unit 4). The types of proteins expressed in cells and the functions performed by these specialized proteins dictate the characteristics of the tissues containing that particular type of cell (Figure 1-7) (Unit 3, Unit 5).

Cells and Genes Make Life Possible

To learn more about how genes function to make cells and organisms work, it is important to have a basic understanding of the two major types of cells, eucaryotic cells and prokaryotic cells. These types of cells are very different from each other but are primarily distinguished by the

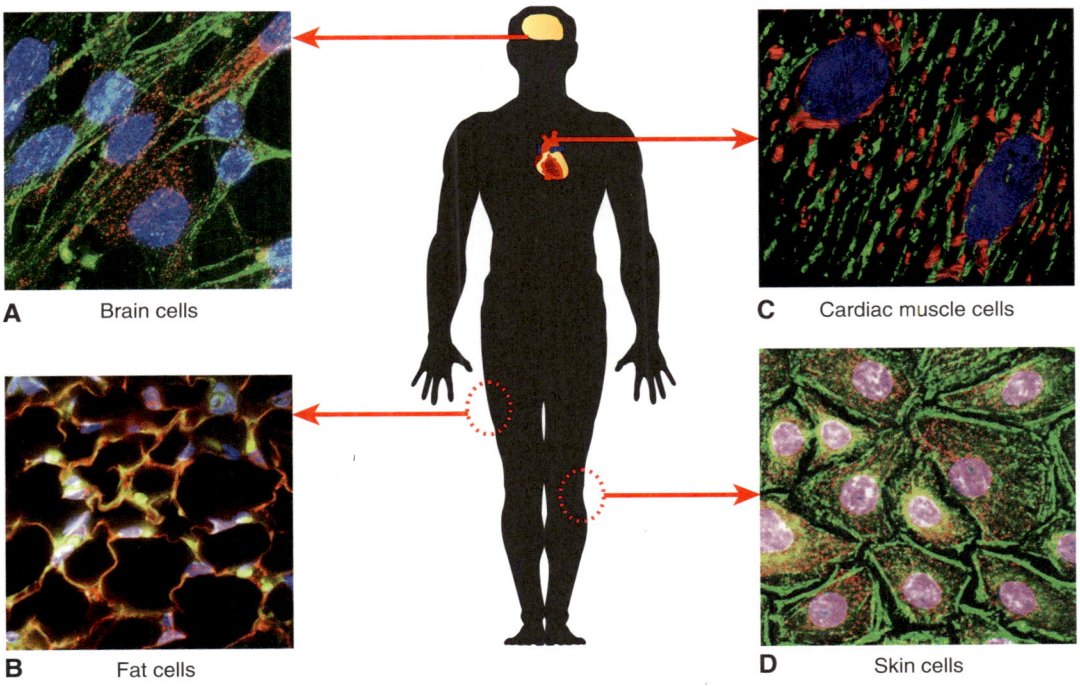

FIGURE 1-7 The human body contains many types of specialized cells The human cells were stained with fluorescent dyes to indicate different cellular components. (A) Cancer cells growing in the brain [nuclei (blue), actin (green)]. (B) Fat cells (adipocytes) have large internal fat droplets (black) with the nuclei (blue) pushed against the cell wall (orange). (C) Heart muscle cells contain contractile proteins including actin (green) as well as calcium channel receptor proteins (red). The influx of calcium into the cell triggers muscle cell contraction. (D) Keratinocytes are epithelial skin cells containing filamentous f-actin proteins (green) surrounding a central nucleus (pink).

presence or absence of a membrane-bound nucleus, a critically important difference that gives eukaryotic cells certain advantages in life.

Humans, other animals, insects, and plants are made up of eukaryotic cells containing nuclei (Figure 1-8), whereas bacteria and most other single-celled microbes are prokaryotes. Exceptions include single-celled eukaryotic organisms such as budding and fission yeasts (*S. pombe* and *S. cerevisiae*, respectively), which have nuclei containing linear double-stranded DNA chromosomes and undergo cell division just like multi-cellular eukaryotic cells.

The compartments inside cells such as nuclei are separated from the outside environment by biological membranes, which contain two layers (bilayer) of fat-like molecules called phospholipids (Figure 1-8) (Unit 3). The "fatty" components of the inner layer of the membrane do not dissolve in water, making the membrane a very effective wall that separates the watery (aqueous) contents inside the cell from the watery environment outside the cell. The eukaryotic nucleus is surrounded by a special double membrane, the nuclear envelope, which separates the chromosomes and the contents of the nucleus from the rest of the cell (Figure 1-9). The nucleus allows eukaryotic cells to use sophisticated mechanisms to control gene expression because the nuclear

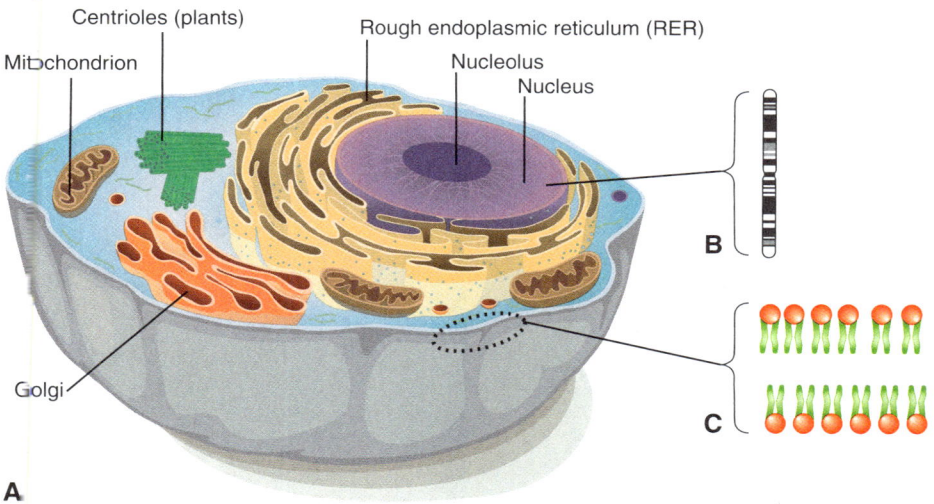

FIGURE 1-8 Eukaryotic cells have nuclei (A) Eukaryotic cells contain membrane-bound organelles including the nucleus, the endoplasmic reticulum and the mitochondria. The nucleolus is the site of ribosomal RNA synthesis and processing inside the nucleus and is not surrounded by its own membrane. Plant cells have centrioles, which function as part of the cell division spindle apparatus. (B) Eukaryotic cells contain linear chromosomes, shown here condensed and with a stained pattern. In living cells the chromosome structure changes with the cell cycle. (C) The cell is surrounded by a bilayer plasma membrane.

membrane separates the two major compartments in the cell and controls communication between compartments.

> **KEY CONCEPT**
> The complex regulation of gene expression and developmental events is necessary to create the numerous tissues, organs, and other biological structures in the human body.

The double membrane surrounding the nucleus contains complicated nuclear pores that control the flow of molecular traffic in and out of the nucleus (Figure 1-9). The nuclear pore structures are composed of hundreds of proteins that regulate the export of RNA molecules from the nucleus and into the cytoplasm for translation into proteins. The nuclear pores also help to control the import of proteins into the nucleus. These proteins are made in the cytoplasm but function in the nucleus in processes such as replication and gene expression. Other organelles found inside eukaryotic cells, such as mitochondria, chloroplasts (in plant cells), and the Golgi apparatus, are also surrounded by membranes (Unit 3).

Eucaryotic and prokaryotic cells both contain DNA genomes, although the chromosomes in eukaryotic and prokaryotic cells have very different structures. Each human chromosome contains a linear, double-stranded DNA helix molecule extending from one end to the other end of the chromosome

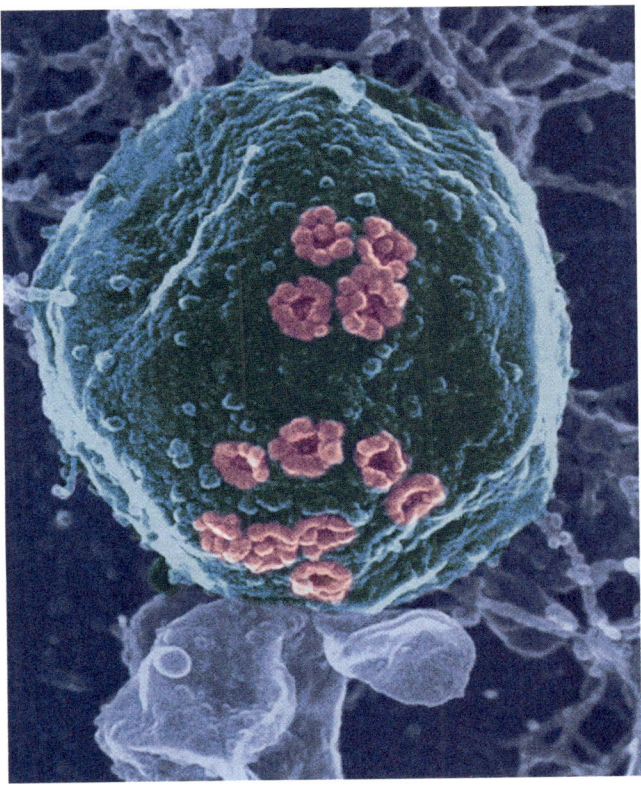

FIGURE 1-9 Nuclear pores in the nuclear membrane of a yeast cell. This scanning electron micrograph (SEM) shows the nucleus (blue) of a budding yeast cell (*Saccharomyces cerevisiae*) containing nuclear pore complexes (pink). Nuclear pores regulate the transport of molecules in and out of the nucleus in eukaryotic cells.

(Figure 1-2), while the much smaller genomes of prokaryotic cells are circular, double-stranded DNA molecules (Figure 1-10). Prokaryotic cells lack membrane bound nuclei but the circular DNA chromosome is contained in the nucleoid region inside the bacterial cell. Procaryotic cells are typically much smaller than most human cells, and unlike eukaryotic cells that exist in a wide range of shapes and sizes, most bacterial cells are shaped like spheres or rods (Figure 1-10).

A close look inside an *E. coli* cell reveals that bacteria contain many of the same cellular components that are usually present in eukaryotic cells (except for a nucleus), including ribosomes, enzymes, a chromosome, tRNAs, and mRNAs. The membrane surrounding the *E. coli* cell contains the motor proteins that anchor and drive the flagellum that allow the cell to move. Some prokaryotic organisms live in the human intestinal tract, including the usually harmless bacterium, *E. coli* (Figure 1-10). Recently a new strain of *E. coli* became famous when the U. S. Centers for Disease Control (CDC) announced that a toxic strain of *E. coli* bacteria could be lethal if consumed by humans. This toxic strain of *E. coli* bacteria is responsible for the outbreak of cases of lethal food poisoning in the U.S. in the past decade and killed many people in Germany in summer 2011.

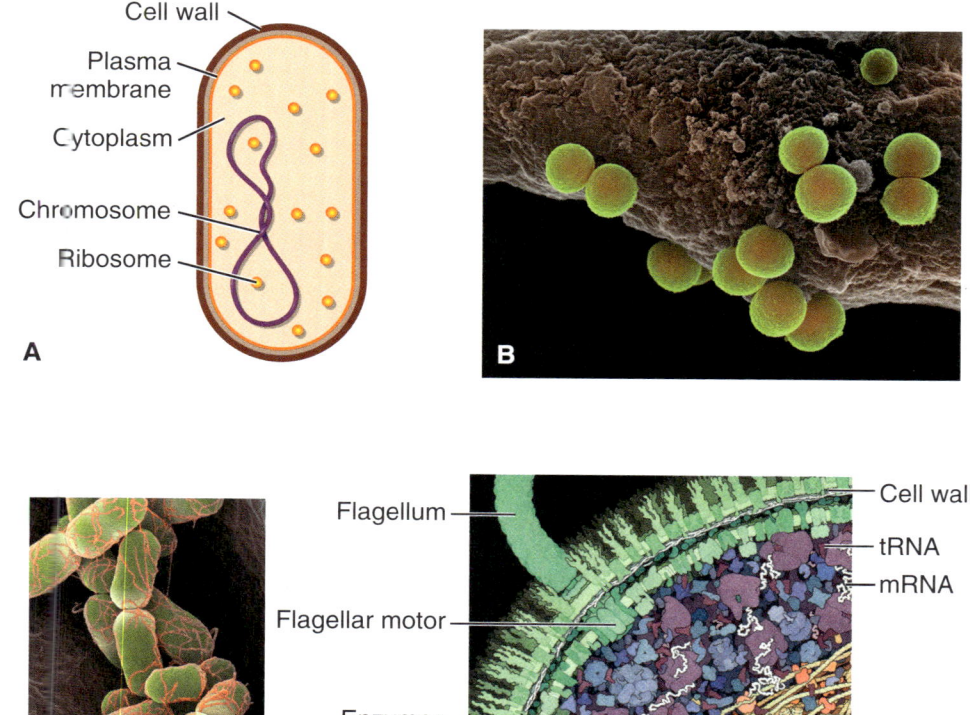

FIGURE 1-10 Prokaryotic cells do not have nuclei (A) Bacteria are prokaryotic cells and do not have nuclei. The circular, double-stranded DNA chromosomes reside in a nucleoid structure without membranes. The ribosomes are located in the cytoplasm along with the DNA. (B) *Staphylococcus aureus* bacterial cells are shaped like spheres. (C) The familiar *E. coli* bacteria are rod-shaped. (D) The internal contents of this bacterial cell are surrounded by a cell wall (green). The two bilayer membranes contain the motor proteins that anchor and drive the flagellum, which allows the cell to move. The cytoplasm also contains tRNA molecules (maroon) and mRNA strands (white) involved in protein synthesis at the ribosomes (purple). The chromosome is carried in the nucleoid (yellow and orange).

Gene Expression Directs Protein Production in Cells

Human genes are a major influence on human traits such as personality, physical strength, biological characteristics, blood type, health risks (heart disease, stroke, alcoholism, Alzheimer's, etc.), genetic diseases (sickle cell, hemophilia, cancer, etc.) and behavior. However, genes do not act alone, scientists now know that the environment also plays a large role in human development and can have a strong influence on the outcome of genetic inheritance (Unit 5, Unit 6).

Most people know that genes determine individual human traits, such as eye color and height, but few people have had the chance to learn about the biological connections between genes and traits, and the important roles played by DNA and proteins. Human characteristics result from the functions of proteins made in response to instructions from the genes, and the influence of the environment, which not only affects gene expression in the organism but can also alter the status of the genes transmitted to future offspring. Recent research has shown that both the chromosome DNA and the chromosome packaging proteins are modified in a mechanism involved in transmitting changes in genetic information over many generations (Unit 6).

Protein expression in the cell begins with gene expression, when the DNA letters (bases) coding for a specific protein are copied into an RNA strand, a message that transfers the genetic information from the DNA gene to the protein product (Figure 1-11) (Unit 3). Special proteins bind to the RNA copies (transcripts) and transport them through the nuclear pores and into the cytoplasm where the genetic information in the RNA is translated into a protein by a ribosome. This fundamental pathway of gene expression: DNA to RNA to Protein was once called "Central Dogma" because scientists thought that the gene expression pathway could function in only one direction. However, we now know that the DNA to RNA to protein gene expression pathway can operate in the reverse direction, as evidenced by natural exceptions to the "Central Dogma" pathway where RNA is copied into DNA (Figure 1-11).

Proteins Do The Work in The Cells

The function of the human body depends on trillions of proteins that perform an astonishing array of intricate jobs in many different cells. The human genome DNA encodes genetic information, but it is the proteins that actually carry out genetic instructions and perform the necessary physical and biochemical work in the cells. All the proteins in eukaryotic and prokaryotic cells are made up of various combinations of only 20 different naturally occurring amino acids, and the genes in living cells are encoded by only four

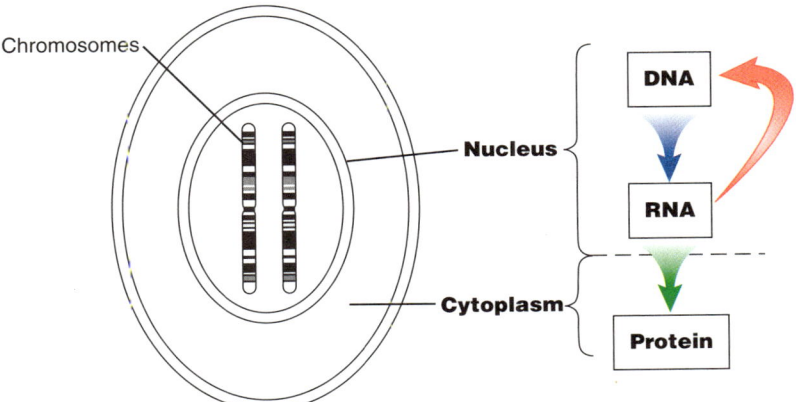

FIGURE 1-11 Central Dogma is reversible! The transmission of genetic information in a cell begins with gene expression, the process of copying DNA into RNA (blue arrow). In eukaryotic cells this process occurs in the nucleus. The RNA strand carrying the genetic information is transported from the nucleus to the cytoplasm and translated into protein (green arrow). This Central Dogma pathway transmits the genetic information from DNA to RNA to the protein product. However, in some rare cases the pathway is reversible and RNA is copied into DNA (pink arrow).

DNA letters. Yet this seemingly small amount of genetic information is sufficient for cells to generate a myriad of proteins with different shapes, sizes, and structures that perform a wide range of essential biological functions.

The gene expression pathway begins with a gene that stores and transmits genetic information, but it is the protein products expressed by the genes that carry out a wide range of different jobs in cells. For example, some proteins are enzymes that greatly increase the speed (catalyze) of the biochemical reactions that are involved in many cell functions. Without enzymes to catalyze essential biochemical reactions, the cells will die because they cannot convert fuel molecules into energy fast enough or duplicate chromosomes before the cell finishes division.

Some types of specialized cells in the human body make proteins that are secreted from the cells and are transported to distant locations in the body through the bloodstream. Insulin is a well known hormone protein made in the mammalian pancreas that is absolutely essential for life because insulin regulates the uptake of sugar from the blood into the cells. The hormone is produced in the pancreas as an inactive pre-hormone protein that must be cleaved to produce the active insulin hormone. Diabetes occurs if the pancreas fails to make enough insulin or if the cells cannot respond to the signals generated by the insulin. Diabetes is a metabolic disease that prevents the body from properly regulating sugar levels in the blood. Type II diabetes, which has now reached an epidemic level in the United States and is on the rise world-wide is caused in part by an increase in childhood obesity, a diet with lots of fat and sugar, and poor lifestyle habits without much physical exercise.

> **KEY CONCEPT**
> The environment plays a large role in human development and influences the results of genetic inheritance.

Close-Up Look Inside a Eukaryotic Cell

Students who are studying human molecular and cellular biology often have difficulty trying to visualize the molecular structures of components such as chromosomes and proteins inside the cells. The drawings of cells usually simplify the structures inside the cell in order to focus on the components under discussion and do not provide an accurate idea of the environment inside a eukaryotic cell. A more realistic picture of the components inside a pancreas cell is provided by dual-axis electron microscope (EM) tomography, which portrays a 3D reconstruction of the internal contents of the cell, showing the many components involved in actively producing and secreting many copies of the insulin protein (Figure 1-12).

Scientists do not yet know exactly how the DNA helix is packaged by proteins into a dynamic chromosome structure that protects the DNA but also

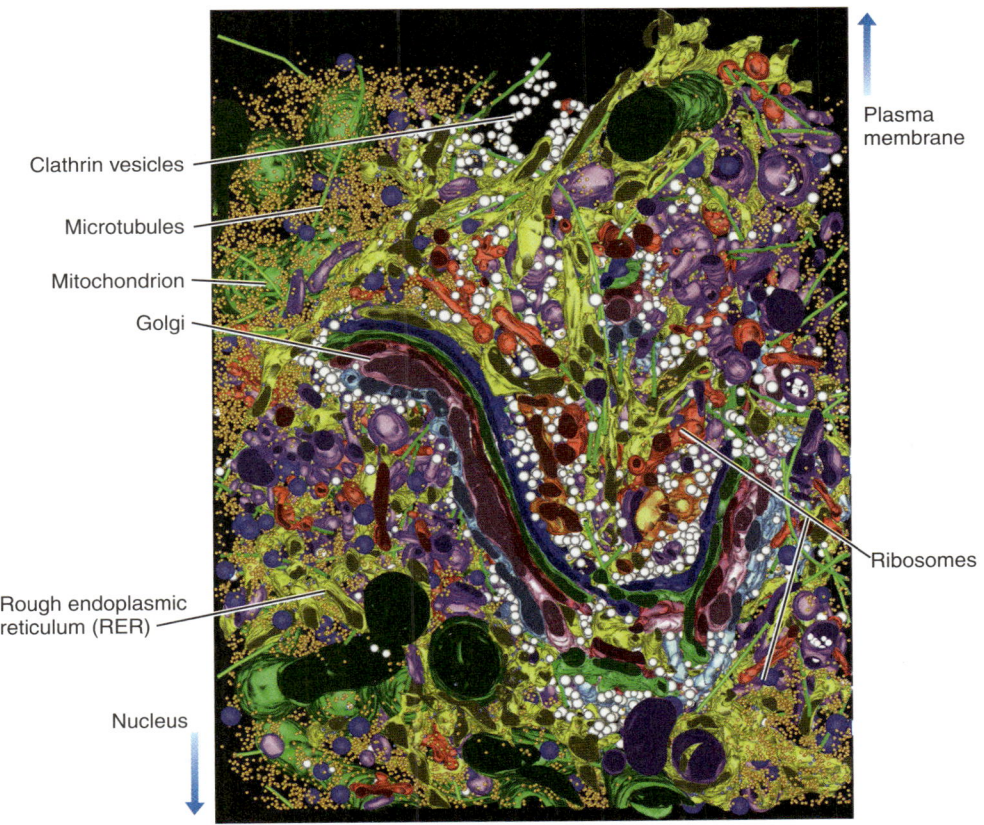

FIGURE 1-12 A pancreas cell produces insulin hormone Cells contain many active components and intricate structures. This region of the cytoplasm in this pancreas cell was analyzed using high resolution electron tomography to visualize the components involved in making insulin proteins for secretion from the cell. The clathrin proteins (white) on coated vesicles allow specific proteins to move across the membrane to exit the cell (at top). The microtubule fibers (green fibers) are a primary part of the cell's cytoskeleton, and the mitochondria (green circles) generate energy for the cell. Proteins to be secreted are translated on ribosomes (red) on the Rough Endoplasmic Reticulum (RER) (yellow) and are modified in the Golgi apparatus (maroon). The nucleus of this cell is located off of the bottom of the figure.

changes structure throughout the different phases of the cell cycle. Each human chromosome contains a very long double-stranded DNA molecule that is packaged by specialized proteins into highly condensed chromatin that allows the chromosome DNA molecules to fit inside the space available in the cell nucleus. When the cell prepares to divide to make two cells, the chromosomes become very compact, so they can be moved easily and segregated to the two new nuclei of the two new cells.

Human cells spend the most time in interphase, the part of the cell cycle when the cell is devoted to the specialized functions directed by the expression of the genes encoded by the chromosome DNA. When a gene is expressed (or turned on), the DNA encoded by the gene is copied into RNA; but to accomplish this feat, the chromosome structure must first be rearranged and less compact, thereby allowing the enzyme access to the DNA template to copy the gene into RNA.

> **KEY CONCEPT**
> Human genes are located in the nucleus of the cell. When a gene is expressed, the DNA in the gene is copied into RNA. The mRNA is transported through the nuclear membrane and is translated into protein in the cytoplasm.

Human Chromosomes have Small DNA Differences

All mammals including humans share a common body plan featuring two arms, two legs (or four legs), one head, one nose, two eyes, and no tail, characteristics that distinguish mammals from creatures such as snakes, beetles, and lizards (Figure 1-13). These observations suggest that humans have genes that specify the general organization of the human body. Super imposed on this general body plan are the many physical traits (phenotypes) that vary widely among individuals, in part resulting from the expression of different versions of the same inherited genes called alleles.

Comparing the outward appearances of people in a crowd (or students in a classroom) makes it clear that people are both similar to each other and different from each other. Except for genetically identical twins, each person has a unique combination of different physical features such as height, weight, eye color, body type, and personality traits that are specific to the individual. However, further comparison shows that humans are much more similar to other humans than humans are to primates, plants, insects or other animals.

Human traits can be categorized as either simple or complex, but only a few human traits are actually controlled by the expression of a single gene. These include tongue-rolling, a widow's peak, extra fingers and toes (polydactyly), the Rh

FIGURE 1-13 All mammals including humans have a common body plan Humans have a common body plan featuring two arms, two legs (or four legs), one head, one nose, two eyes, and no tail, which is distinct from other creatures such as snakes, birds, fish and lizards.

factor, gender, some genetic diseases such as Huntington's disease and the ability to taste bitterness. This sensation is controlled by the PTC (TAS2R38) gene, which encodes a taste receptor protein located in the taste bud cells on the human tongue.

Research shows that most human eye color (blue, brown, green, etc.) results from inheriting different combinations of parental gene alleles. These polygenic characteristics determine the pigmentation of the iris, the colored part of the human eye that surrounds the black pupil in the center. Human skin and hair color result from the types of melanin pigment that accumulates in the pigment cells, including eumelanin (brown-black) and phaeomelanin (red-brown). Red hair is the least frequent naturally occurring human hair color; it appears when the cells make more phaeomelanin pigment than eumelanin pigment. The combination of red hair, fair skin, and freckles are characteristic traits of a person who has inherited two mutant MC1R genes (alleles) and as a result can produce only defective mutant melanocortin 1 (MC1R) receptor proteins, causing the mutant pigment cells to accumulate extra phaeomelanin, a red-brown pigment.

People who inherit one normal (wild-type) MC1R gene allele and one mutant MC1R gene have a 25% chance of conceiving a redhead (Unit 4). Anyone with limited skin pigmentation, including albinos and redheads, must use extreme caution in sunlight. These people are at a significantly higher risk of developing a serious skin cancer called melanoma because of the low levels of melanin in their skin cells. Melanin is a natural protection against cancer. Recently researchers found that the ultraviolet radiation (UV light) from the sun interacts with the phaeomelanin protein, which in turn can damage the chromosome DNA, and promote the development of cancer cells (Unit 4).

> **KEY CONCEPT**
> Expressing different genes leads to changes in the molecular components inside the cell, which in turn alters cell function and traits.

Discovering The Structure of The DNA Molecule

In biological organisms the genetic information for simple and complicated traits is encoded in genes and written in the universal DNA language, which contains four DNA letters: A, G, C, and T. These four DNA letters, and the DNA language, make up the genes of all living things, from microbes, bacteria, and most viruses to all human cells. The unique biological and chemical properties of the DNA molecule allow the DNA chromosome to carry genetic information in every cell and to be inherited by every generation. The secret behind how DNA stores, reproduces, and transmits genetic information is actually hidden in the molecular structure of the DNA molecule. To laypeople, molecules often appear to be complicated structures made up of an apparently random array of balls and sticks (atoms and chemical bonds), but the patterns of atoms and bonds in a protein molecule actually represent the amino acids that make up the protein structure. It is important to learn about the key parts of each molecular structure that are essential to the biological function of the molecule.

The discovery of the structure of the DNA helix in the early 1950s marked the beginning of the molecular age of DNA science. This work revealed key structural features of the DNA helix molecule that are important to understand how DNA functions to store, copy and transmit genetic information. When a cell is rapidly dividing and reproducing, the chromosome DNA must be replicated and the chromosomes duplicated each time the cell divides. Based on the efficient enzyme reactions used by cells to copy DNA, scientists developed DNA sequencing technologies that were used to determine the DNA sequence of the entire human genome. The Human Genome Project explained many mysteries about the arrangement of DNA sequences in the human genome, identified the locations and functions of many human genes, and revealed the mystery surrounding the vast stretches of non-coding DNA sequences in the human genome.

The structure of the DNA helix is very important to understand because the DNA molecule carries and expresses genetic information in the cell. The order or sequence of the DNA bases in a gene code for the protein product. The DNA sequencing technique is used to determine the base sequence, the linear order of the bases, in a given segment of DNA. Analyzing the overlapping DNA sequences allows scientists to assemble very long regions of DNA sequence, eventually compiling the sequence of entire chromosomes (Unit 2). DNA sequencing is now used routinely in many research and clinical labs. Scientists can perform DNA sequencing "by hand" in the lab or send DNA samples to commercial services that determine DNA sequences for a price and return the results electronically. As the practice of personalized medicine becomes more widespread, the DNA sequences of individual human genomes will soon be a necessary component of healthcare decisions made to determine the best medical treatment or drug for an individual patient (Unit 4).

The race to determine the 3D structure of the DNA molecule was scientifically and politically important and was a common topic of public discourse in 1950. At the time James Watson was an American graduate student who went to Cambridge University in London to study with Francis Crick, a prominent British biochemist (Figure 1-14). Together, Watson and Crick tried unconventional approaches to solve the problem of the structure of DNA. They gathered all the known facts published about DNA and tried to fit the facts into possible molecular structures, much like solving a 3D puzzle. For example, Watson and Crick knew that the amount of the "A" chemical in DNA is equal to the amount of "T" chemical in DNA and the amount of "G" chemical in DNA is equal to the amount of "C" chemical in DNA. Watson and Crick made cardboard cut outs of the chemical shapes of the four bases and used them to try to build a model of a DNA molecule.

At the same time another young scientist Rosalind Franklin was working with John Randall and Maurice Wilkins at King's College in London (Figure 1-14). Randall's lab was studying the structure of DNA using X-ray diffraction, a method that is still commonly used to determine the 3D shapes of biological molecules such as proteins based on their crystal structure. Even now, this method continues to be a fundamental way to understand the relationship between the 3D shapes of proteins and their biological functions in the cell. The protein sample is converted into a uniform crystal, which is placed in the path of the X-ray beam (Figure 1-15). When the X-ray beam is scattered (diffracted) by the atoms in the crystal, they form a specific diffraction pattern on film that reflects the position of each atom in the protein crystal. The X-ray diffraction information in the pattern is analyzed further to generate the accurate 3D shape of the protein.

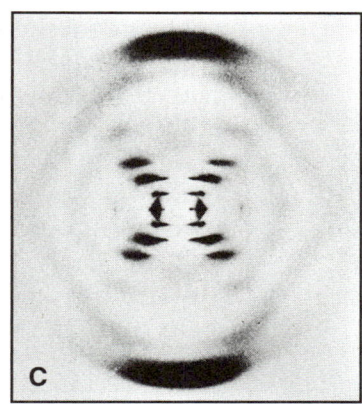

FIGURE 1-14 Meet the people who solved the structure of DNA (A) James Watson and Francis Crick built a six foot tall model of the DNA helix structure. (B) Rosalind Franklin's x-ray crystallography research provided essential information about DNA structure. (C) This x-ray diffraction picture of DNA was made by Rosalind Franklin and provided essential information to solve the structure of the DNA helix.

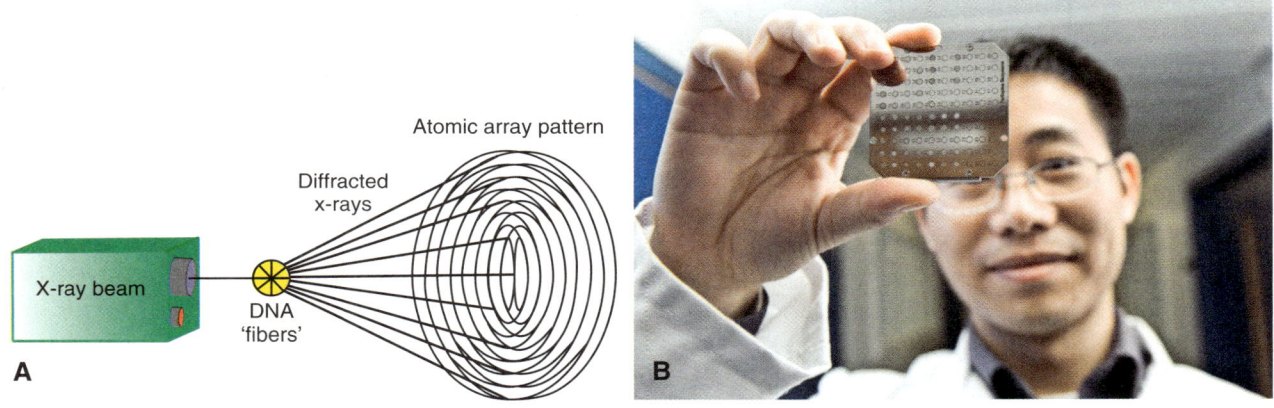

FIGURE 1-15 Methods used to determine the 3D shape of a molecule (A) An x-ray diffraction instrument sends an x-ray beam through a protein crystal. The atoms in the crystal alter (diffract) the path of the x-rays, which creates a diffraction pattern on the film revealing structural information about the sample. (B) Protein researchers also use MALDI-TOF (matrix-assisted laser desorption/ionization time-of-flight) a new form of mass spectrometry that can identify and characterize biological macromolecules including large proteins. This technician has prepared a target plate that contains a series of protein samples that are ready for MALDI-TOF analysis.

Franklin's X-ray Diffraction Image Helped to Solve The DNA Puzzle

By the early 1950s, the efforts to discover the molecular structure of DNA using X-ray diffraction of DNA crystals had failed, even in the Randall lab. Then Rosalind Franklin devised a method to make uniform DNA fibers instead of DNA crystals. When she analyzed the DNA fibers by X-ray diffraction, Franklin saw for the first time the beautiful X-ray diffraction pattern that revealed the 3D structure of the DNA helix molecule (Figure 1-14).

Watson and Crick reported the basic structural features of the DNA helix molecule for the first time in a famous paper they published in *Nature* in 1953. This historic one-page paper contained a sketch depicting a putative DNA helix structure that was supported by available scientific evidence, although at the time, Watson and Crick lacked their own independent experimental data to support their DNA helix structure. In 1962 James Watson, Francis Crick and Maurice Wilkins were awarded the Nobel Prize for solving the structure of the DNA helix. Unfortunately, Rosalind Franklin was not included in the Nobel Prize because she died of ovarian cancer four years earlier at age 37, and the Nobel Prize is not awarded posthumously. Recently the Franklin Medal was created in Britain to honor Rosalind Franklin's important contribution to the discovery of DNA double helix with an award each year to the best female scientist in the U.K.

Researchers Determine The Composition and Shapes of Biomolecules

In addition to X-ray diffraction, scientists also use nuclear magnetic resonance (NMR) and mass spectrometry (MS) to determine the molecular composition and 3D shapes of proteins and other biological molecules. The "mass spec" became familiar to the public as Abby Sciuto's (played by Pauley Perrette) favorite crime-solving machine in her forensics lab on the CBS TV show *NCIS*. Researchers also use a new form of mass spectrometry called MALDI-TOF (matrix-assisted laser desorption/ionization time-of-flight), which permits fragile biomolecules and biopolymers, including proteins, sugars, and large macromolecules, to be analyzed without damaging the sample. MALDI is often used in large-scale protein studies to identify proteins that were separated by one-dimensional gel electrophoresis and by two-dimensional gel electrophoresis (Figure 1-15).

Computers use the data from X-ray diffraction and mass spectrometry to create accurate 3D images of cellular components. The ability to visualize the activities going on inside living cells including the 3D images of different biological components has a dramatic impact on how well people remember information about cells and molecules. The use of computer images makes it much easier for people to distinguish between DNA, RNA, and proteins and to remember which of these fundamental components encodes genes (DNA) and which acts as a messenger to carry genetic information from the nucleus to the cytoplasm (RNA).

Computer animations are particularly helpful to visualize biological activities such as the function of a restriction enzyme protein that binds to the DNA helix and breaks chemical bonds to cleave the DNA helix into two DNA helices. Tutorials, videos, and simulations are also very useful to learn more about many cellular processes such as chromosome duplication, DNA replication, cell division, and gene expression. Excellent computer animations featuring DNA, RNA and proteins are now available from several sources.

Molecules Contain Atoms Held Together by Chemical Bonds

DNA and protein molecules are made up of a set of "organic" atoms, including carbon (C), hydrogen (H), oxygen (O), nitrogen (N), and phosphorus (P), which are commonly found in living organisms (Figure 1-16). Many types of atoms are necessary for the cell to perform biological functions. For example, calcium

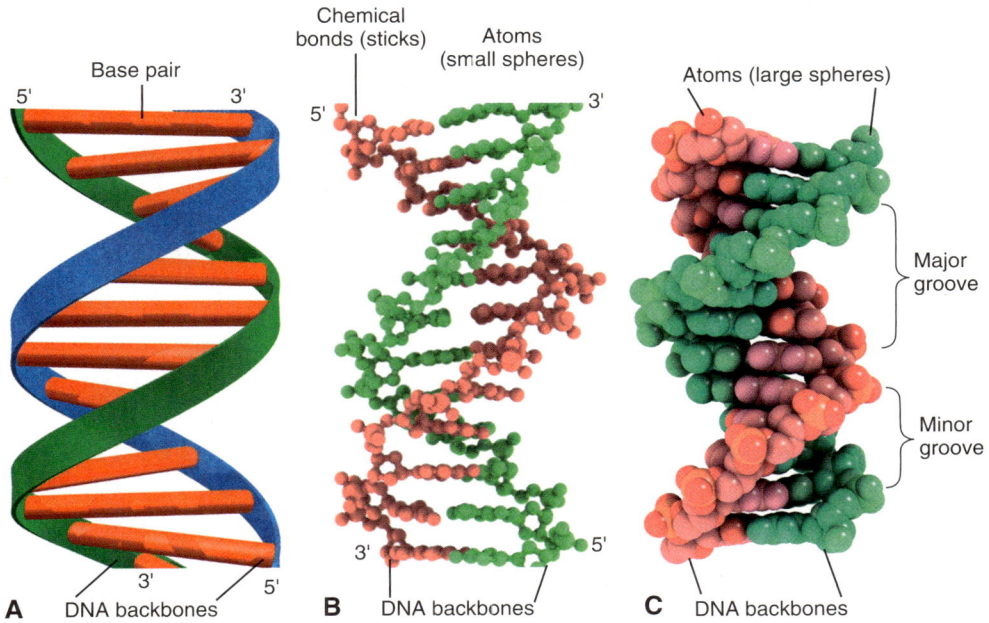

FIGURE 1-16 Different ways to represent a DNA helix The molecular structure of a DNA helix can be represented in different ways, including molecular structures and depictions with little molecular detail. (A) Ribbon DNA helix drawn without specific molecular details but including the backbone strands and base pairs. (B) This ball and stick DNA model with the atoms (balls) that are connected together by chemical bonds (sticks) is a more accurate way to depict a DNA helix. (C) Also more accurate is this space-fill DNA helix model, which contains spheres that represent the atoms and also encompass the space occupied by the chemical bonds. Note the major and minor grooves in the helix.

(Ca) is needed to build bones, and the food we eat provides nutrients such as amino acids to make proteins, as well as fat and various vitamins, for our cells. Energy is stored in the human body in the form of carbohydrate molecules.

The atoms in a molecule are connected to each other by three main types of chemical bonds: covalent, hydrogen (H-bonds), and ionic bonds. Covalent bonds make the strongest connections between atoms, while hydrogen bonds (H-bonds) are weak chemical bonds. The ionic bonds are based on the familiar attraction between opposite charges (positive and negative charges attract).

Becoming familiar with the structure of the DNA molecule is essential in order to be able to understand how DNA functions to express genes in cells. Fortunately, the overall structure of the DNA helix is aesthetically pleasing and the most important features are easily remembered by most people. The DNA helix molecule is symmetrical and made up of repeating units that are essential to the role of the molecule in storing, transmitting and expressing genetic information.

The three DNA helices in Figure 1-16 represent the same DNA molecule but depicted in different ways. The ribbon DNA model shows just the major structural features of the helix, the two DNA backbone strands, and the paired bases in between the strands (Figure 1-16). The ball and stick DNA molecular model represents the atoms as balls that are connected together by the covalent chemical bonds represented by sticks. An atom in the spacefill molecular model of the DNA helix is represented by larger spheres that encompass the space filled by both the chemical bond and the atom (Figure 1-16).

The atoms in a DNA molecule are connected together by strong covalent bonds, and the DNA molecules do not fall apart under the physiological conditions in the body with a temperature of 37°C (98.6°F). The covalent bonds that connect the atoms in the sugar-phosphate backbone of the DNA molecule ensure that these extremely long DNA strands do not break apart into shorter strands in the cell (Figure 1-16, Figure 1-17, Figure 1-18).

The weak H-bonds in the DNA, RNA, and protein molecules are also extremely important for these molecules to function in cells. The H-bonds between the base pairs in the center of the DNA helix are responsible for the highly specific base pairing interactions that are essential to transmit the correct genetic information. H-bonds also have important functions in interactions between proteins and between proteins and DNA. The formation of the correct hydrogen bonds between the base pairs in a DNA helix ensures that *only complementary bases can form base pairs*. The "A" base forms H-bonds with the "T" base in an A-T base pair, and the "C" base forms H-bonds with the "G" base in a C-G base pair (Figure 1-19).

> **KEY CONCEPT**
> Don't confuse molecules and atoms. Molecules are made up of atoms that are held together by chemical bonds. Some molecules have only two atoms, such as oxygen gas, which is a molecule with two atoms (O_2) held together by a chemical bond that has a different structure than a single oxygen (O).

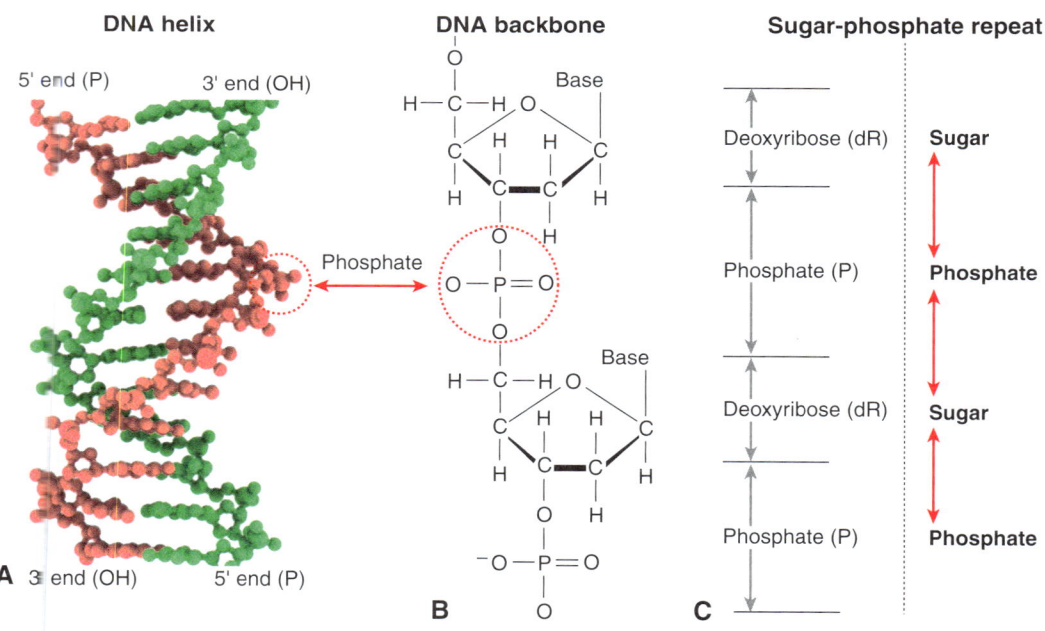

FIGURE 1-17 The DNA backbone strand has sugar-phosphate repeating units (A) Each DNA strand has a phosphate on the 5' end and an OH group on the 3' end. In the DNA helix the two DNA strands are anti-parallel; one strand is 5' to 3' while the opposite DNA strand is arranged 3' to 5'. (B) Both single-stranded and double-stranded DNA molecules contains strong covalent bonds. (C) The backbones of the DNA helix molecule are comprised of sugar-phosphate repeating units.

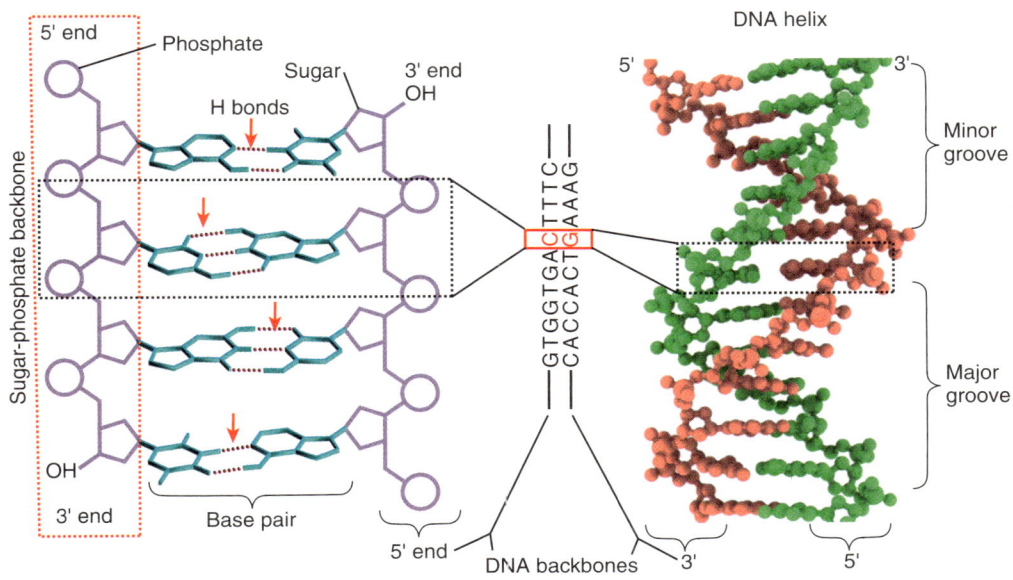

FIGURE 1-18 DNA base pairs written in English letters The inside of a DNA helix can be better visualized by flattening the helix to reveal the two DNA backbone strands and the A-T and G-C base pairs (left). The sugar-phosphate backbones (purple) and the base pairs (cyan) are shown with the hydrogen bonds indicated with red arrows. The DNA sequence is indicated in English letters with one base pair shown in red (center). The molecular DNA helix structure is shown (right) to relate the components in the DNA molecule with the letters used when DNA sequence is written as English letters.

Understanding the basic structure of the DNA helix molecule is central to understanding how genes are expressed. It is important to become familiar with the parts of the DNA molecule that have key roles in the cellular mechanisms involved in gene expression, including the DNA backbone, the DNA base pairs and the different ends of a DNA strand.

Each DNA helix contains two DNA backbone strands that wrap around each other on the outside of the helix and form alternating major groove and minor grooves in the DNA helix (Figure 1-16, Figure 1-18). A DNA backbone strand is made up of repeating phosphate and sugar chemical units that are linked together by strong covalent phosphodiester bonds (Figure 1-17). It is important to note that each DNA strand has different chemical structures at the two ends, a phosphate (P) on the 5′ end of the DNA strand and a hydroxyl group (OH) on the 3′ end (Figure 1-16, Figure 1-17, Figure 1-18). The two backbone strands in the DNA helix molecule are positioned in opposite (anti-parallel) directions such that one of the DNA strands is arranged 5′ to 3′ in the helix and the other DNA strand is arranged 3′ to 5′ in the same DNA helix (Figure 1-18). These key features of the DNA molecule will continue to play important roles in the biological mechanisms involving DNA and proteins in the cell.

The base pairs in the center of the DNA helix are an essential structural feature that permits the DNA molecule to carry genetic information. The DNA bases are connected to the DNA backbones by strong covalent bonds. In the center of the DNA helix, the bases are connected to each other by weak H-bonds. Opening and flattening the DNA helix reveals the two parallel

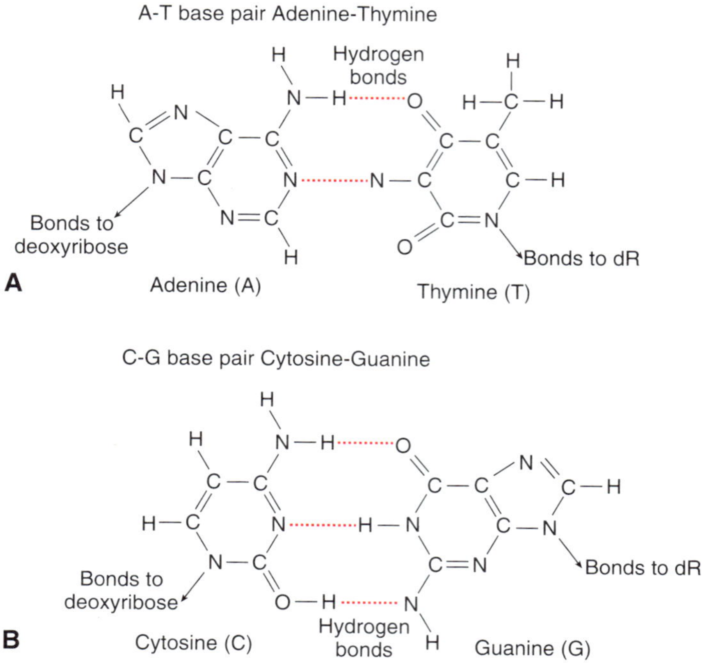

FIGURE 1-19 Specific H-bonds connect the base pairs in the center of the DNA helix The A-T and G-C base pairs have specific 3D chemical shapes that permit only complementary base pairs to fit together inside the diameter of the DNA helix molecule. The A-T and G-C base pairs are held together by weak hydrogen bonds (H-bonds) (red dotted lines). (A) Each A-T base pair has 2 H-bonds. (B) Each G-C base pair has 3 H-bonds. dR represents DNA backbone.

backbone strands and the A-T and G-C base pairs that fit together inside the DNA helix like 3D puzzle pieces (Figure 1-18). The A-T and G-C base pairs are held together by weak H-bonds, A-T base pairs have 2 H-bonds and G-C base pairs have 3 H-bonds (Figure 1-19). The H-bonds involved in forming even a short stretch of several complementary A-T and G-C base pairs make extremely specific interactions that prevent the formation of incorrect base pairs such as an A base paired with a G or a T base paired to a C. Base pairing interactions between complementary DNA strands are extremely specific because of the formation of many H bonds that cooperate over a length of several base pairs.

The ability of two single-stranded DNA molecules to correctly base pair together to make a double-stranded DNA helix is essential for the biological functions of DNA molecules, including gene expression and DNA replication. This feature of the DNA helix is necessary for cells to store, copy, and transmit genetic information when cells divide and grow. The genetic information is carried by the base sequence, which is represented by the linear order of the bases along one backbone strand of a DNA helix. In fact, the sequence of the bases along one DNA strand of the helix automatically reveals the sequence of the bases in the opposite DNA strand. The complementary DNA strands in the helix obey the universal base-pairing rules (A base pairs with T and C base pairs with G), which is made possible by the formation of many H-bonds between a region containing several complementary bases (Figure 1-19).

Life is possible because the weak H-bonds between complementary DNA base pairs can be broken by a moderate increase in temperature (in the lab) or by the action of certain protein enzymes (in the cell). This means that the annealing reaction that brings two complementary DNA strands together to make a DNA helix is reversible. The H-bonds between the bases in the base pairs can be broken, causing the DNA strands in the helix to separate into two single DNA strands (Figure 1-20). Inside living cells, specific enzymes help to bring together complementary DNA strands to promote annealing and form a double-stranded DNA helix containing A-T and G-C base pairs. The chromosome DNA in cells is always covered with bound proteins and is never completely naked. However, in the research lab, under conditions where the proteins have been removed from the DNA, the physical state of the DNA strands can be manipulated by changes in temperature. Decreased heat (temperature) causes complementary DNA strands to anneal together into a DNA helix and increased heat will separate the helix into two single DNA strands (Figure 1-20). This structural characteristic of the DNA helix molecule is required for the cell to express and transmit genetic information.

> **KEY CONCEPT**
> Complementary DNA strands can form the highly specific hydrogen bonds needed to make A-T and G-C base pairs. The ability of a DNA helix to naturally transition from a double-stranded helical molecule into two separate complementary single strands of DNA is a remarkable and essential feature of the DNA molecule. Just as important is the ability of two complementary single strands of DNA to re-anneal and form a DNA helix.

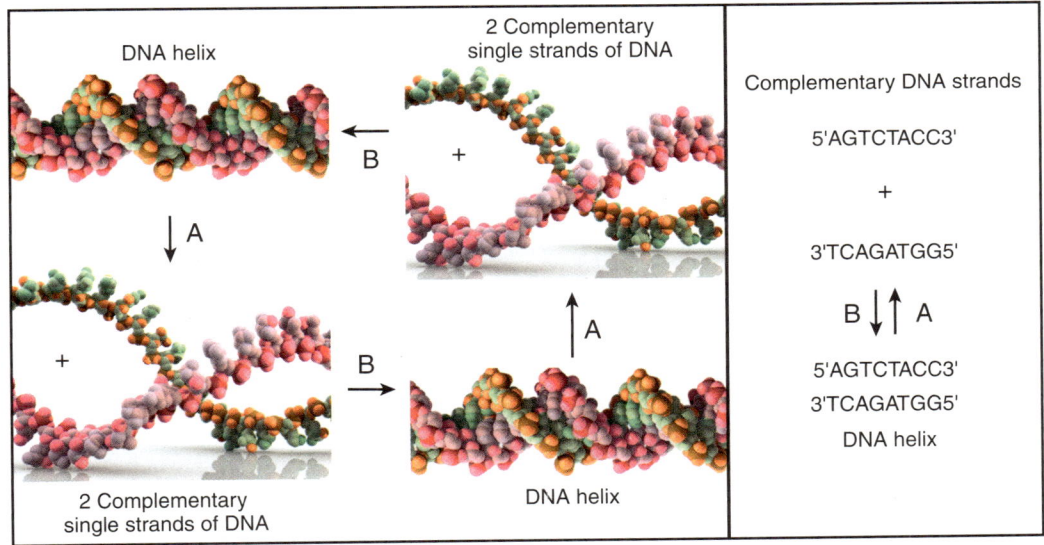

FIGURE 1-20 Complementary DNA strands can anneal and denature in a reversible process (A) When the H-bonds between the base pairs in a DNA helix are broken, the DNA strands in the helix separate (denature) and become two complementary single strands of DNA. (B) Complementary DNA strands can base pair together again by annealing (or hybridization) to form a DNA helix. The formation of a DNA helix from two complementary DNA strands is a reversible process.

Gene Expression Starts When DNA is Copied Into RNA

The basic transcription process in human cells involves the transfer of genetic information from the DNA gene in the chromosome to an RNA molecule. Copying a DNA gene into an RNA strand requires that the RNA polymerase enzyme "read" the order of the DNA bases in the gene and produce an RNA copy written in the RNA language. The DNA language used in genes is different from the RNA alphabet (Figure 1-21). The DNA letters (bases) A, G, C, and T are chemically similar to the letters in RNA but the DNA letters are clearly different from RNA letters (bases) A, G, C, and U. Most important, the RNA alphabet includes the letter "U" but not the letter "T", and the RNA "U" letter base-pairs to the RNA "A" letter (base). When a DNA strand is copied into an RNA strand, DNA-RNA base pairs form between the DNA template and the RNA transcript (Figure 1-21). Like DNA base pairs, RNA base pairs and DNA-RNA base pairs are also held together by weak H-bonds.

> **KEY CONCEPT**
> RNA and DNA molecules carry genetic information but the two languages use different alphabets. The DNA letters (bases): A, G, C, and T have different chemical structures than the RNA letters (bases): A, G, C, and U.

When genetic information is transferred from a DNA gene to an RNA, the letters (base sequence) in the template strand of the DNA helix must be available so that the RNA polymerase enzyme can read the order of the bases

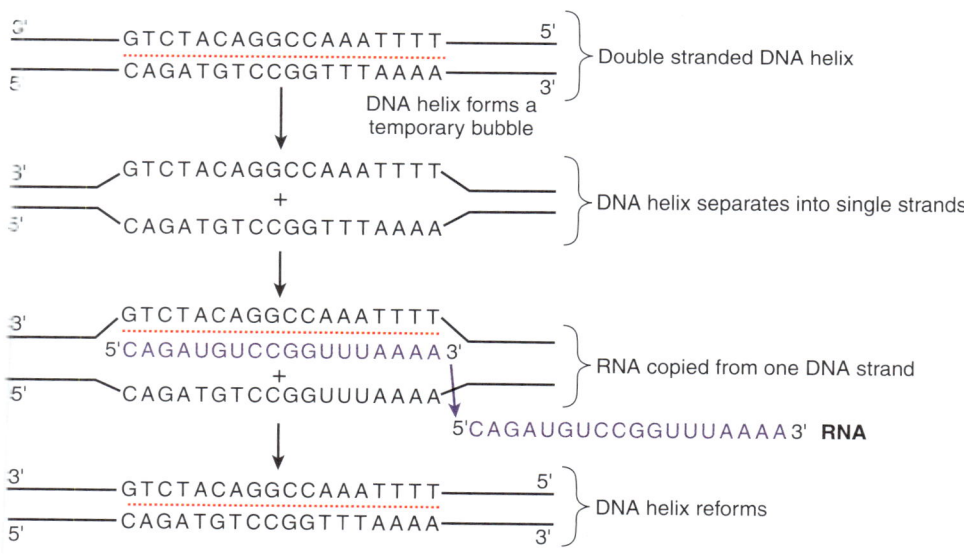

FIGURE 1-21 During gene expression the DNA helix strands separate to reveal the base sequence Genes are located in small regions of the very long DNA molecules in eukaryotic chromosomes. In the region of the DNA helix encoding the gene, the H-bonds break and the DNA strands temporarily separate to allow access to the DNA template strand to be copied into RNA.

along the DNA template strand (Figure 1-22). In living cells this feat involves rearranging many chromatin proteins that are involved in packaging and condensing the chromosome structures to fit inside the nucleus. A short localized region of the DNA helix unwinds, breaking the H-bonds and physically separating a region of the two DNA strands to form a temporary single-stranded "bubble" in an otherwise double-stranded DNA helix (Figure 1-22). This region of single-stranded DNA acts as a template for the enzyme to copy the DNA bases into an RNA strand that is then released from the DNA template, the DNA helix reforms, and the chromatin proteins bind to the DNA again. In human cells, the very long RNA strands (pre-mRNAs) that are copied from genes must be processed in the nucleus before the mature messenger RNAs (mRNAs) can be transported to the cytoplasm to make proteins (Unit 3).

> **KEY CONCEPT**
> For DNA helix strands to separate into two DNA single-strands, the H-bonds between the bases must break. The same H-bonds are formed when the complementary single DNA strands base-pair together again and reform the DNA helix.

DNA "Language" Codes for Genes in The DNA Helix

The DNA language is written in a linear order of A, G, C, and T letters called bases that are contained in a specific order in the template DNA strand. In living cells, the template DNA strand is always part of the double-stranded chromosome DNA and does not exist as a separate single-stranded DNA template

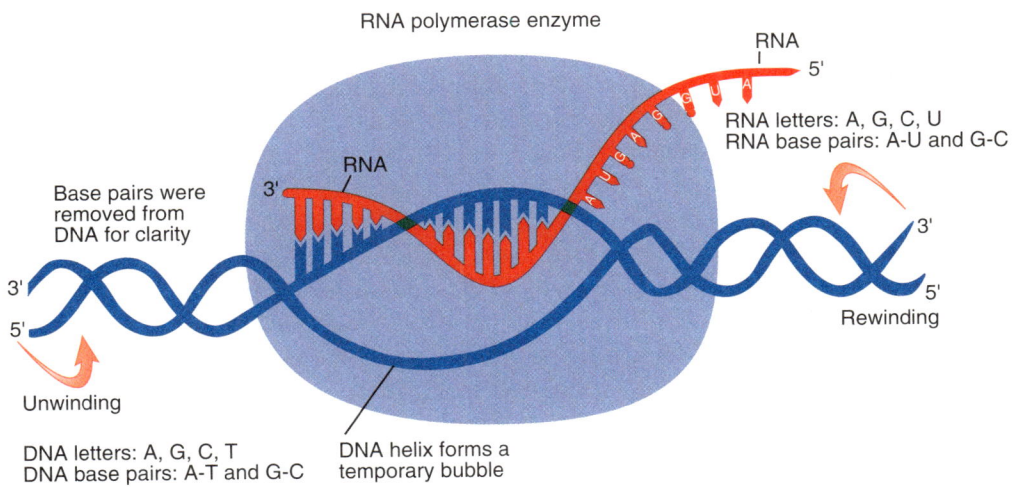

FIGURE 1-22 Gene expression occurs in many regions along the chromosome DNA This picture shows how a localized region of the DNA helix becomes temporarily single-stranded in a DNA bubble. This allows the polymerase enzymes to have access to the single-stranded DNA template to copy DNA into RNA. The white arrows at either end of the temporary bubble indicate that the DNA helix must wind and unwind during this process. The DNA base pairs have been removed for clarity.

(Figure 1-23). For this reason the strand of the double-stranded DNA helix containing the template sequence must temporarily separate into a single-stranded region in order for the enzyme to correctly read the base sequence from the template strand of the gene.

In the research lab it is often necessary to identify the template DNA strand for a specific gene by reading the double-stranded DNA sequence information that is available from DNA sequencing studies of the human genome. Each human chromosome carries one double-stranded DNA helix molecule extending from one end of the chromosome to the other end (telomere to telomere) and contains a large number of genes. Each gene is encoded by a local section of the chromosome DNA that contains the coding region of the gene. Only one of the DNA strands in this particular region of the chromosome DNA helix is the template strand for this gene and is copied into an RNA strand. The opposite, complementary DNA strand in this region of the chromosome is the non-template strand. The polymerase enzyme does *not* switch from one DNA strand to the other DNA strand in the middle of a gene. Follow the series of figures to observe how the polymerase enzyme would read the template strand of the DNA helix to produce an RNA copy. This DNA strand reads 3'-TAGAGATCC-5' (Figure 1-23).

> **KEY CONCEPT**
> The DNA molecule encodes genetic information in a gene sequence and transmits the information by copying the gene DNA into RNA (transcription). When a gene in the chromosome is expressed, only one strand of the DNA helix acts as the template and codes for the genetic information that is copied into RNA.

Single-Stranded RNAs Form Secondary Structures in Cells

DNA and RNA molecules have similar chemical structures but in living cells these two macro-molecules have different, essential roles in gene expression. The chemical name for RNA, ribonucleic acid, lacks the "deoxy" part of the DNA name, *deoxy*ribonucleic acid, because the chemical structure of RNA contains oxygen atoms that DNA lacks. This tiny molecular difference between these molecules strongly influences the chemical properties of DNA and RNA and determines their distinct biological functions in the cell.

Although the DNA and RNA molecules both carry and transmit genetic information, they have very different jobs in living cells. The chromosomes protect, duplicate, and express the genome DNA and reside within the nucleus throughout the cell cycle, even when the nuclear membrane disassembles (and reassembles) during cell division (mitosis). The DNA molecules in living cells exist exclusively as very stable, double-stranded DNA helices, either linear molecules as in eukaryotic cells or as circular double-stranded molecules in mitochondria (mtDNA) and in prokaryotic cells such as bacteria. Whether linear or circular, the DNA molecules must be copied, packaged, and distributed correctly when cells divide.

UNIT 1 Genes are Written in DNA Language

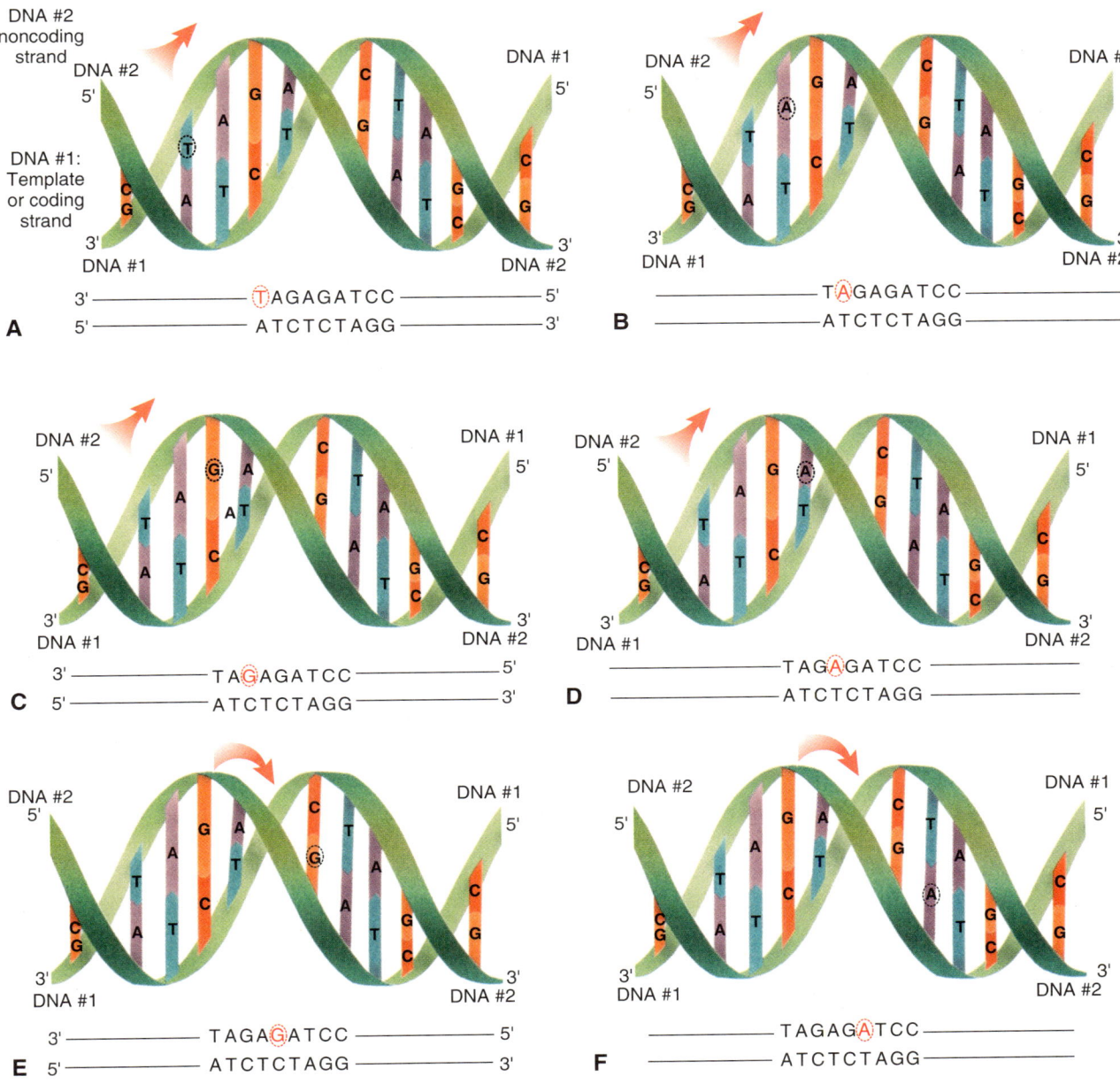

FIGURE 1-23 How to read the DNA "language" Each gene is encoded by a local region of an extremely long DNA helix in a eukaryotic chromosome. In this region of the DNA helix only one DNA strand is the template strand and is copied into an RNA strand. The opposite, complementary DNA strand is the non-coding strand. The polymerase enzyme cannot switch from one DNA strand to the other strand in the middle of copying a gene into RNA. This series of figures indicates how the enzyme would read the DNA template strand when making an RNA copy of the gene. The DNA sequence read from the DNA helix is: 3′ TAGAGATCC- 5′.

Continued

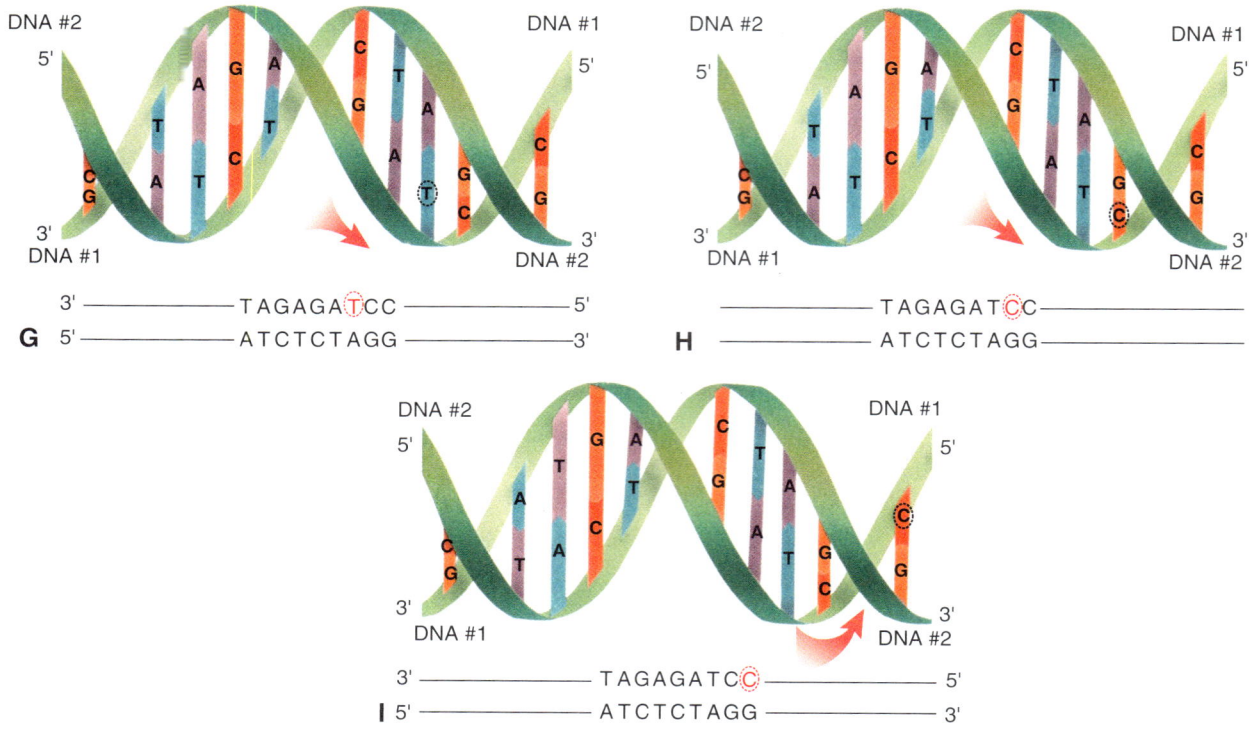

FIGURE 1-23 How to read the DNA "language" (con't).

The oxygen atoms in the molecular structure of RNA molecules cause RNA to be inherently much less stable than DNA molecules, which reflects the more temporary function of the RNAs in cells. The messenger RNA molecules (mRNA) copied from genes are sent to the cytoplasm to transfer genetic information and produce specific proteins, but after the RNA has been translated a prescribed number of times, the mRNAs are degraded. In cells, the RNAs exist primarily as single-stranded molecules that are easily degraded and discarded. Gene expression can be controlled by destroying an mRNA before it is translated into protein, but it is much more common for the cell to control gene expression at the point where the DNA gene is copied into RNA (Unit 3). A gene can be silent because it is not copied into RNA, or a gene can be expressed at high levels by making many multiple RNA copies of the gene, ensuring that many copies of the protein translated from those mRNAs will be produced in the cell. Many different protein factors and other types of signals control the expression levels of genes, which can change quickly in response to biological signals. Some genes are completely inactive or suppressed and are not copied into RNA, while other genes are expressed at reduced levels and produce only small amounts of certain proteins (Unit 3).

RNA molecules are never naked in the cell; they associate with specific RNA binding proteins as soon as they are made. The single-stranded RNA molecules form intricate secondary RNA structures that contain single-stranded loops and short double-stranded stem regions formed by intrastrand C-G and A-U base pairs (Figure 1-24). For example, the 18S ribosomal

UNIT 1 Genes are Written in DNA Language

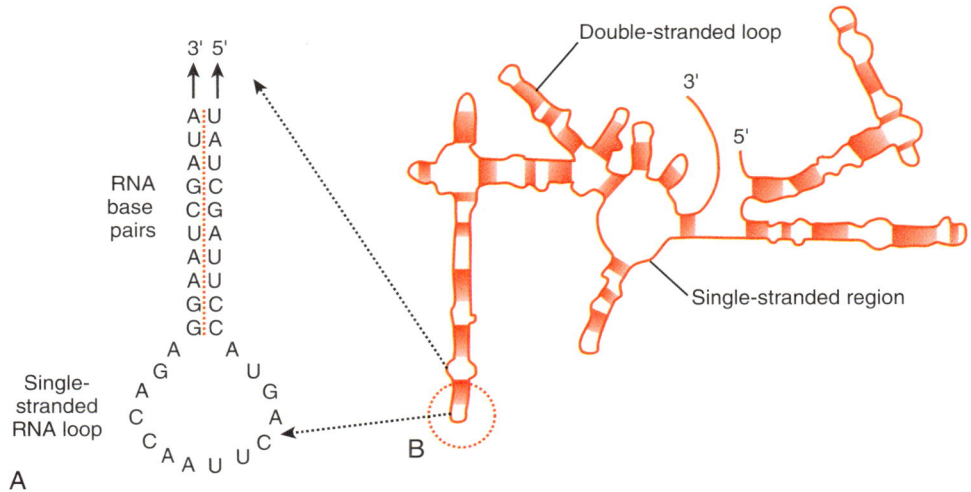

FIGURE 1-24 RNA strands fold into secondary structures by forming RNA:RNA base pairs
(A) In living cells the RNA molecules are predominantly single-stranded but they usually fold into secondary structures by forming base pairs within the same RNA strand. The loops, stems and bubbles formed in the folded RNAs are binding sites for proteins with specialized functions. (B) This single strand of RNA is drawn from 5' end to the 3' end and folded as it might appear in the cell. The RNA folds by base pairing between complementary RNA bases within the RNA strand.

RNA (rRNA), functions as a fundamental component of the small 40S ribosomal subunit found in eukaryotic cells. The 18S rRNA is about 1,900 bases long and folds into an intricate secondary structure that plays a key role in translation by the ribosome. The ribosomal proteins bind specifically to the double-stranded stems and single-stranded loops of the folded 18S rRNA. The "S" designation in 18S rRNA refers to Svedberg, which is the unit used to determine the rate of sedimentation of a macromolecule in an ultracentrifuge. An ultracentrifuge is a type of centrifuge that spins tubes very fast to separate and analyze cellular components. Prokaryotic ribosomes contain a 16S rRNA that is very similar in RNA sequence to the 18S rRNA molecule in eukaryotic ribosomes. The rRNA sequences exhibit a slow rate of change in RNA sequence over time, which allows scientists to study the divergence of the rRNA sequences observed when the sequences of the ancient and modern species are compared. The evolutionary history of different organisms can be determined by tracing the changes in the rRNA sequences over time.

> **KEY CONCEPT**
> The intra-strand A-U and C-G base pairs in the folded RNA molecules form biologically important secondary structures such as stems and loops. Some of these RNA structures are often specific binding sites for special proteins that perform essential roles in key processes such as transcription, RNA processing, RNA splicing, translation, and RNA transport.

Gel Electrophoresis is a Key Technique to Separate DNA Molecules

Scientists routinely analyze DNA and RNA molecules using gel electrophoresis, a method that separates these linear molecules by length. How DNA is manipulated in the lab depends on the physical characteristics of the DNA under study. The relatively short, circular double-stranded DNA plasmids found in bacteria are easy to manipulate in the lab compared to the very long double-stranded DNA molecules isolated from eukaryotic chromosomes. Once the proteins have been removed from eukaryotic chromosomes, the very long DNA molecules are released and break easily. In preparation for gel electrophoresis scientists can physically shear the long chromosome DNA molecules into shorter DNA fragments or cut the DNA molecules with a restriction enzyme to produce DNA fragments of shorter, varying lengths.

To analyze the DNA products of a restriction enzyme reaction, the DNA fragments are separated by electrophoresis on either an agarose or a polyacrylamide gel (Figure 1-25). The term "gel" refers to a stiff Jello-like matrix that supports the DNA molecules as they migrate through the gel matrix in response to an electric current (Unit 2). Depending on the requirements of the experiment, a researcher can adjust the composition of the gel to achieve conditions that will most effectively resolve DNA or RNA molecules within a range of lengths. Agarose gels are typically used to separate DNA fragments that are longer than about 10–20 kilobase pairs (kb), while polyacrylamide gels are used to separate DNA fragments that are shorter than 10 kb in length. Double-stranded DNA fragments tend to migrate as rod shapes through native gels. However, single-stranded DNA or RNA molecules tend to form various secondary structures due to intra-strand base pairing. Polyacrylamide gels containing a denaturing reagent are used to keep the DNA molecules in a single-stranded form and prevent intra-strand base-pairing while the DNA (or RNA) migrates through the gel. The denaturing agent breaks the H-bonds between the internal base pairs that form in the DNA strand in order to maintain the single-stranded form required to resolve the short DNA strands produced by DNA sequencing reactions that can differ by only a single base in length (Unit 2). Gels containing denaturing reagents are also necessary when separating RNA molecules by electrophoresis in order to keep the RNA molecules single-stranded and to prevent the formation of intra-strand base pairs in the RNA.

Most research labs have simple plastic gel electrophoresis equipment with a tray that supports the agarose gel slab in a horizontal position when the gel is poured (Figure 1-25). During electrophoresis the tray is used to position the gel so that both ends of the slab are in contact with the liquid buffer and the electrodes can send an electrical current through the gel slab. Agarose gels can be made in a range of concentrations and in various sizes with different combs and different plastic molds to cast the melted agarose into solid gels. To prepare an agarose gel, the appropriate weight of agarose powder is mixed with a predetermined amount of liquid buffer, and heated almost to boiling to melt the agarose powder. After the agarose has completely dissolved and cooled a little, the liquid is poured into the plastic gel tray and the comb with teeth is

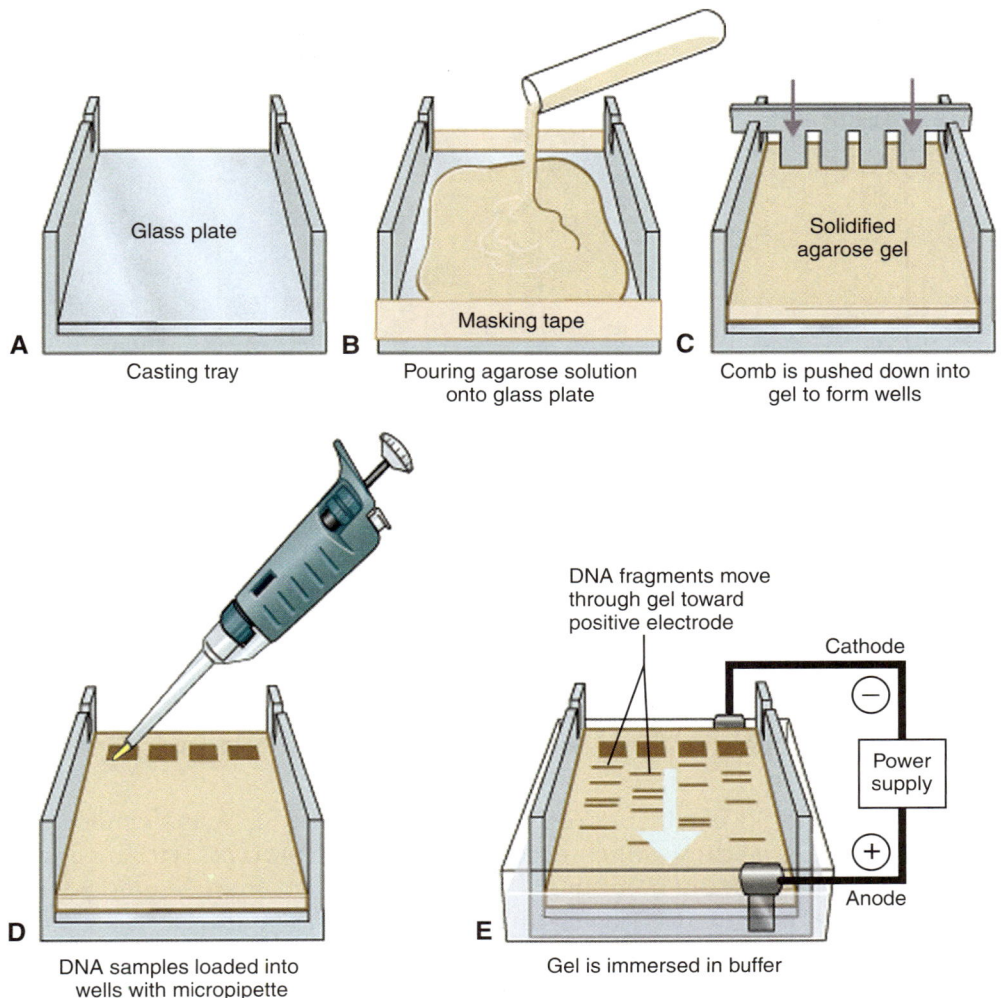

FIGURE 1-25 Agarose gels are used to separate DNA molecules using an electric current
DNA fragments are analyzed by gel electrophoresis using a Jello-like solid matrix, a gel, that supports the DNA molecules as they move. A comb with teeth is used to make wells as the gel solidifies. Later the wells are used to load the samples on to the gel. The gel apparatus is designed to pass the electrical current through the gel during the experiment, causing the DNA strands to migrate through the gel.

inserted into the gel and positioned near one end of the gel mold. When the agarose has solidified and the comb is removed, the teeth have made a line of indentations called wells that are parallel to the top of the gel.

The solid gel slab is placed in the gel apparatus in contact with the liquid buffer. The DNA samples to be analyzed contain a dense loading buffer so that when the DNA samples are loaded onto the gel they are layered into the wells underneath the liquid buffer (Figure 1-26). The DNA samples will remain temporarily at the bottom of the wells until the electric field is turned on and the negatively charged DNA molecules start moving in the gel. Too long a delay will allow the DNA or RNA samples to diffuse out of the wells, ruining the experiment.

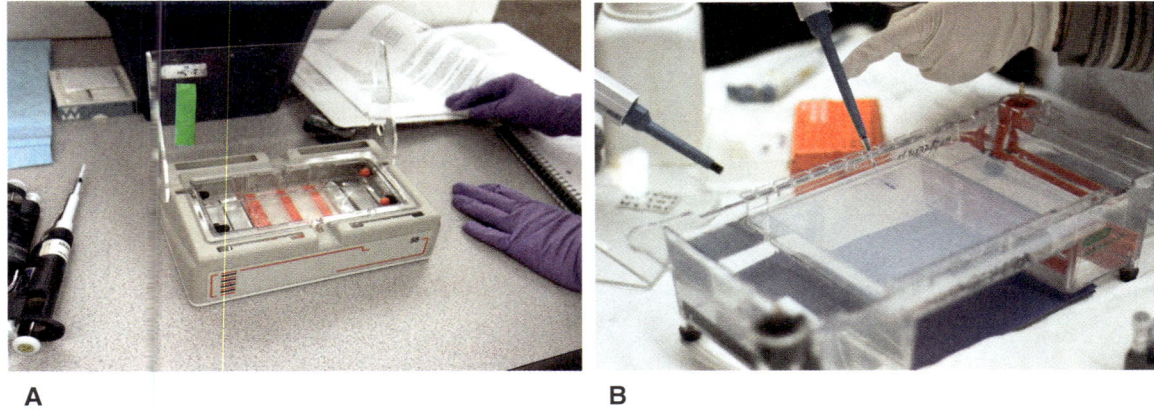

FIGURE 1-26 DNA samples are loaded into the wells in the gel (A) The gel is supported in an apparatus that supplies the electric current. (B) The DNA samples are mixed with a dense buffer so the samples can be layered into the bottom of the well underneath the buffer. DNA molecules have an overall negative charge due to the phosphate DNA backbone. In an electric current, the DNA molecules will migrate in the gel toward the positive (+) electrode. The DNA samples also contain a blue-purple dye that migrates in the gel with the DNA.

Interpreting The Meaning of DNA Bands in a Gel

During gel electrophoresis double-stranded DNA molecules adopt rod-like shapes as they migrate through the gel matrix, with the shortest DNA strands moving more quickly through the gel than the longer DNA strands, which move more slowly through the matrix (Figure 1-27). At the end of the experiment the longer DNA strands are located in the gel closer to the wells at the top of the gel, while the shorter DNA strands migrate closer to the bottom of the gel (Figure 1-27). The distances that the new DNA molecules have migrated in the gel are compared to the distances migrated by the DNA standards in the same gel, revealing the lengths of the unknown DNA fragments. The results obtained for the unknown DNA samples are compared with the DNA standards used as experimental controls to determine the length of the unknown DNA fragments.

When the electric current is turned on, the gel is subjected to an electric field (volts) with the negative electrode (−) at the top of the gel closest to the wells and the positive electrode (+) at the bottom of the gel. The DNA molecule has an overall negative charge due to the phosphate backbone so that the DNA strands migrate through the gel toward the positive (+) electrode at the bottom of the gel. Hooking up the electric leads in the wrong orientation will cause the DNA molecules to move in the opposite (wrong) direction, leave the gel, and dive into the buffer chamber! The longer DNA molecules migrate more slowly through the gel than the shorter DNA molecules, which migrate more rapidly through the gel. The information derived from a gel electrophoresis experiment depends on the distance the DNA fragment migrates down the gel relative to the loading wells.

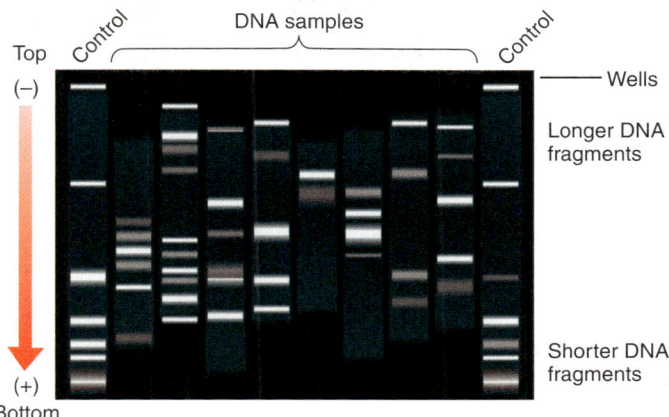

FIGURE 1-27 DNA molecules are separated by length The DNA bands in the gel are easily visualized by staining the DNA with ethidium bromide, which makes the DNA molecules fluoresce when illuminated with ultra violet light (UV). The DNA shows up as white bands in the figure. The longer DNA molecules migrate more slowly through the gel than the shorter DNA molecules. DNA molecules have an overall negative charge due to the phosphate backbone. In an electric current, the DNA molecules migrate toward the positive (+) electrode.

At the end of the experiment, the locations of the DNA bands on the gel are detected in different ways depending on the experiment. For example, when the DNA in the gel contains a radioactive, bioluminescent, or fluorescent tag, then the DNA bands in the gel are visualized by exposing the gel to X-ray film (Unit 2). DNA or RNA molecules are usually labeled when the DNA or RNA molecules are present at such low levels they would not be detected without the signal amplification provided by labeling the molecules.

When the amount of DNA used in a restriction enzyme reaction is sufficient the DNA bands in the gels can be easily detected by staining the DNA in the gel with ethidium bromide (EthBr) (Figure 1-27). EthBr is a planar molecule that binds in between the base pairs of the DNA helix and causes the DNA molecules to fluoresce when illuminated with ultra violet light (UV). Each DNA band visualized by ethidium bromide staining represents millions of DNA fragments of the same length and as a result the DNA fragments migrate to the same approximate position in the gel.

> **KEY CONCEPT**
> When analyzing unknown DNA by electrophoresis, each sample is loaded into separate wells on the same gel so that the DNA bands can be directly compared at the end of the experiment. The lengths of DNA fragments are determined by comparing the distances that the unknown DNA fragments have migrated in the gel compared with the control DNA bands run in a separate lane on the same gel.

Restriction Enzymes Cut DNA at Specific Base Sequences

Restriction enzymes are essential protein tools used to manipulate DNA molecules in the lab. Hundreds of different restriction enzymes with known DNA cleavage sites are now commercially available. These restriction enzymes recognize and cut only at specific base pair sequences in the DNA helix, and produce DNA fragments with different types of DNA ends.

Restriction enzymes are naturally produced in bacterial cells where these enzymes are components of a restriction-modification system that is used by bacterial cells to defend against invading bacteriophage DNA. These bacterial viruses inject their DNA chromosomes into the cells where the bacteriophage DNA replicates and eventually kills the bacterial cell. The restriction enzymes have evolved to recognize and cut the "foreign" invading DNA, but they do not cut the cell's chromosome DNA, because the cell's "host" DNA has been modified by the addition of small chemical methyl groups (Unit 6).

Once the lambda (λ) DNA enters the cell the 12 bases of cohesive sticky ends (*cos*) on either end of the lambda (λ) DNA base pair to each other to form a circular λ genome that replicates to generate λ DNA to be packaged into the new λ phage that kill the cell and release new λ phage to infect surrounding bacteria (Figure 1-28). Alternatively, the λ DNA can insert into the bacterial DNA and remain integrated as a prophage in the bacterial chromosome, and the λ prophage DNA replicates with the chromosome DNA each time the cell divides.

Restriction enzymes are especially important in DNA research because these enzymes cut double-stranded DNA only at specific cleavage sites

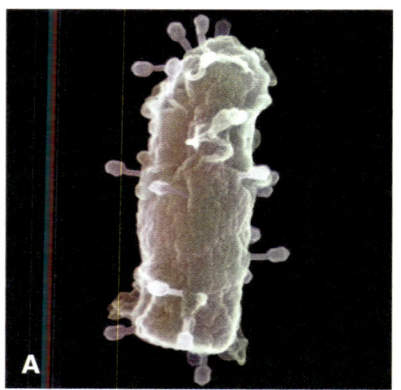
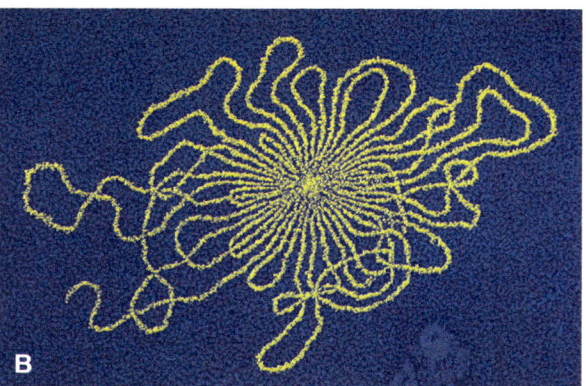

FIGURE 1-28 Bacteriophage infect bacterial cells (A) The T4 bacteriophage attaches to the outside wall of an E. coli cell by its tail and injects the phage DNA into the cell. The T4 DNA makes T4 proteins that take over the basic functions of the bacterial cell and make hundreds of T4 phage viruses, which kill the bacterium. (B) This figure shows the lambda (λ) DNA when released from the phage capsid in the lab. To infect a bacterium the lambda (λ) phage injects its double-stranded DNA genome into the cell. The λ DNA can either direct the production of many λ phage that kill and lyse the cell, or the λ DNA genome can follow a different pathway and insert into the bacterial chromosome so that the phage DNA will replicate with the chromosome DNA each time the cell divides. When the λ cos sequences on both ends of the λ DNA chromosome base pair together, the λ DNA forms a circular molecule that is necessary for part of the phage life cycle.

that are composed of specific base pair sequences. The DNA sequences cut by restriction enzymes are typically 4-8 bases in length (but can be longer) and are usually palindromes, the base sequences on both DNA strands are the same when the DNA is read in the 5' to 3' direction on both strands (Figure 1-29). For example, the HindIII restriction enzyme makes a staggered cut across both strands of the DNA helix between the 5'-A-A-3' bases in the sequence, 5'-AAGCTT-3', and the BamHI restriction enzyme cuts across the DNA helix between the 5'-G-A-3' bases in the sequence, 5'-GGATCC-3' (Figure 1-29). The BamHI cleavage site is the same sequence when either the top or the bottom DNA strand is read in the 5' to 3' direction: 5'-GGATCC-3'/5'-GGATCC-3' (Figure 1-29).

Restriction Enzymes can Produce "Sticky Ends" on DNA

The HindIII and EcoRI restriction enzymes both leave short 5' single-stranded ends on the DNA fragments, but the ends have different base sequences (Figure 1-29). Sticky ends on DNA fragments provide a technical advantage in DNA cloning experiments because the single-stranded bases at the ends of the fragments can base pair together, stabilizing the link between the two DNA fragments that are connected by the four base pairs at the ends of each DNA fragment (Figure 1-30). Another type of restriction enzymes such as PstI, cut the DNA helix to produce DNA fragments with specific single-stranded 3' ends (Figure 1-29). The DNA fragments with 3'

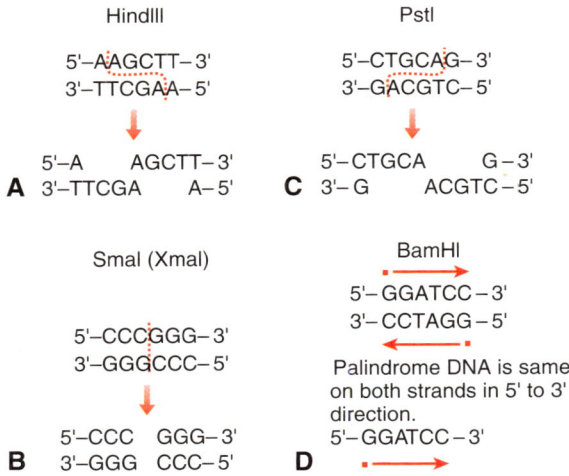

FIGURE 1-29 Restriction enzymes cut DNA at specific sequences Restriction enzymes cut DNA only at specific base pair sequences. (A, D) HindIII are examples of restriction enzymes that cut the DNA and leave short 5' single-stranded ends on the DNA fragments. (C) PstI leaves short 3' single-stranded ends. (B) SmaI cuts directly across the DNA helix and produces DNA fragments with blunt ends (no single-stranded extensions). (D) Most restriction enzyme cleavage sites are palindromes, both strands have the same sequence when read in the 5' to 3' direction: 5'-GGATCC-3' (BamHI).

single-stranded ends made by PstI cannot base pair to DNA fragments with either 5′ single-stranded ends or blunt ended fragments produced by restriction enzymes such as SmaI (Figure 1-29).

HindIII and a large number of restriction enzymes, including EcoRI and BamHI, cut DNA to produce fragments with short, 5′ single-stranded ends (Figure 1-30). The four bases in the single-stranded ends on these DNA fragments can base-pair with four complementary bases in the 5′ single-stranded ends of other DNA fragments generated using the same restriction enzyme (Figure 1-30).

How to Analyze DNA by Restriction Enzyme Digestion

In this example two DNA fragments, DNA #1 and DNA #2, were isolated from the *same region of the same chromosome* from two different people and cut separately with the HindIII restriction enzyme. This is an important point because we now know that the same chromosomes (for example chromosome #7) from different people will have almost the same DNA sequences. The products of cleavage by HindIII in the two reactions were analyzed by gel electrophoresis to determine the lengths of the DNA fragments produced (Figure 1-31). HindIII cleavage of DNA #1 produced two DNA fragments that are 3,000 bp and 2,000 bp in length, as determined by comparing the distance migrated in the gel by the unknown DNA bands with the migration

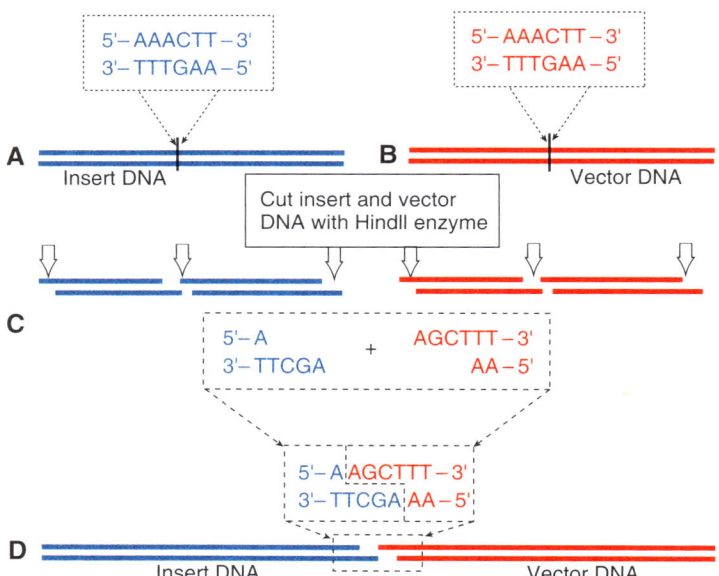

FIGURE 1-30 HindIII sticky ends are used to clone a DNA fragment (A, B) HindIII is used to cut both the vector DNA and the insert DNA to make complementary sticky ends on both the vector DNA and the DNA fragments. (C, D) Sticky ends are often an advantage in DNA cloning because the complementary single-stranded bases at the ends of the DNA fragments can base pair together to temporarily stabilize the link between the two DNA fragments.

of DNA fragments of known length (DNA Markers) (Figure 1-31). However, when the DNA #2 fragment was cut with HindIII, the enzyme reaction produced only one DNA product 5,000 bp in length, also determined by comparing the migration of the unknown DNA band in the gel with the migration of known DNA fragments in the gel (Figure 1-31).

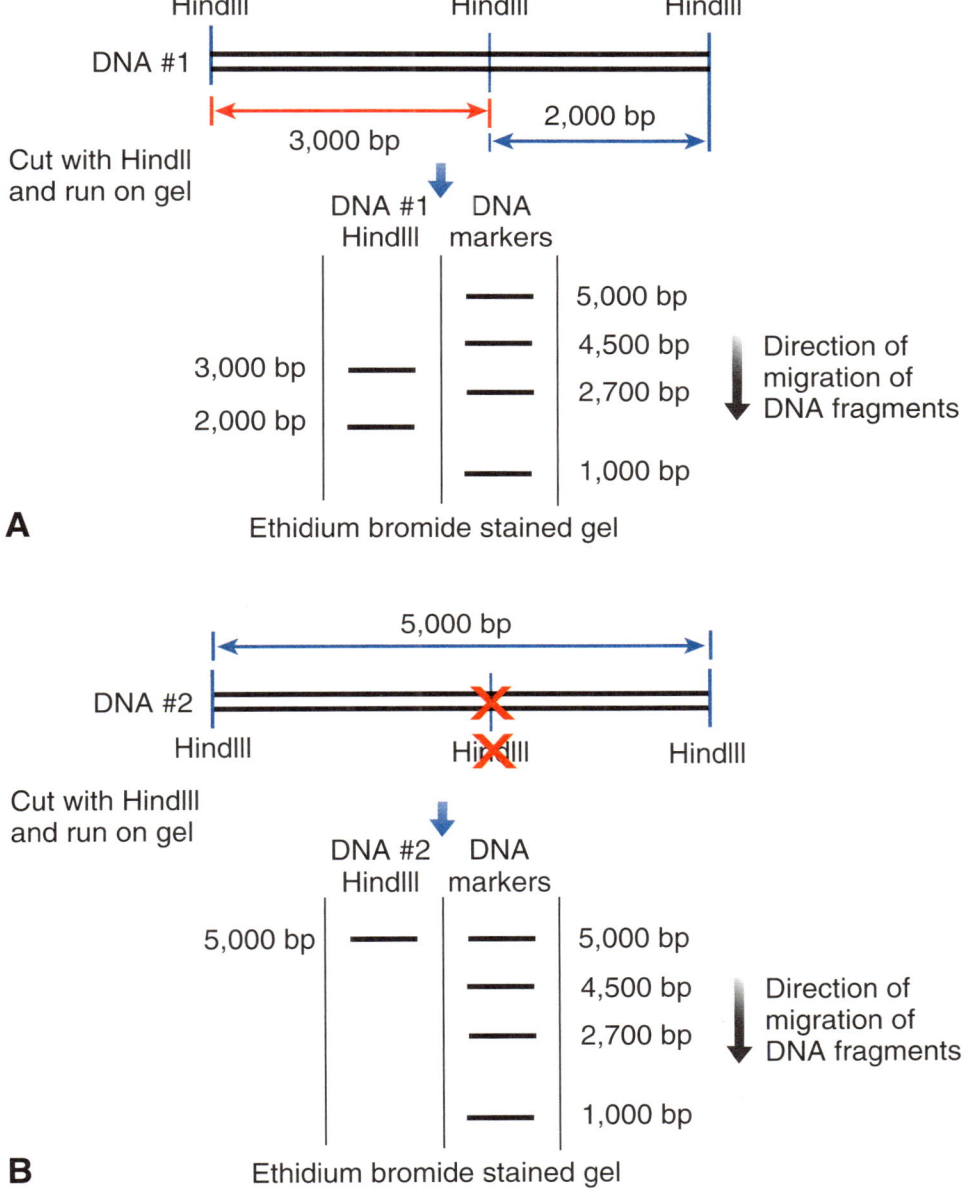

FIGURE 1-31 Restriction enzyme analysis of DNA In this experiment the two DNA fragments DNA #1 and DNA #2, were cut with the HindIII restriction enzyme. The products of the two HindIII reactions were analyzed separately by gel electrophoresis to determine the lengths of the DNA products. (A) HindIII cleavage of DNA #1 produced two DNA products. The lengths of these DNA products are determined by migration in the gel. (B) HindIII cleavage of DNA #2 produced no new DNA products indicating that there is no HindIII site in DNA #2.

Continued

The HindIII restriction enzyme will cut double-stranded DNA only at this base sequence:

5'– AAGCTT – 3'
3'– TTCGAA – 5'

Any mutation that changes even a single base pair in the HindIII cleavage site will prevent HindIII from cleaving the DNA.

HindIII will NOT cut:

C
5'– AAACTT – 3'   5'– AAGCGT – 3'
3'– TTTGAA – 5'   3'– TTCGCA – 5'

FIGURE 1-31 Restriction enzyme analysis of DNA (con't). (C) A mutation that alters a single base pair in the HindIII restriction enzyme cleavage sequence (5'AAGCTT3') will prevent the enzyme from recognizing and cutting the DNA.

Cutting the two DNA fragments (#1 and #2) with HindIII produced different DNA fragments that revealed important information about the characteristics of the two DNA fragments (#1 and #2). The results of the experiment show that DNA fragment #1 contains an intact HindIII site, since digestion of DNA fragment #1 with HindIII produces two DNA fragments (Figure 1-31). The two DNA fragments (#1 and #2) could produce different fragments when cut by HindIII if a mutation has altered the base sequence of the HindIII cleavage site in DNA #2 compared to DNA #1. Even a mutation as small as a single base pair change that alters the HindIII recognition sequence will prevent HindIII from cutting the DNA. The idea that the genomes from different individuals differ at many positions in the DNA sequence is the basis of the DNA fingerprinting technology used in forensic DNA testing (Unit 6). The techniques used to analyze single nucleotide polymorphisms (SNPs) in the human genome rely on the use of DNA probes designed to detect the SNP sites in human DNA (Unit 6). This is more efficient than earlier methods that rely on random mutations that by chance change the base sequence of a HindIII cleavage site (Figure 1-31).

Scientists working on the human genome found that the genome sequences of unrelated people are 99.9% identical (Human Genome Project, Unit 2). In practical terms this means that humans have almost identical DNA genomes, with some exceptions, including about 10 million single nucleotide polymorphisms that naturally occur in each genome (Unit 2). These tiny sequence differences among human genomes have become extremely important for many biomedical applications of this DNA technology.

DNA fingerprinting can be used to identify individual people because the technology is sensitive enough to detect the relatively rare DNA base pair differences among human genomes (Unit 2 and Unit 6). SNPs and short repeated DNA sequences that occur naturally in the human genome are analyzed in the DNA fingerprinting tests currently used in the United States by the legal system and law enforcement, including the FBI (Unit 6).

In cases where a DNA change (polymorphism) alters the cleavage sequence of a specific restriction enzyme, the change in sequence will prevent the enzyme from cutting at the altered DNA sites in the human genomes, which can be detected by Restriction Fragment Length Polymorphism (RFLP) technology (Unit 6).

Recombinant DNA Cloning and DNA Sequencing Strategies

Over the past three decades numerous DNA cloning and sequencing strategies have been developed to manipulate the cloned genes and DNA fragments and maximize the amount of information obtained from each DNA sequencing reaction (Figure 1-32). Before single-stranded DNA primers and Polymerase Chain Reaction (PCR) DNA amplification became routine, shotgun cloning strategies were essential to prepare recombinant DNA clones carrying the thousands of different DNA inserts needed to sequence an entire genome.

Recombinant DNA cloning strategies typically start by cutting long double-stranded DNA molecules with a restriction enzyme that cleaves the DNA at a specific base sequence that occurs many times in the genome. The restriction enzyme cuts across both backbones of the DNA helix, cleaving the long double-stranded chromosome DNA into millions of shorter double stranded DNA fragments that can be cloned into a DNA vector (Figure 1-32). Plasmid DNA vectors are short segments of circular DNA that contain specific marker genes to permit the vectors to be located in cells by genetic tests. The vectors are also modified to contain 'cloning enzyme cut sites' where the vector DNA can be cut with specific restriction enzymes to accommodate the insertion of the foreign DNA fragments into the vector DNA. The DNA vectors containing inserted fragments of foreign DNA can replicate inside the bacterial cells and produce thousands of identical copies of the cloned recombinant DNA vector in each cell. The DNA vectors containing fragments of foreign DNA are referred to as recombinant DNA clones.

The strategies used to make recombinant DNA molecules involve creating permanent recombinant DNA molecules by joining together DNA fragments from two or more different sources. For this reason, an approach that involves producing DNA fragments and cut vectors with sticky ends is an advantage for DNA cloning experiments. DNA fragments with sticky ends are easily made by digesting human genome DNA with EcoRI for example. These fragments can base pair with the complementary sticky ends on vector DNA molecules that were also generated by cutting with the EcoRI restriction enzyme. DNA fragments with sticky ends made by digestion with HindIII for example, can base pair with sticky DNA ends that were` generated by cutting with the HindIII restriction enzyme, but cannot base pair with sticky ends made by cutting the DNA with EcoRI or most other restriction enzymes except HindIII.

The temporary recombinant DNA molecules, which are vector DNA molecules base paired to insert DNA fragments, are stabilized only by the H-bonds between the base-paired sticky ends. These temporary connections are converted into permanent DNA helices by the DNA ligase enzymes, which seal the DNA backbone strands with covalent bonds (Unit 1, Unit 2). The foreign DNA fragments that are permanently ligated to the

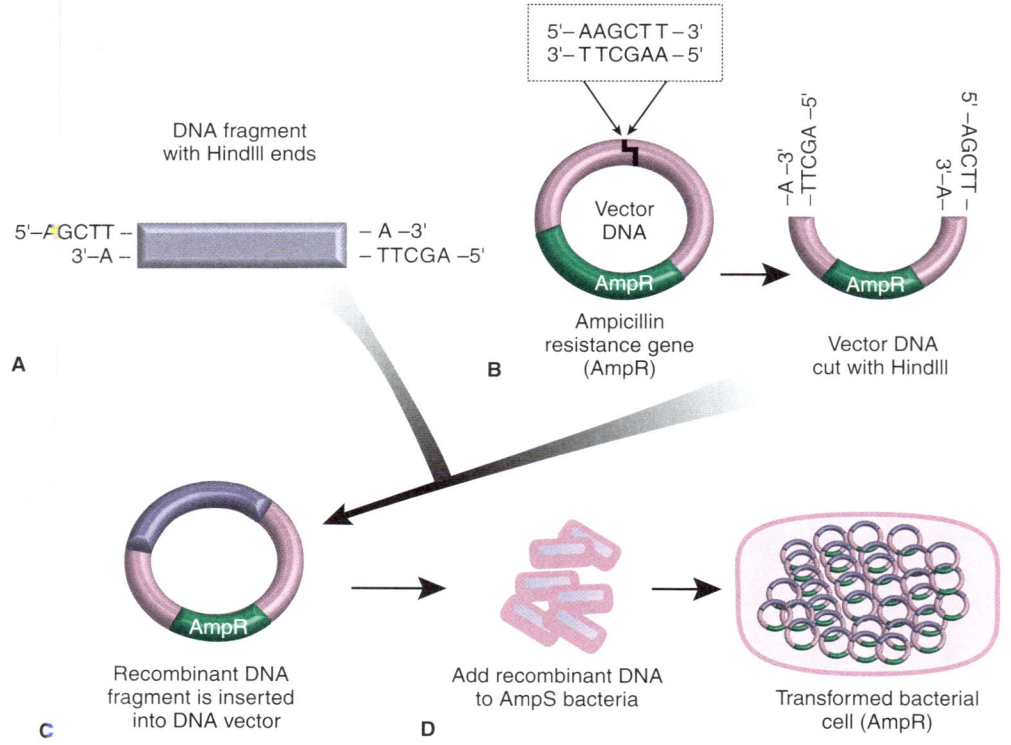

FIGURE 1-32 Recombinant DNA cloning strategy (A) The DNA to be cloned (blue) is cut with the HindIII restriction enzyme, which leaves 5' sticky ends on the DNA products (B) The vector DNA containing an Ampicilin resistant (AmpR) gene (green) is cut with HindIII. (C) The cut DNA fragments and the cut vector DNA are combined and connected together by a ligase enzyme to make recombinant molecules. (D) The ligated DNA molecules are used to transform ampicillin sensitive (AmpS) bacterial cells. Cells that have picked up a recombinant DNA molecule are ampicillin resistant (AmpR) and can grow on media containing ampicillin.

DNA vectors will replicate with the recombinant vector DNA, producing many copies of identical recombinant DNA clones in the bacterial cells (Figure 1-32).

Once the ligation reaction is finished the reaction mixture containing the recombinant DNA is introduced into bacterial cells by a method called transformation, which is followed by a genetic selection designed to identify the rare transformed cells that have picked up the recombinant DNA molecules. The bacterial cells that pick up DNA will acquire only one recombinant DNA molecule per cell. These transformed cells continue to divide and generate more identical cloned cells that carry copies of the recombinant vectors with DNA inserts. Most of the bacterial cells exposed to the DNA reaction mixture do not pick up any recombinant DNA and as a result they die during the antibiotic selection part of the experiment.

To identify the transformed recombinant bacterial cells among the untransformed cell population, the recombinant DNA vector carries a gene that confers a specific trait on the transformed cells that is not exhibited by the untransformed cells. In many transformation experiments the untransformed bacterial host cells are sensitive to an antibiotic such as ampicillin (AmpS), which means that the AmpS cells will die if exposed to ampicillin. However, the transformed cells are resistant to ampicillin (AmpR) because they have acquired recombinant DNA vectors carrying a gene that codes for resistance to ampicillin-(Figure 1-32). The transformed cells divide and grow into a pure, cloned population of identical ampicillin-resistant bacteria carrying the recombinant vector with a cloned DNA insert. The cloned DNA insert can then be analyzed by restriction enzyme digestion, subjected to DNA sequencing or inserted into vectors engineered to promote expression of the cloned DNA in certain cell types (Unit 2).

After the PCR method became routine to amplify DNA fragments in lab without the process of cloning and transformation, it became less common for researchers to grow large quantities of the cloned bacterial cells with the purpose of obtaining large quantities of cloned recombinant DNA. However, recombinant DNA cloning still has many important uses to manipulate foreign genes and other DNA fragments in the lab. The transformation protocols used to introduce recombinant DNA molecules into host cells vary depending on the transformation conditions required to induce a particular type of host cell to pick up and internalize the recombinant DNA molecules. Specialized recombinant vectors have been developed for hundreds of cloning applications, including protein expression vectors that are made to express the products of specific foreign genes in many different types of cells in addition to bacteria, including human cells. This approach continues to be an important way to study the proteins expressed in different types of cells in the human body (Unit 3).

Unit 1 Questions

1. Which of these DNA base pairs are allowed in the DNA helix?
   a. A-G and C-T
   b. T-G and C-A
   c. A-T and G-C
   d. All of the above
   e. None of the above

2. Genes are made up of DNA letters called bases that:
   a. Encode genetic information
   b. Encode the amino acid sequence of a protein
   c. Are copied into RNA strands carrying genetic information
   d. All of the above
   e. None of the above

3. The process of copying DNA into RNA in the cell is called:
   a. Translation
   b. Transcription
   c. Replication
   d. Xeroxing

4. The DNA helix contains how many backbone DNA strands?
   a. 2 strands
   b. 4 strands
   c. 6 strands
   d. 8 strands

5. Which of these components are present in eukaryotic cells but not in prokaryotic cells?
   a. Ribosomes
   b. Cell membranes
   c. Chromosomes
   d. Nuclei

6. Which of these choices shows the order of the gene expression pathway known as "Central Dogma"?
   a. DNA to RNA to protein
   b. RNA to DNA to protein
   c. Protein to RNA to DNA
   d. DNA to protein to RNA

7. Proteins are made up of various combinations of:
   a. 20 different genes
   b. 20 different amino acids
   c. 20 different cells
   d. 20 different base pairs

8. Which type of chemical bonds hold the DNA bases together in base pairs?
   a. Hydrophobic bonds
   b. Covalent bonds
   c. Hydrogen bonds (H-bonds)
   d. Hydrophonic bonds

9. Which of these is the strongest type of chemical bond?
   a. A hydrophobic bond
   b. A covalent bond
   c. A hydrogen (H-bond)
   d. A hydrophonic bond

10. The hemoglobin complex in the red blood cells is made up of:
    a. 4 separate proteins
    b. 4 different amino acids
    c. 4 different base pairs
    d. 4 small RNAs and proteins

UNIT 2

DNA Replication and Human Genome Sequencing

Introducing Human Chromosomes

A human karyotype is a display of the condensed human chromosomes taken from the nucleus of a human cell before cell division. Recall that most cells in the human body contain 23 pairs of chromosomes or a total of 46 chromatids. In the karyotype shown in Figure 2-1, the 46 chromosomes have duplicated in preparation for cell division for a total of 92 DNA molecules. Human

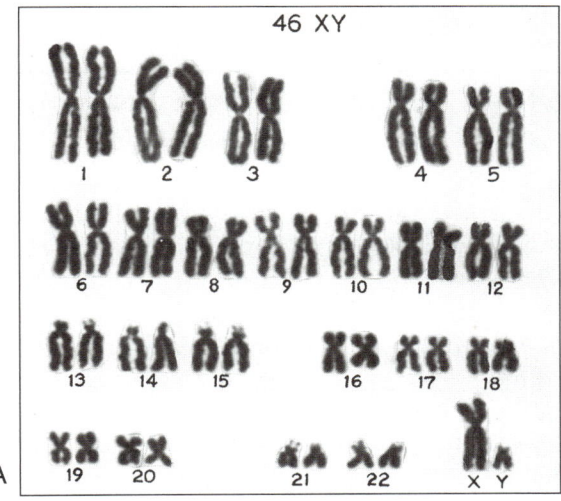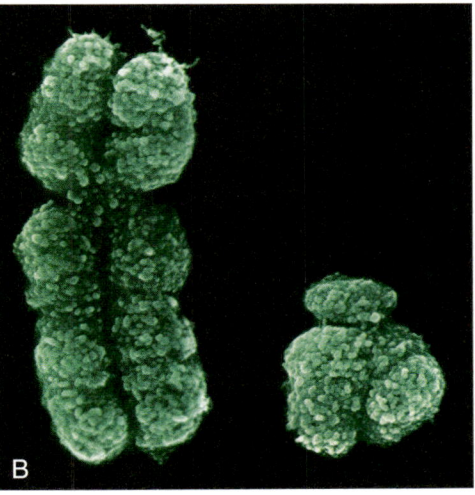

FIGURE 2-1 Replicated human chromosomes are carried in 46 chromosome pairs (A) This karyotype displays 46 chromosomes that have duplicated in preparation for cell division; each chromosome is represented by duplicated chromosome pairs for a total of 92 DNA helices. The human chromosomes are arranged by overall length, the length of the chromosome arms and the position of the centromere relative to the ends of the chromosome as well as banding pattern (not shown here) and are numbered from 1 to 22. The human sex chromosomes are the 23rd chromosome pair: X, Y (male) and X, X (female). (B) This is a scanning electron microscope (SEM) picture showing the rough surface topology of the human X (left) and Y (right) sex chromosomes.

chromosomes are arranged in a karyotype display according to the overall length of each chromosome, starting with the longest human chromosome, chromosome 1, and arranged in decreasing length to the shortest, chromosome 22, followed by the X and Y sex chromosomes.

Duplicated chromosomes contain two 'sister' chromatids that appear to be attached to each other at the centromere, which is morphologically defined as the pinched region that divides a chromosome into the long and short chromosome arms. The chromosome DNA helix molecule passes through the centromere region of the chromosome unbroken. The position of the centromere relative to the ends of the chromosome is a consistent feature of chromosome structure that is one of the key characteristics used to identify individual human chromosomes. A karyotype reveals that chromosomes 13, 14 and 15 are examples that have centromeres located very close to one end of the chromosome (telocentric) (Figure 2-1). When stained with cytological dyes, the condensed human chromosome arms exhibit unique patterns of light and dark bands that are specific to each chromosome.

> **KEY CONCEPT**
> In addition to the banding patterns observed on stained chromosomes, each human chromosome has different physical characteristics that make it possible to identify human chromosomes by visual karyotype analysis. These structural features include overall chromosome length, the lengths of the long and short chromosomes arms and the position of the centromere relative to the chromosome ends.

A DNA Helix Extends from End to End in a Eukaryotic Chromosome

Scientists now have an accurate picture of the overall arrangement of human genes based on the fact that each human chromosome contains one double-stranded DNA molecule extending from one end to the other end of the same chromosome. The results of the human genome DNA sequence provided a more detailed understanding of how genes are organized on human chromosomes and revealed surprising information about the structure and function of noncoding DNA sequences in human chromosomes.

It is very difficult to visualize how such an extremely long DNA molecule can be efficiently packaged into a DNA-protein chromatin fiber in each chromosome without the DNA strands becoming hopelessly tangled. Scientists still have much to learn to completely understand the hierarchy of the DNA–protein interactions that package human genes into dynamic chromosome structures. However, we do know that the process starts with the nucleosome, a DNA-protein complex that represents the fundamental repeating unit of chromatin fibers in all eukaryotic cells (Unit 6). A scanning electron micrograph (SEM) shows that the condensed human sex chromosomes appear to have a very bumpy surface topology because of the densely packed DNA-protein chromatin fibers that make up the condensed chromosome (Figure 2-1).

The human genome is the set of hereditary DNA instructions needed to build and maintain a person, which is written in the DNA language and carried on chromosomes in human cells. Figuring out the DNA sequences of the human genes was a primary goal of the scientists who worked on the Human Genome Project and sequenced the entire human genome. This accomplishment enabled researchers to better understand the functions of all the proteins in the human body, including the proteins that control development and proteins that cause genetic diseases (Units 4 and 5).

DNA Replication and DNA Sequencing Reactions are very Similar

Whether performed by hand in the lab or automatically by a DNA sequencing machine, the biochemical reactions used to determine the sequence of DNA in the lab are very similar to the biochemical reactions that replicate chromosome DNA in the cells. The DNA replication reactions and DNA sequencing reactions also involve similar components, making it helpful to learn about DNA replication and DNA sequencing at the same time. Both processes involve copying single-stranded DNA templates into double stranded DNA, and require single-stranded DNA primers with a 3'OH group to initiate DNA synthesis (Figure 2-2). The strands of the DNA helix must be able to separate into a localized single stranded region of the DNA helix. The reverse process of DNA annealing occurs when the two complementary DNA strands base-pair to each other to form a DNA helix. The processes of

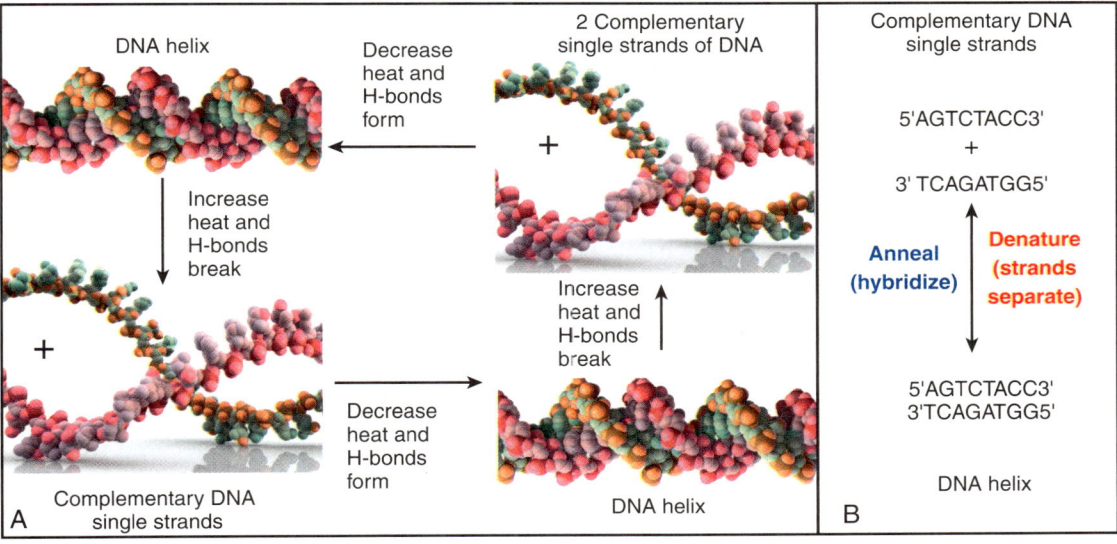

FIGURE 2-2 DNA strand separation and annealing is reversible (A) Increased heat breaks the H-bonds between the base pairs in the DNA helix, causing the DNA strands in the helix to separate (denature) and become two complementary single strands of DNA. Decreasing the temperature allows the complementary DNA strands to base pair (anneal or hybridize) together to form a DNA helix. (B) The formation of a DNA helix from two complementary DNA strands is a reversible process as shown here with DNA letters.

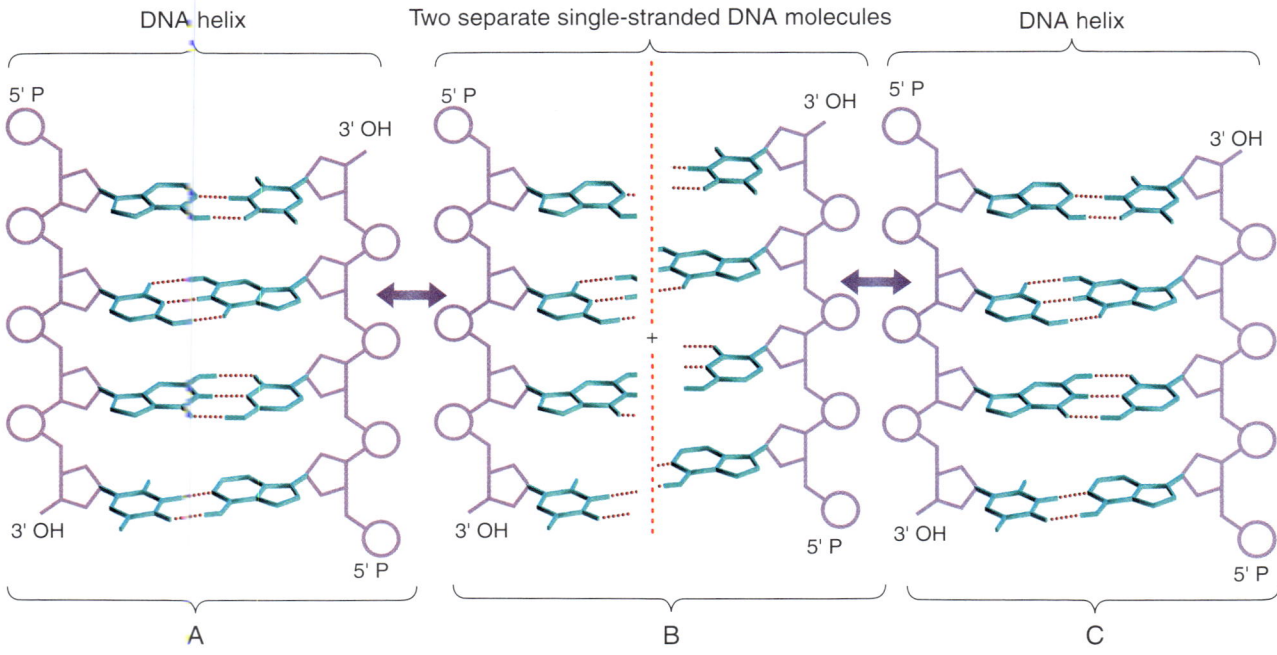

FIGURE 2-3 A look inside the DNA helix during strand separation (A) The chemical units in the center of the DNA helix molecule are shown, including the H-bonds between the base pairs. The sugar-phosphate backbones (purple) and the base pairs (cyan) are shown. (B) When the H-bonds break, the two complementary single DNA strands separate. However, the covalent bonds do not break and the DNA single strands are quite stable. (C) The two complementary DNA strands follow base pairing rules and the DNA strands base pair together by forming the appropriate H-bonds between the complementary bases (A-T and G-C).

DNA strand separation and annealing are fundamental, reversible functions of the DNA helix structure that are essential for many cellular processes, including gene expression and DNA replication (Figure 2-3) (Unit 1).

Comparing the DNA replication process as it occurs in cells with the DNA sequencing reactions conducted in the lab and considering the similarities and differences between them provides a better understanding of these biochemical reactions. The DNA replication and the DNA sequencing reactions are catalyzed by proteins called DNA polymerase enzymes, which copy the DNA template strand into a complementary DNA strand. Different types of DNA polymerases perform different DNA synthesis reactions in the cell; for example, one type of DNA polymerase enzyme is necessary to repair damaged DNA, while a different DNA polymerase enzyme replicates the chromosome DNA. A special type of DNA polymerase is now used in most DNA sequencing reactions that also involve the PCR to amplify the template DNA.

In the cell, DNA replication begins at specific regions in the chromosome DNA where the DNA helix becomes temporarily single-stranded to allow the DNA polymerase enzyme access to the DNA template. To start DNA replication, the DNA polymerase enzyme requires a short primer that base-pairs to the DNA template (Figure 2-4). All DNA polymerases require the free

3′ OH group on the end of a base-paired primer to begin the DNA synthesis reaction. The DNA polymerase enzyme then extends the new DNA strand by adding a new base (dNTP) to the free 3′ OH group at the end of the primer (Figure 2-4). To be incorporated into the growing DNA strand, the new DNA base must be able to base pair with the base in the template DNA. The DNA polymerase continues to copy the template DNA strand into a new, complementary DNA strand, producing a DNA helix. At the completion of DNA replication, the 'old' double-stranded DNA helix has made two new identical double-stranded DNA helices. Each new DNA helix contains one 'new' DNA strand and one 'old' DNA strand.

The accuracy of chromosome DNA replication is very important for the function of cells and for the fidelity of the genes passed to future generations. The importance of the accuracy of DNA replication is emphasized by the fact

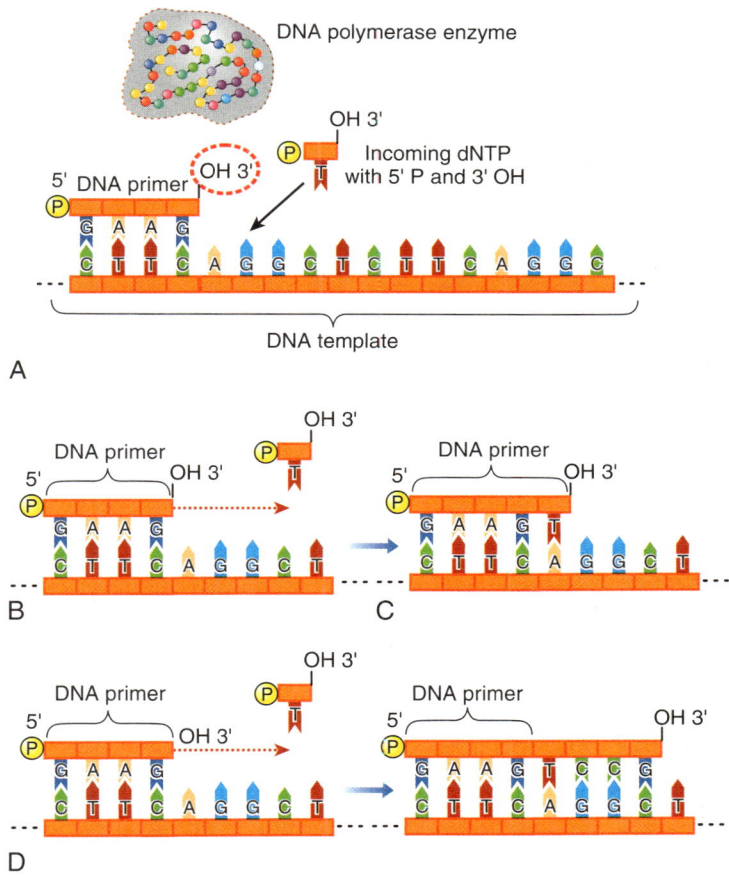

FIGURE 2-4 DNA polymerase needs a primer with a 3′OH group to start replication
(A) The DNA helix strands separate to permit DNA polymerase access to a single-stranded region of the DNA template. To initiate DNA replication the DNA polymerase requires a short primer with a 3′ OH group base paired to the template DNA strand. (B) The DNA polymerase requires a primer (with a free 3′ OH group) to start DNA replication. (C) The 3′ OH makes a covalent bond with the incoming dTTP, which is incorporated into the new DNA strand. (D) Once DNA replication has started, the DNA polymerase continues to the end of the template strand.

that the eukaryotic DNA polymerase enzyme has a "proofreading" function that can locate and correct replication errors. As the DNA polymerase enzyme copies the DNA template, it simultaneously checks the base sequence of the newly made DNA strand against the sequence of the old DNA strand, searching for any mismatched base pairs. When this multifunctional enzyme finds a base-pairing mistake, it deletes a short section of DNA bases on one strand to remove the mismatched base. The DNA polymerase enzyme then initiates DNA synthesis at the 3′ OH end of the gap in the double-stranded DNA and copies the strand to repair the double-stranded DNA helix (Figure 2-5).

In addition to the DNA polymerase enzymes many proteins are involved in DNA replication (Figure 2-6). Some of these proteins help the DNA polymerase to function more efficiently. For example, as the DNA polymerase enzyme copies the DNA template, the enzyme tends to fall off of the template DNA, making the DNA synthesis reaction very inefficient. Cells convert DNA replication into a more efficient process by making a sliding clamp protein called PCNA, which assembles around the DNA helix (Figure 2-6). As the DNA polymerase moves along the DNA helix, it binds to the clamp proteins, which help to secure the enzyme to the DNA template.

The Action at The DNA Replication Forks

DNA polymerase enzymes initiate DNA synthesis at many locations on the chromosome DNA called origins of DNA replication (ori). Once replication has begun, the two replication forks are bidirectional and travel away from the origin on the same DNA helix (Figure 2-7). Specialized proteins move along the helix in front of the DNA polymerase enzyme to unwind the DNA strands, allowing the DNA helix to create a temporary single-stranded region for replication to take place. Replication forks contain the DNA polymerase enzymes, accessory proteins and other components needed to synthesize both

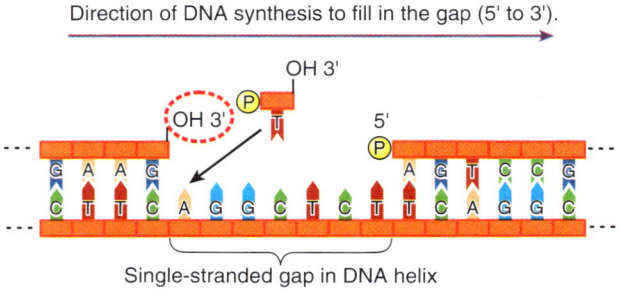

FIGURE 2-5 DNA polymerase proofreads new DNA and repairs replication mistakes
Eukaryotic DNA polymerase has a "proofreading" function that permits the enzyme to locate and repair replication errors. The DNA polymerase enzyme checks the base sequence of the newly made DNA strand against the sequence of the old DNA strand, searching for mismatched base pairs. The enzyme removes the strand of the DNA helix containing the mistake and repairs the gap by initiating DNA synthesis at the 3′ OH end of the gap, synthesizing the base pairs needed to repair the DNA into a double-stranded DNA helix.

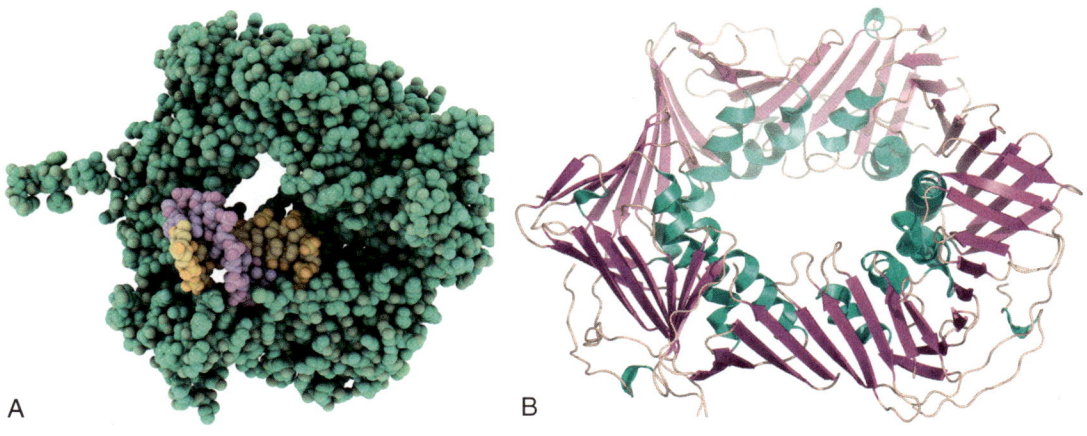

FIGURE 2-6 DNA polymerase enzyme replicates DNA in the cell (A) DNA polymerase enzyme (green) catalyzes the DNA synthesis reactions in the cell, copying the DNA template strand (yellow) into a new, complementary DNA strand (purple). (B) Three proliferating cell nuclear antigen (PCNA) proteins with domains defined by beta sheets (pink) and alpha helices (blue) form a complex that secures the moving DNA polymerase enzyme to the DNA template.

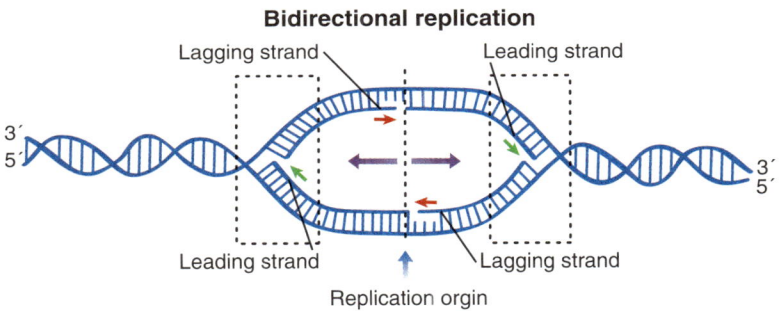

FIGURE 2-7 Bidirectional DNA replication forks The DNA replication forks begin at a specific site on the DNA helix called the origin (Ori). The forks move on the DNA helix in both directions away from the Origin. As the replication forks move away from the origin, each fork carries polymerase enzymes and many accessory proteins that help to replicate the old DNA helix, and simultaneously generate the two new DNA helices. Each new DNA helix contains one new DNA strand base-paired to one old DNA strand.

strands of DNA. The DNA strands are called the leading and the lagging strands because somewhat different replication mechanisms are required to copy the DNA on the leading and lagging strands (Figure 2-7). The replication of the two different DNA strands must be closely coordinated as the replication forks move apart.

When initiating DNA synthesis, the DNA polymerase adds the first dNTP to the 3′ OH group on the DNA primer and makes a covalent bond to seal the backbone of the new DNA strand. On the leading DNA strand, the DNA polymerase enzyme requires one DNA primer to initiate the 5' to 3' synthesis of a long new DNA strand. However, to copy the lagging strand DNA in the required direction, the lagging strand makes a loop as the helix exits and re-enters the complex of proteins at the replication fork. In addition, the DNA polymerase enzyme must initiate DNA synthesis at multiple sites on the template strand and requires multiple primers. Surprisingly, the primers used to initiate DNA replication on the lagging strand are actually short RNA strands made by the RNA primase enzyme in the cell. Like the DNA primers on the leading strand, the RNA primers base-pair to the lagging strand template and DNA polymerase uses the 3' OH groups on the RNA primers to initiate synthesis in the 5' to 3' direction.

On the lagging strand the RNA primers are extended by DNA polymerase into short DNA copies of the lagging strand called Okazaki fragments (Figure 2-8). The RNA primer sequences are then removed by the RNase H enzyme, which digests just the RNA bases in the RNA-DNA heteroduplex molecule, a DNA strand that is base paired to a complementary RNA sequence. This leaves a short single-stranded gap in the DNA helix, which is repaired by DNA polymerase. At the end of DNA replication, the chromosome DNA in the eucaryotic cell does not contain any RNA sequences remaining from the RNA primers used during replication.

The DNA replication fork is a carefully choreographed dance of moving, looping DNA strands and numerous moving components that replicate the old DNA helix and generate two new DNA helices (Figure 2-8). The activities on the DNA strands at each replication fork are closely coordinated in time

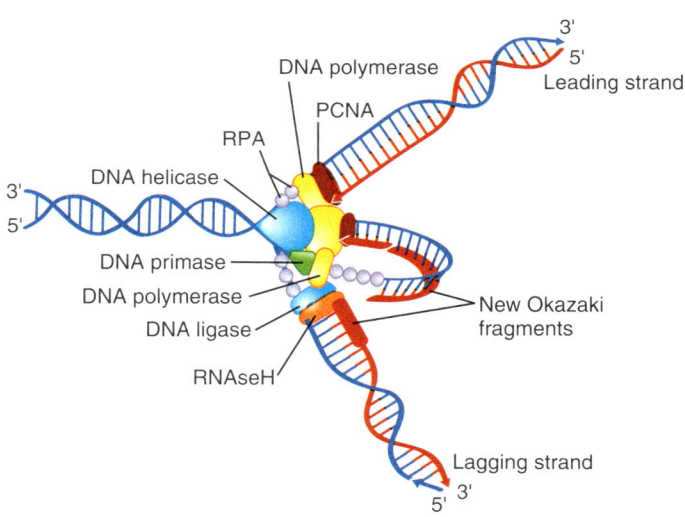

FIGURE 2-8 Molecular choreography at a DNA replication fork The replication fork is an amazing choreographed array of moving DNA strands and proteins that simultaneously replicate the old DNA helix and generate two new DNA helices. This is accomplished efficiently though two similar DNA replication mechanisms operating on the two different DNA strands, leading and lagging, so that both strands are copied in the 5' to 3' direction only. The protein activities on the DNA strands at each replication fork are closely coordinated in time and space so that the leading and lagging DNA strands can be copied as the replication fork moves along the DNA helix.

and space to simultaneously replicate the leading and lagging DNA strands as each replication fork moves along the DNA helix (Figure 2-8). This amazing process occurs every time a cell prepares to undergo cell division and produce two new cells, each with a perfect copy of the chromosome DNA.

DNA Sequencing Reactions Determine the Order of the DNA Bases

In contrast to the complexities of chromosome DNA replication in living cells, the biochemical reactions that are used to sequence DNA are much less complicated. Thirty years ago the early methods of sequencing DNA were very time consuming and required the use of hazardous chemicals to cut the DNA strands at specific sequences including exposure to radioactive isotopes used to radioactively label the DNA sequencing products for detection. Current DNA sequencing methods use DNA polymerase enzymes because scientists discovered that when DNA polymerase was isolated from cells it could replicate DNA molecules in the lab, even in the absence of accessory proteins.

Today the most commonly used DNA sequencing protocol is based on the Sanger chain termination (dideoxy-chain termination) method developed by Dr. Fred Sanger, winner of two Nobel Prizes! The Sanger chain termination method is currently the most common way to sequence DNA for research purposes and has also been adapted for use in automated DNA sequencing machines. The Sanger chain termination method is well named, because the major feature involves the controlled termination of growing DNA strands (chains) that in the end will reveal the base sequence of the template DNA.

Fred Sanger knew that the chemical structures of the dideoxynucleotides (ddNTPs: ddATP, ddCTP, ddGTP, and ddTTP) are identical to the chemical structures of the normal deoxy-nucleotides (dNTPs) used in DNA replication in cells, except that the dideoxy-nucleotides (ddNTPs) do not have a 3′ OH group (hence the name, dideoxy). Sanger used this information to design a totally novel method of sequencing DNA that relies on the occasional incorporation of a terminator dideoxy-nucleotide (ddNTP) into the growing DNA strand in place of a normal deoxy-nucleotide (dNTP). The 3′ OH group on the end of the primer that is central to DNA replication is also key to the molecular mechanism behind the chain termination DNA sequencing method (Figure 2-9). When a dNTP is incorporated into the growing strand, the 3′ OH group is used by DNA polymerase to continue to copy the template DNA. However, if a ddNTP is incorporated into the growing strand, the lack of a 3′ OH group will prevent further DNA synthesis and the DNA strand (chain) must terminate (Figure 2-9).

> **KEY CONCEPT**
> The newly synthesized DNA strand grows only in the 5′ to 3′ direction because the incoming dNTP must make a covalent bond with the 3′ OH group on the end of the growing DNA strand and base pair to the base in the template strand. The dideoxy-nucleotides lack free 3′ OH groups, causing controlled strand termination in the Sanger chain termination DNA sequencing reactions.

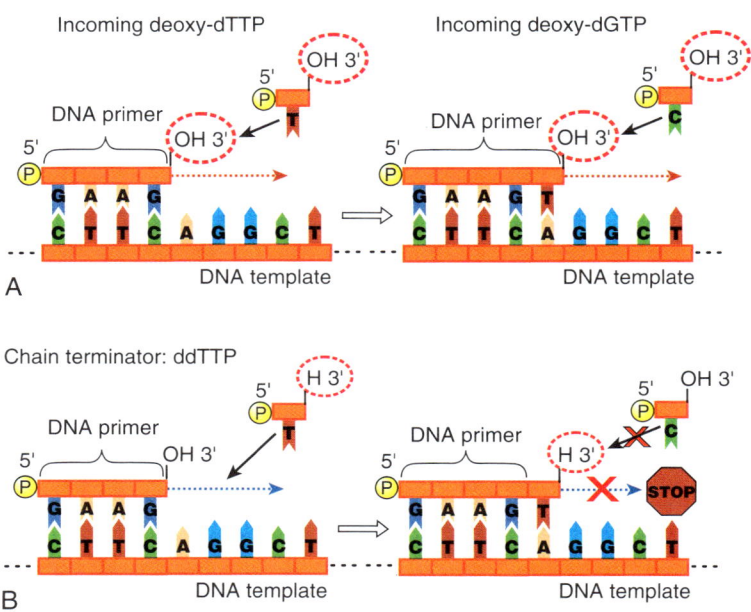

FIGURE 2-9 Chain terminators (ddNTPs) block DNA replication (A) Normally DNA polymerase adds a deoxy-nucleotide such as dTTP to the 3'OH end of the primer, which makes a covalent bond with the 5' phosphate and adds the new base to the new strand. (B) When an incoming dideoxy-GTP (ddGTP) is added to the 3' end of the primer, it replaces the normal free 3'OH group with a 3' H group. Without the 3'OH on the primer, the DNA polymerase stops DNA replication because the next base cannot be added to the growing strand. This chain termination mechanism was adapted for the Sanger Chain Termination DNA Sequencing method.

Sanger Chain Termination DNA Sequencing Reactions

To understand how the Sanger chain termination method of DNA sequencing works, it is important to examine the actual steps of the DNA sequencing method (Figure 2-10, Figure 2-11).

Step 1. The DNA sequencing reaction begins in one tube. The reagents needed for a set of four (A, G, C, and T) DNA sequencing reactions are mixed together in a tiny amount of liquid buffer in a small plastic microfuge tube as follows (Figure 2-10):

- Template DNA (single-stranded or double-stranded DNA)
- Deoxy-nucleotides (dNTPs: dATPs, dTTPs, dGTPs, dCTPs)
- DNA polymerase enzyme
- Primer DNA (single-stranded DNA 10–20 bases in length that base-pairs to one site in the template DNA)

Step 2. Next the reaction mixture is heated to separate double-stranded DNA molecules into single strands, ensuring that the template and primer are single-stranded. Then the temperature of the reaction is slowly decreased, allowing the single-stranded DNA primers to base-pair (anneal) to a unique complementary sequence on each DNA template strand. The DNA polymerase enzymes use the 3' OH groups on the annealed primers to initiate the DNA synthesis reactions on the DNA template strands (Figure 2-10).

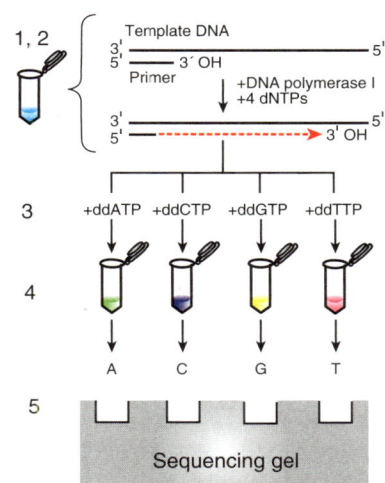

FIGURE 2-10 Flowchart of the chain termination DNA sequencing method The protocol for DNA sequencing a DNA molecule starts by combining reagents in a plastic tube. Step 1: The template/primer mix contains liquid buffer, dGTP, dTTP, dCTP, dATP, DNA polymerase and primer DNA. Step 2: DNA primers base pair to the target site on the single-stranded template DNA molecules. Step 3: The template/primer mix is divided into four tubes labeled A, C, G, and T. Each tube contains a limiting amount of one dideoxynucleotide, either ddATP, ddCTP, ddGTP, ddTTP, as labeled on the tube. Step 4: DNA polymerase copies the DNA template to produce a set of nested DNA strands that all start at the same 5′ base, but have 3′ ends terminating in a dideoxy-base. Step 5: The DNA products made by the four DNA sequencing reactions are analyzed using a denaturing polyacrylamide gel that resolves single-stranded DNA molecules differing by only a single base in length. This method has been adapted to automated DNA sequencing machines.

Step 3. To successfully determine any DNA sequence it is necessary to determine the order of all four bases. The reaction mix containing the annealed primers, enzyme and dNTPs is divided equally into four separate microfuge tubes, labeled A, C, G, and T (Figure 2-10, Figure 2-11). Each tube has been pre-loaded with a small amount of *one* of the four different dideoxy-nucleotides, A (ddATP), C (ddCTP), G (ddGTP), and T (ddTTP) according to the label on the tube.

The chain termination DNA sequencing reactions take place in all four tubes (A, G, C, and T) using the same primer-template DNA (Figure 2-10). A close-up shows that the "T" tube contains the "T" sequencing mix, the ddTTPs, and the DNA polymerase. When the enzyme needs to incorporate a T base into the new DNA strand to base-pair with an A base in the template, the enzyme will most likely incorporate a "normal" deoxy-nucleotide (dTTP) into the new DNA strand because there are many more deoxy-nucleotides (dTTPs) in the tube than there are dideoxy-nucleotides (ddTTPs). When the DNA polymerase enzyme incorporates a dTTP into the growing DNA chain, DNA synthesis continues. However, on the rare occasion that the DNA polymerase incorporates a ddTTP into the growing DNA strand, the lack of a 3′ OH group will permanently terminate that DNA strand (Figure 2-10, Figure 2-11).

As a result the DNA sequencing reactions in the "T" tube produce only DNA strands that terminate with various ddT ends base-paired to different A bases in the DNA template (Figure 2-10). At the end of the experiment,

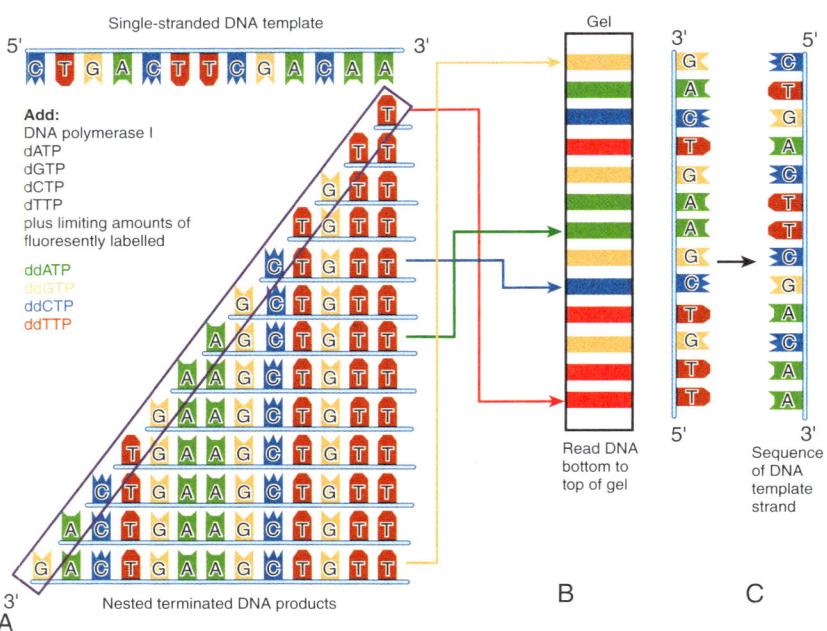

FIGURE 2-11 ddNTPs stop DNA strand growth at specific sites on DNA template (A) Taken together the four different DNA sequencing reactions produce a set of nested DNA strands that all start at the same 5′ base, with the 3′ ends of the strands terminating with the dideoxy-base indicated by the label on the tube. (B) The DNA products of the four sequencing reactions are resolved on a sequencing gel. Each colored band represents a specific DNA strand in the set of nested strands produced by the sequencing reactions. The shortest DNA strands are located near the bottom of the gel and the longer DNA strands are located closer to the top of the gel. (C) The DNA sequence is deduced from the sequencing gel by reading the DNA bands from the bottom to the top of the film.

these prematurely terminated DNA strands will reveal the sequence (order) of the A bases in the template. The long DNA strands that do not terminate do not contribute to the DNA sequence information and are not detected.

Different DNA sequencing reactions take place in the other tubes, labeled G, C, and A. For example, the "G" tube contains the "G" sequencing mix, including the ddGTPs. The DNA polymerase copies the template in the "G" tube, and when it needs to incorporate a G base into the new DNA strand to base-pair with a C base in the template, the enzyme will most likely incorporate a "normal" deoxy-nucleotide (dGTP) into the new DNA strand because there are more deoxy-nucleotides (dGTPs) in the tube than there are dideoxy-nucleotides (ddGTPs). When the DNA polymerase enzyme incorporates a dGTP into the growing DNA chain, DNA synthesis continues. However, on the rare occasions that the DNA polymerase incorporates a ddGTP into the growing DNA strand, the lack of a 3′ OH group will permanently terminate that DNA strand (Figure 2-10).

The DNA sequencing reactions in the "G" tube produce only DNA strands that terminate with the various ddG ends base-paired to the different C bases in the DNA template. At the end of the experiment, these prematurely terminated DNA strands will reveal the sequence of the C bases in the template.

As a result, the DNA sequencing reactions in the G tube produce only DNA strands that terminate with the various ddG ends base-paired to the different C bases in the DNA template. Similar reactions occur in the C and A tubes as described for the T and G tubes (Figure 2-10, Figure 2-11).

Step 4. Analyzing the products of the DNA sequencing reactions will reveal the sequence of the DNA template. The key to the Sanger chain termination sequencing method lies in the different DNA strands produced in the four different reaction tubes (Figure 2-10, Figure 2-11):

- In the T tube, the only terminated DNA strands will have ddT ends that are base-paired to the A bases in the DNA template.
- In the A tube, the only terminated DNA strands will have ddA ends that are base-paired to the T bases in the DNA template.
- In the C tube, the only terminated DNA strands will have ddC ends that are base-paired to the G bases in the DNA template.
- In the G tube, the only terminated DNA strands will have ddG ends that are base-paired to the C bases in the DNA template.

At the finish of the DNA sequencing reactions, each tube contains a collection of DNA strands of different lengths. Taken together the DNA strands in all four reaction tubes have the same 5' ends with different nested 3' ends located at every position in the base sequence (Figure 2-10, Figure 2-11). The DNA strands produced by the sequencing reactions are separated by length usually by electrophoresis on a special polyacrylamide gel that can resolve DNA strands that differ in length by a single base. The four separate products of the DNA sequencing reactions, A, G, C, and T, are loaded in four adjacent wells on the gel so that the DNA bands in the four adjacent lanes can be directly compared at the end of the experiment. The locations of the DNA strands are visualized by exposing the gel to an X-ray film (Figure 2-12).

DNA migrates through polyacrylamide gels much as described previously for agarose gels (Unit 1), the shortest DNA strands migrate more quickly through the gel than the longer DNA strands. At the end of the electrophoresis the DNA strands are distributed throughout the length of the gel, with the longer DNA strands closest to the wells near the top of the gel and the shorter DNA strands closest to the bottom of the gel.

Comparing the positions of the dark bands relative to each other on the final X-ray film will identify the lengths of the DNA strands produced by the four DNA sequencing reactions. The DNA sequence is read directly from the X-ray film, starting with the dark DNA band closest to the bottom of the gel (dotted red box) and is: 5'-TACAAAATTGTCACATTTGG-3' (Figure 2-12). Although the DNA sequence can be read from the exposed x-ray film by eye, long DNA sequences are routinely read by the computer to reduce human error. Automated DNA sequencing machines have the advantage that the results of the DNA sequencing reactions are sent directly to a computer to compile and analyze the DNA sequence information.

The earliest strategies used to sequence large DNA genomes entailed cloning the DNA of interest into vectors and growing the recombinant clones in bacteria to harvest enough DNA to perform DNA sequencing reactions (Unit 1). These approaches changed when it became routine for researchers to purchase commercially available custom designed DNA primers. Soon it

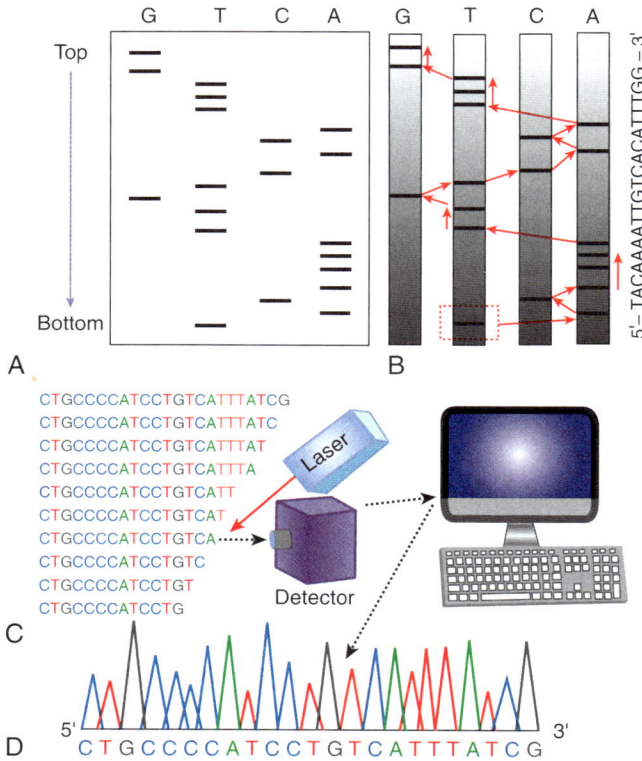

FIGURE 2-12 The DNA sequence is read from DNA bands or peaks (A) The dark bands on the X-ray film represent the DNA strands of different lengths produced by the DNA sequencing reactions. The DNA sequencing reactions were analyzed by loading the reaction products in four adjacent wells (GTCA) on the same gel. (B) The DNA sequence is read from the gel starting at the bottom of the exposed x-ray film and following the red arrows: 5'-TACAAAATTGTCACATTTGG-3'. (C) Automated DNA sequencing machines use chain termination reactions to analyze DNA sequence. The single-stranded DNA products are analyzed by a laser detector and the sequence is reported to the computer, in this case: 5'-CTGCCCCATCCTGTCATTTATCG-3' (D) The automated DNA sequencing machines report the sequence as a string of colored peaks that represent the linear order of the four different DNA bases in the DNA sequenced.

was feasible for scientists to make large amounts of inexpensive DNA primers for use in the Sanger chain termination DNA sequencing reactions.

A big advantage of the Sanger chain termination method of DNA sequencing is that the sequencing reactions require very little template DNA compared to the methods of DNA sequencing used 30 years ago. Also the Sanger method is easily combined with PCR to amplify the available template DNA and provide a sufficient amount of template DNA for the chain termination reactions (see PCR below). Although it was no longer necessary to clone the DNA fragments to be sequenced into vectors and grow them in bacteria in order to obtain sufficient DNA, recombinant DNA cloning continues to be a very important approach used in many applications such as protein expression experiments. In addition, foreign DNA that has been cloned into a vector can be easily sequenced using the chain termination method. The vector DNA primer base-pairs to a unique single site on the vector DNA located immediately adjacent to the inserted cloned DNA. In the sequencing reactions, the

3' OH group on the vector primer is used by the DNA polymerase enzyme to extend the primer DNA in the chain termination sequencing reactions.

Each set of four sequencing reactions can determine the sequence of only about 300 bases to <1000 bases (in automated sequencing machines), because the method is limited by the ability to physically separate DNA strands that differ in length by only a single base. It is much easier to separate short DNA strands that differ in length by only a single base (100 bases from 101 bases) than it is to separate longer DNA strands that differ in length by only a single base (1000 bases from 1001 bases). Now that inexpensive DNA primers are available, researchers can obtain longer stretches of sequence using a series of multiple primers. After the first few hundred base pairs of new DNA sequence is determined using the vector primer, the new DNA sequence is used to design a new primer that will base pair to a unique target site on the template DNA located a considerable distance away from the position of the vector primer site. This new DNA primer is used to initiate the next set of DNA sequencing reactions, which reveals new DNA sequences that overlap and extend the sequence information obtained with the first vector DNA primer. This sequencing strategy permits scientists to use successive DNA primers and sequencing reactions to determine the sequence of very long regions of DNA.

> **KEY CONCEPT**
> At the end of the four DNA sequencing reactions, each tube contains a collection of nested, single-stranded DNA with different lengths, common 5′ ends, and nested 3′ ends.

Automated DNA Sequencing Machines

Automated DNA sequencing machines typically use a version of the Sanger chain termination method that employs four different dideoxy-nucleotide (ddNTP) derivatives containing fluorescent tags that emit four different colors when excited by a laser (Figure 2-12). The four differently colored ddNTPs can be distinguished from each other, making it convenient to mix the four dideoxy-nucleotide sequencing reactions (A, G, C, and T) together in one tube. The DNA strands produced by the sequencing reactions are separated by length and then pass by a detector, which interprets the different fluorescent colors and reports the order of the four bases in the DNA sequence as a series of four differently colored peaks (Figure 2-12).

The DNA sequence data generated by the DNA sequencing machine are automatically sent to the computer, where the DNA sequences are stored and analyzed. The computer aligns the different DNA sequences generated using different primers and identifies regions of DNA sequence overlap between the newly sequenced stretches of DNA. The overlapping sequences are assembled by the computer into extremely long stretches of DNA sequence. The sequences of large regions of chromosome DNA have been determined using this strategy, although several other approaches were required to complete the entire chromosome DNA sequence (Figure 2-13).

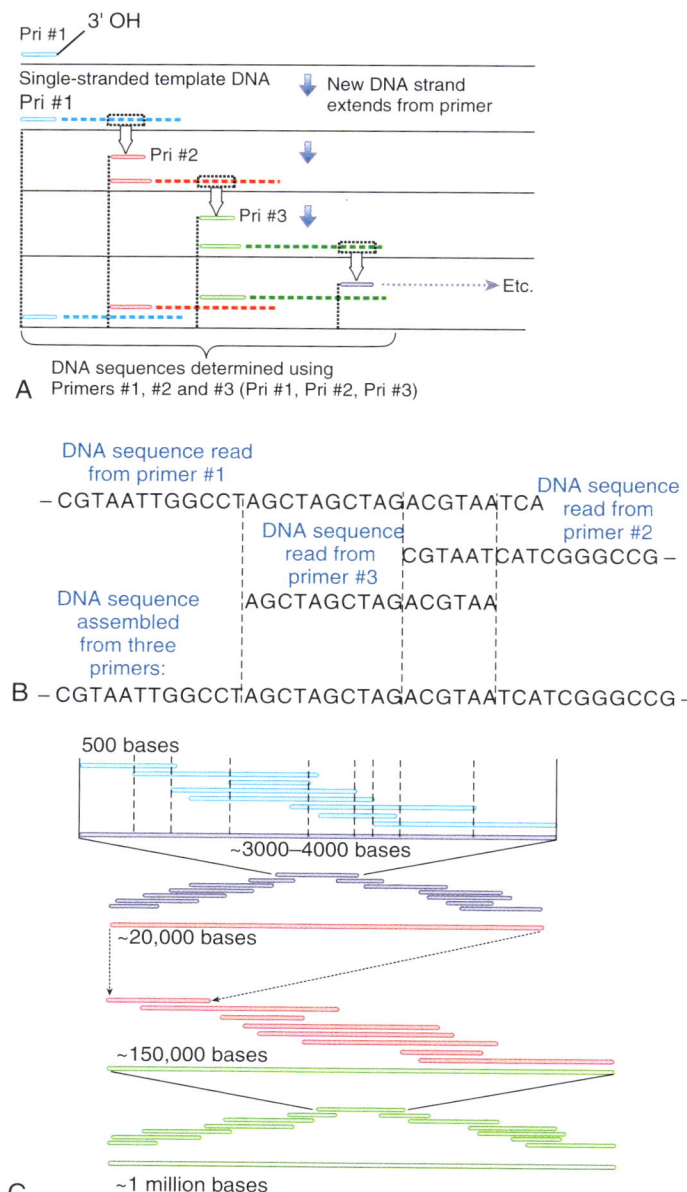

FIGURE 2-13 Overlapping DNA sequences are assembled into very long stretches of new DNA sequence (A) DNA sequence information determined from sequencing reactions using the first primer is used to design a new primer that anneals to a distant region on the DNA template. Additional DNA sequence data obtained from reactions using the second DNA primer is used to design a third primer for sequencing a more distant region of DNA. (B) The DNA sequences using several different primers are compared by computer to identify regions of sequence overlap. (C) By aligning the overlapping DNA sequences, it is possible to assemble shorter DNA sequences into longer stretches of accurate DNA sequence.

The Polymerase Chain Reaction Amplifies DNA

Kary Mullis developed polymerase chain reaction (PCR) technology while he was a scientist working at Cetus Corporation in the early 1980s. Mullis wanted to create an enzyme system that would produce large numbers of identical DNA copies in a tube, without relying on recombinant DNA cloning and bacterial transformation, using cells to grow large quantities of identical, cloned DNA fragments. Mullis applied his understanding of DNA replication in cells as a foundation to design his DNA amplification strategy. Mullis started with chromosome DNA containing a target sequence to be amplified. He separated the DNA strands and then annealed the long chromosome DNA to two single-stranded DNA primers (Figure 2-14). The DNA region between the two different DNA primers represents the target DNA sequence that will be copied and eventually amplified.

After the two DNA primers have annealed (base-paired) to the complementary sites on the DNA template strand, the DNA polymerase enzyme is added to copy the target DNA located between the base-paired primers (Figure 2-14). The temperature is raised to separate the double-stranded DNA molecules into single strands, including the DNA primers and the copied template DNA. When the temperature is decreased, the complementary DNA primers anneal

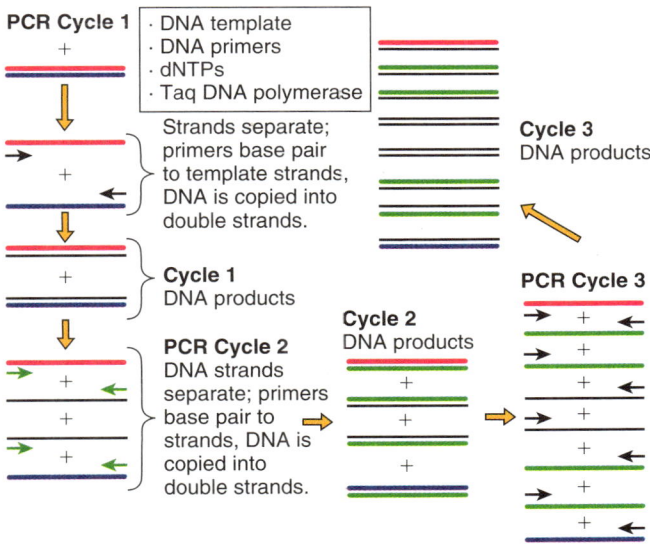

FIGURE 2-14 Polymerase Chain Reaction amplifies identical copies of DNA molecules The Polymerase Chain Reaction (PCR) method is an efficient way to make unlimited copies of a DNA molecule by copying the DNA over and over again in cycles. A new PCR cycle starts when the strands of the DNA helix separate into two single-stranded templates and the primers (small arrows) base pair to the template strands. In each PCR cycle, after the DNA strands separate in high heat, the mixture cools, the DNA primers base pair to two sites on the template strands and the DNA is copied into double-stranded DNA by a heat tolerant DNA polymerase. Then the reaction is heated again at the start of the next PCR cycle to separate the DNA strands, cooled to permit the primers to base pair to the templates, and the DNA is copied into by the DNA polymerase. Depending on the requirements of the experiment the PCR can continue for 15 to 20 cycles.

to the two unique sites on the DNA template strands. The DNA polymerase initiates DNA synthesis and once again copies the target DNA sequences between the primer sites. The temperature is raised once again, separating the DNA strands and beginning another PCR cycle (Figure 2-14). Sufficient DNA amplification by PCR typically requires 15 or more PCR cycles.

For the PCR method to work effectively, Mullis needed to heat the reaction to 90°C after each round of replication in order to separate the DNA into single-strands. However, this posed a problem since the high temperatures irreversibly inactivated the DNA polymerase enzyme, requiring Mullis to add fresh DNA polymerase enzyme for each new PCR cycle. Amazing thermophilic organisms that live in very hot environments provided the answer in the form of heat-tolerant polymerase enzymes. One such enzyme, *TaqL* DNA polymerase, was isolated from *Thermus aquaticus,* a thermophilic bacterium (Figure 2-15). *TaqL* polymerase is a typical protein enzyme, yet it can withstand very high temperatures and still synthesize DNA at 72°C. Other thermostable DNA polymerase enzymes have also been studied and are used in DNA sequencing reactions involving PCR amplified template DNA including the *Pfu* DNA polymerase enzyme, which was isolated from the thermophilic archaeon *Pyrococcus furiosus*.

In 1983 Kary Mullis received the Nobel Prize for his work on PCR. The PCR method is not just another DNA technique. PCR is now used routinely in every molecular biology lab on Earth and has been adapted to amplify DNA for many kinds of applications in basic research, clinical and genetic testing, and disease diagnosis (Unit 4). PCR amplification has had a huge impact on the use of DNA as a molecular detection tool. The ability of PCR

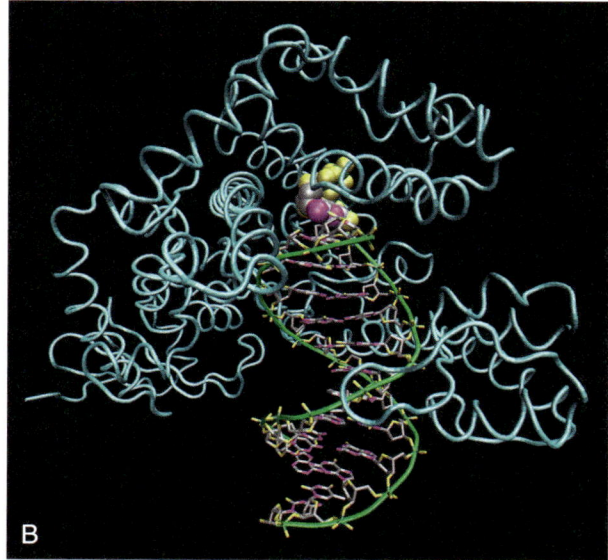

FIGURE 2-15 PCR is performed by a heat stable DNA polymerase (A) Dr. Tom Brock isolated a new type of DNA polymerase enzyme from the thermo-stable bacterium, *Thermus aquaticus*. This heat stable DNA polymerase, *Taql* is commonly used to amplify DNA *in vitro* by PCR. (B) The *Taql* polymerase enzyme (blue) is bound to the two DNA strands (green) being replicated, with the atoms in the nucleotide bases colored grey (carbon), pink (nitrogen), yellow (oxygen) and green (phosphate).

to amplify extremely small amounts of DNA significantly increases the sensitivity of DNA detection methods. PCR-based advances in the forensic analysis of DNA evidence, especially in DNA fingerprint testing, has had significant effects on the criminal justice system (Unit 6). Advances in DNA testing have also provided an important tool for people who use DNA science to challenge the mistakes made by the legal system and exonerate wrongly convicted people (Unit 6).

> **KEY CONCEPT**
> PCR can produce virtually unlimited amounts of identical DNA molecules starting with a tiny amount of the DNA template. This is an extremely powerful DNA technology with applications in many areas of science.

RT-PCR is a Way to Amplify mRNA

The reverse transcriptase PCR (RT-PCR) method is designed to amplify a specific RNA molecule that is present at low copy number in a population of many other RNA molecules. PCR technology is used to copy specific mRNA strands into complementary DNA (cDNA) strands (Figure 2-16). Under normal conditions the mRNAs transcribed from protein-coding genes are present

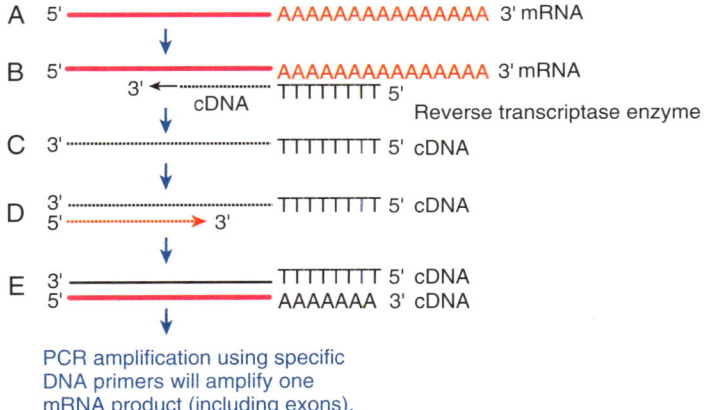

FIGURE 2-16 Reverse transcriptase and PCR are combined to amplify low copy mRNAs
The RT-PCR is a combination of reverse transcriptase treatment and PCR used to amplify a low copy mRNA molecule in a population of many other RNAs. The mRNAs copied from protein-coding genes are usually present at very low numbers compared to the tRNAs and rRNAs. (A) The total RNA molecules isolated from the cells are mixed with short strands of oligo(dT) (5'TTTTTTTTT3'), which base pair to the poly(A) tails on the mRNAs. The RNAs without poly(A) tails do not base pair with the oligo(dT). (B) The oligo(dT) that base pairs to the poly(A) tail provides a primer with a free 3'OH, which the reverse transcriptase enzyme uses to copy the mRNA into a complementary DNA (cDNA) strand. (C) The RNA strand is removed from the RNA:DNA heteroduplex by RNAseH enzyme. (D) The DNA polymerase copies the cDNA strand into a double-stranded DNA molecule. (E) The resulting DNA can be inserted into a vector for cloning or amplified so that this DNA product to use as a very specific DNA probe that will detect the very low copy number mRNA under study.

at low copy number in the cell and are greatly outnumbered by the tRNAs and rRNAs made in large quantities because the cell is growing. RT-PCR relies on the fact that the mRNAs of interest have long tails of A bases added on to the 3' end called poly(A) tails, and the tRNAs and rRNAs do not.

To perform RT-PCR the researcher harvests RNA molecules from the cells (total RNA) and then mixes the RNAs with short strands of oligo(dT) (5'-TTTTTTTTT-3'), which base-pair to the poly(A) tails on the mRNAs. The tRNAs and rRNAs do not have poly(A) tails and do not base pair with the oligo(dT). This distinguishes the relatively rare mRNAs with poly(A) tails from the much more abundant tRNAs and rRNAs that do not have poly(A) tails. The oligo(dT) strands base-paired to the poly(A) tails provide primers with 3' OH for the reverse transcriptase enzyme to use to initiate DNA synthesis and copy the mRNA strand into a complementary DNA (cDNA) strand (Figure 2-16). In the next step the DNA polymerase enzyme copies the cDNA strand into a double-stranded DNA product with the sequence of one of the mRNAs in the population.

The products of RT-PCR have many potential applications. Specific DNA primers can be used to amplify and identify one specific cDNA product from among the population of cDNAs. This PCR product can be used as a powerful DNA probe that detects only one specific mRNA for study, the product of copying a specific gene in the cell. In certain controlled conditions, RT-PCR can be used to indirectly measure the number of specific mRNA molecules produced by the expression of a single gene, providing important information about the activity of the gene being studied.

RT-PCR is very useful for studying eukaryotic gene expression, because it is a highly sensitive method that permits the detection of RNA molecules that are present at very low copy numbers in cells. The intron sequences that are present in the eukaryotic genome DNA but are not present in the mature processed mRNA are not represented in the RT-PCR product (Figure 2-16) (Unit 3). In most eukaryotic cells the pre-mRNA strands copied from the DNA and containing the introns and exons are processed very rapidly and are difficult to study. However, the final mature mRNAs in the cytoplasm can be detected even at relatively low levels using RT-PCR (Figure 2-16).

The majority of eukaryotic protein-coding genes express pre-mRNAs containing introns and exons that cannot be processed or spliced properly in prokaryotic cells like bacteria. For these reasons, the genomic copies of most eukaryotic genes cannot be expressed in prokaryotic cells even if the vector provides the eukaryotic gene with an appropriate promoter to permit transcription in prokaryotic cells (Unit 3). However, cloning the cDNA of a specific mRNA into a recombinant expression vector is a routine way to express specific eukaryotic proteins for production and purification in host cells, including prokaryotic cells. The cDNA copy of a eukaryotic mRNA can be expressed in prokaryotic cells if the appropriate prokaryotic promoter is included before the cDNA coding region in the vector. When the cDNA copy of a eukaryotic gene is transcribed in prokaryotic cells, the mRNA is directly translated into protein without requiring RNA splicing or other RNA processing events (Unit 3). In other applications the RT-PCR method is used in clinical labs to diagnosis certain genetic diseases (Unit 4). RT-PCR is also commonly used to study viruses with RNA genomes, including influenza A virus and retroviruses such as HIV.

Scientists have developed quantitative versions of the RT-PCR method with various names including real-time polymerase chain reaction, quantitative real-time polymerase chain reaction (Q-PCR/qPCR/qrt-PCR), and kinetic polymerase chain reaction (KPCR). These PCR modifications quantify the specific amount of target DNA amplified by PCR, by measuring the amount of specific RNA transcripts made in the cells. This is one way that scientists can measure the expression of a specific gene by quantifying transcription of the gene. The key feature of quantitative real time PCR is that the amplified DNA products generated in each PCR cycle are detected and reported in real time as the reactions proceed, which differs from standard PCR, which reports the reaction products only at the end of all the PCR cycles. Real-time PCR technology requires a way to detect the DNA products as they are produced by the reactions. The two most common approaches include the use of fluorescent dyes that nonspecifically stain any double-stranded DNA, such as ethidium bromide, or use of sequence-specific DNA probes that fluoresce and are visualized only after the DNA probe has base-paired (hybridized) to its complementary DNA target.

The Human Genome Project

The human genome is the set of hereditary instructions necessary to build and maintain a human body, which is written in the DNA language and carried on 46 chromosomes in human diploid cells. Knowing the human genome DNA sequence was an important advance in understanding the functions of all the proteins in the human body.

The Human Genome Project (HGP) officially began in 1990 as a joint project funded by the U.S. Department of Energy (DOE) and the National Institutes of Health (NIH), with these ambitious goals:

- To identify all the genes in human DNA
- To determine the order of the base pairs in the human genome
- To store DNA sequence information in public databases
- To make the DNA information available free to the public
- To address the ethical, legal, and social issues of the Human Genome Project (ELSI)

At that time, the U.S. Congress allocated $17.4 million for the HGP, and Dr. James Watson (co-discoverer of the structure of the DNA double helix), was named the first Director of the Office of Human Genome Research. Watson made sure that a percentage of the allocated federal funds would be forever reserved to support research on the ethical, legal, and social issues (ELSI) that rise from the release of the human genome sequence information. Work on the human genome sequence in the United States became known as the public HGP to distinguish the research supported by the U.S. taxpayers from the genome work conducted by private companies, called the private HGP. The next public HGP director, Dr. Francis Collins, worked with hundreds of investigators from all over the world who shared the common goal of sequencing all the individual human chromosomes. However, there was controversy over this approach to solving the problem of sequencing the human genome. Some scientists including Dr. J. Craig Venter wanted to begin by sequencing the regions of the human chromosome DNAs known to code for protein products. This alternative strategy was based on evidence at the time that very large regions of the human genome DNA do not code for proteins.

Scientists began by determining the genome DNA sequences of small eukaryotes, including the genomes of the single-celled yeast *S. cerevisiae* (1997), the nematode worm *C. elegans* (1998), and the fruit fly *D. melanogaster* (2000), before sequencing the very large human genome. This research provided the foundation for the emerging field of genomics and played a very important role in the study the much larger genomes of multicellular organisms. The information from small genome research was essential for scientists to develop the necessary technologies to sequence and analyze larger genomes such as the human genome. In addition, researchers used the smaller genome sequences to develop the much more sophisticated computer programs required to manipulate and analyze the billions of base pairs of DNA sequence information obtained from scientists all over the world. The public HGP met its goals, and the human DNA sequence information generated by the public HGP was stored in public computer databases, with free public online access to the human DNA sequences and genes.

J. Craig Venter Led the Private Sector HGP

Many scientists involved in sequencing the human genome DNA worked for the private sector, mostly in U.S. biotechnology companies with commercial interests in identifying the human genes that cause human diseases and could be used for drug development. J. Craig Venter had a major impact on genome sequencing technologies when he started The Institute for Genomics Research (TIGR), a company developed the technologies needed to efficiently sequence small genomes. Then in 1998, Venter and biotechnology entrepreneur Mike Hunkapillar started Celera Genomics, a company that developed automated DNA sequencing technologies and DNA computing software for research on the human genome. To try to answer the question about how much human DNA is non-coding and does not code for proteins, Venter developed ways to identify the regions called expressed human sequences that encode genes in the human genome. Celera invested in technology development and built a DNA computer facility the size of a football field, the largest civilian supercomputer in existence at the time, and created a lab with over 300 automated DNA sequencing machines.

The scientific competition between the public HGP and the private HGP to sequence the human genome increased when Venter boasted that Celera Genomics would win the race and finish the human genome sequence by 2001 for only $200 million. Dr. Collins shifted priorities in the public HGP effort and instituted new DNA sequencing technologies to complete a "working draft" of the human genome by 2001.

The international race to sequence the human genome ended in a historic international collaboration between the public and private HGP groups. In February 2001 the competing public and private HGP scientists simultaneously published the first 'working' draft of the human genome DNA sequence in two different prestigious scientific journals. The private HGP, headed by J. Craig Venter with scientists from corporate research labs, published in

Science, and the public HGP, headed by Francis Collins with U.S. and international scientists, the U.S. Department of Energy, and the National Institutes of Health, published its human DNA sequence in *Nature*. The next goal in the amazing HGP was reached when the updated and improved version of the human genome DNA sequence was published in April 2003.

The human genome DNA sequence cost taxpayers almost $4 billion, but the impact of the research, biotechnology and medicine in the rapidly expanding field of human genomics and related industries, including companies that began as a result of the HGP have contributed nearly $800 billion to the U.S. economy and created hundreds of thousands of jobs.

All of the information generated by the human genome sequencing effort is available free to the public thanks to the foresight of an early HGP Director, James Watson. A series of government websites make it possible for anyone to learn more about the HGP and find out everything about every known human gene and protein. Explore your genes for free.

Lessons Learned from the Human Genome DNA Sequence

The HGP revealed that the human genome DNA sequence contains about 3,200,000,000 bp (3.2 billion base pairs) distributed among the 23 DNA helix molecules that make up the 23 individual human chromosomes in each haploid cell. Diploid cells contain two copies of each chromosome; two copies of chromosome #1 and two copies of chromosome #2 etc. When the DNA sequences of two copies of the same chromosome, such as the copy of chromosome #3 from Mom and the copy of chromosome #3 from Dad, are compared with each other, the DNA sequences are very similar, but are not identical DNA sequences. This is an important difference between the chromosome copies inherited from parents and the chromosomes that result from the process of DNA replication. The replication of the DNA helix in a chromosome produces two identical DNA helices, which are called chromatids. The chromatids in replicated chromosomes appear to be connected at the centromeres.

The DNA helix molecule in each chromosome contains a DNA sequence that is specific to that chromosome; the DNA sequence of chromosome #6 is different from the DNA sequence of chromosome #3 or chromosome #9. Regardless of sequence, the DNA helix extends from one end of the chromosome to the other end, passing unbroken (and not branched) through the region containing the centromere constriction. When stained with specific dyes the condensed chromosomes exhibit patterns of dark and light bands that form as a result of the combination of DNA sequences and the associated proteins at each site and therefore can be used along with other features (length of chromosome arms, position of the centromere) to identify and distinguish between individual human chromosomes (Figure 2-17).

A short region of the chromosome DNA is expanded in Figure 2-17 to show how the overall chromosome structure relates to the DNA sequence of the region determined using strategies that include analyzing the overlapping DNA sequences, and identifying the positions of various restriction enzyme cleavage sites in the chromosome DNA. The physical locations of specific genes along each chromosome are determined by combinations of studies including computer analysis of the DNA sequence to identify any possible control DNA

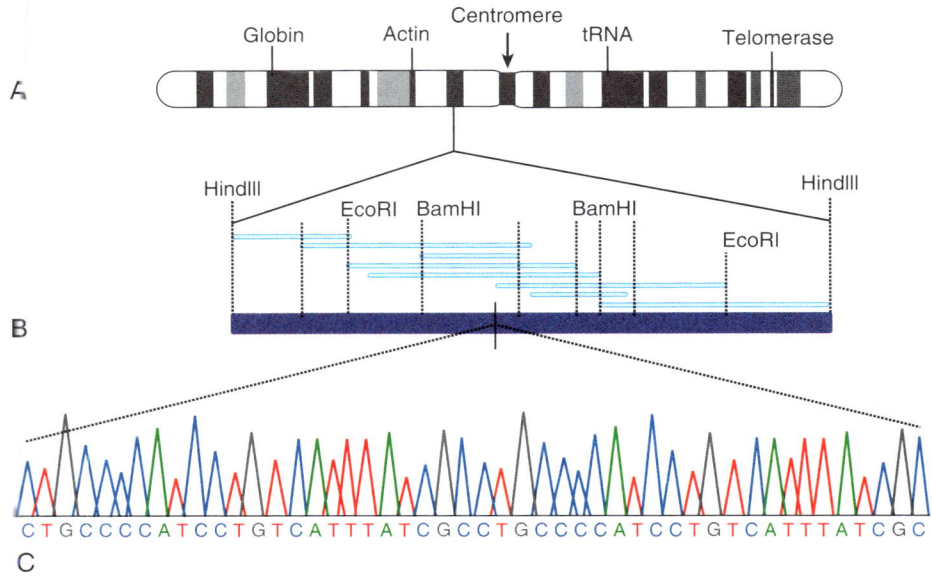

FIGURE 2-17 What did we learn from the human genome DNA sequence? An overall strategy used to sequence the DNA in a human chromosome is shown. (A) The human DNA is carried in chromosomes, extending from one end to the other end of the chromosome. A gene map indicates the positions of the identified genes along the DNA. (B) A close up of a small section of the chromosome shows how overlapping regions of DNA sequence were derived from DNA fragments cloned into vectors or amplified by PCR to provide sufficient DNA for sequence analysis. The positions of a few restriction enzyme cleavage sites are indicated. (C) The linear DNA sequence of a small section of chromosome DNA is shown along the bottom as an array of colored peaks.

regions or open reading frames that code for proteins in the DNA sequence information (Unit 3). Even when the DNA sequence of a region is known, further research is often necessary to identify any genes that might be encoded by that DNA sequence.

Studying the human genome DNA sequences revealed some unexpected and very interesting information about the genome DNA that makes us human. Probably most surprising is that even though we know the entire sequence of the human genome DNA, scientists are still trying to answer the key question, how many human genes are there? Over the past three decades, scientists have made wide ranging predictions about the total number of human protein genes based on the evidence available at the time. Starting at a high of about 140,000 genes, the predicted number of human genes has continually decreased as researchers learned more about how genes are expressed and regulated in eukaryotic cells. Still, many scientists were very surprised to find out that only about *1.5% of the total human DNA sequence (only 48,000,000 base pairs!)* actually codes for protein genes, corresponding to only 18,000 to 22,000 human genes (Unit 3)!

The human genome DNA sequence revealed the positions of thousands of genes encoded along the DNA in the 23 human chromosomes. This research also showed that human genes are not spaced evenly along the DNA molecules but instead are arranged in many gene clusters that are separated by DNA that does not code for proteins (Figure 2-18). Gene clusters contain

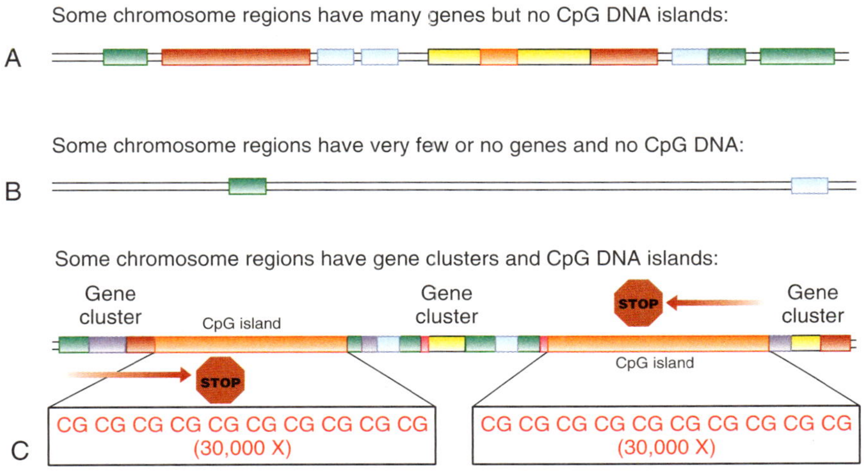

FIGURE 2-18 Where are the genes located in the human chromosomes? Human genes are distributed unevenly along the DNA helix molecule in each chromosome. Many genes are located in gene clusters that are separated from neighboring gene clusters by unusual DNA sequences called CpG islands. The CpG islands contain repeated CG base pairs that extend over tens of thousands of base pairs of chromosome DNA but do not code for proteins. The CpG islands block accidental read-through transcription from an adjacent gene cluster in the genome.

groups of protein-coding genes that are distributed at uneven intervals along the DNA molecules separated by extensive regions encompassing millions of base pairs of human genome DNA that are completely devoid of genes. The clusters of human genes are sometimes separated from each other along the chromosome DNA by regions called CpG islands, which contain repeated CpG base pairs that extend over large regions of chromosome DNA (Figure 2-18). The CpG islands act as transcription barriers between adjacent gene clusters that prevent read-through RNA transcription from neighboring genes.

Most Human Protein-coding Genes Contain Introns and Exons

The human genome sequence information revealed that the majority of human protein-coding genes contain intron and exon sequences in the genome DNA. As a result, like most other vertebrate genes, human genes are actually expressed as pre-mRNA transcripts that are much longer than the gene, and include both introns and exons (Figure 2-19) (Unit 3). A good example is the longest human gene, which encodes the dystrophin protein and encompasses more than 2.4 million bp (megabases, Mb) of chromosome DNA. The 2,000 kb dystrophin pre-mRNA transcripts are processed to generate a 14 kb mRNA encoding 79 exons. In normal cells, this mRNA is translated into the 3,500 amino acid-long dystrophin protein. Mutations in dystrophin cause a common form of muscular dystrophy (Unit 4).

As demonstrated for dystrophin, most human genes are expressed as very long precursor RNA molecules (pre-mRNAs). These mRNA precursors require extensive RNA processing to create a messenger RNA molecule (mRNA) with an intact protein-coding region that can be translated into protein (Unit 3).

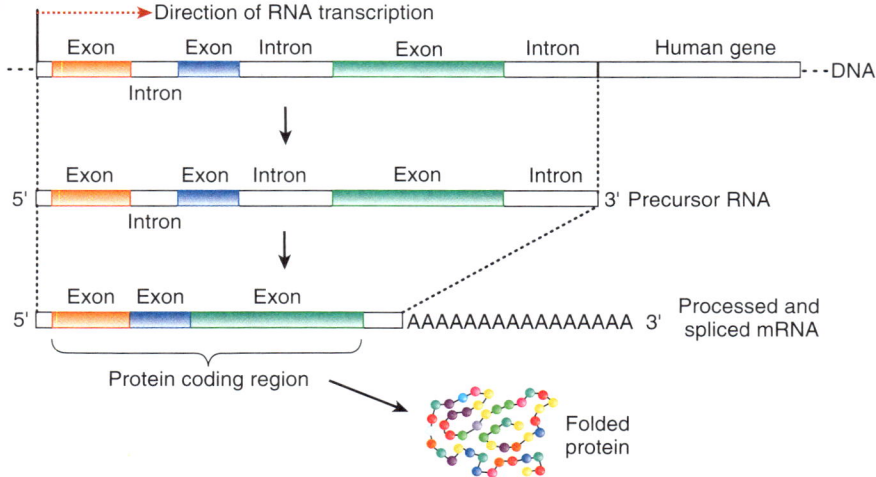

FIGURE 2-19 Introns in human genes are removed by RNA splicing Most human protein-coding genes contain introns and exons in the genome. These human genes are transcribed into very long precursor RNA molecules called pre-mRNAs, which are processed by RNA splicing to remove the introns. The final messenger RNA (mRNA) contains the intact protein coding region and can be translated.

The RNA splicing process is extremely accurate and involves intricate interactions among many specialized proteins and small RNA molecules. RNA splicing is an important regulatory mechanism used to control gene expression in eukaryotic cells (Unit 3).

Humans Usually Inherit two Copies of each Gene

Each human cell has 46 chromosomes (23 chromosome pairs) except for the sperm and egg cells, which have 23 chromosomes each, and mature red blood cells (RBCs), which lose their nuclei during development and don't have DNA. Each person inherits one copy of each chromosome from Mom and inherits a second copy of the same chromosome from Dad (Unit 4). The DNA sequence of the second copy of the chromosome inherited from Dad is very, very similar, but not identical to the DNA sequence of that chromosome inherited from Mom. The fact that the maternal and paternal copies of each chromosome have similar (but *not* identical) DNA sequences is essential to ensure that human children are not genetically identical clones of their parents. This is also why human siblings are usually very different from each other.

> **KEY CONCEPT**
> The two copies of each gene on each chromosome are called alleles, with one allele for each gene inherited from each parent. The different alleles of a gene usually have different DNA sequences.

A child will normally inherit one of three possible combinations of genes:

- Two different alleles of the same gene, for example, a normal allele from Mom and a mutant allele from Dad, or
- Two normal alleles, one from each parent, or
- Two mutant alleles, one from each parent.

Although most human genes are present in two copies per cell (one from Mom and one from Dad), about 10% of the human genome contains sequences that have been altered by copy number variation (CNV). Sometimes genes are inherited in more than one copy per genome as a result of a localized DNA duplication mistake that occurred during DNA replication. The duplicated region of the chromosome DNA can increase the copy number and therefore the expression levels of any encoded genes affected by the duplicated DNA. Typically these CNV differences involve only one gene or a few genes in one region of the genome. However, in some cases large regions of protein-coding DNA sequences are rearranged by CNVs, which typically increases the copy number of the genes and increases the expression levels of the protein products (Figure 2-20). Surprisingly, even the genomes of genetically identical twins can be altered differently by copy number variation because the same

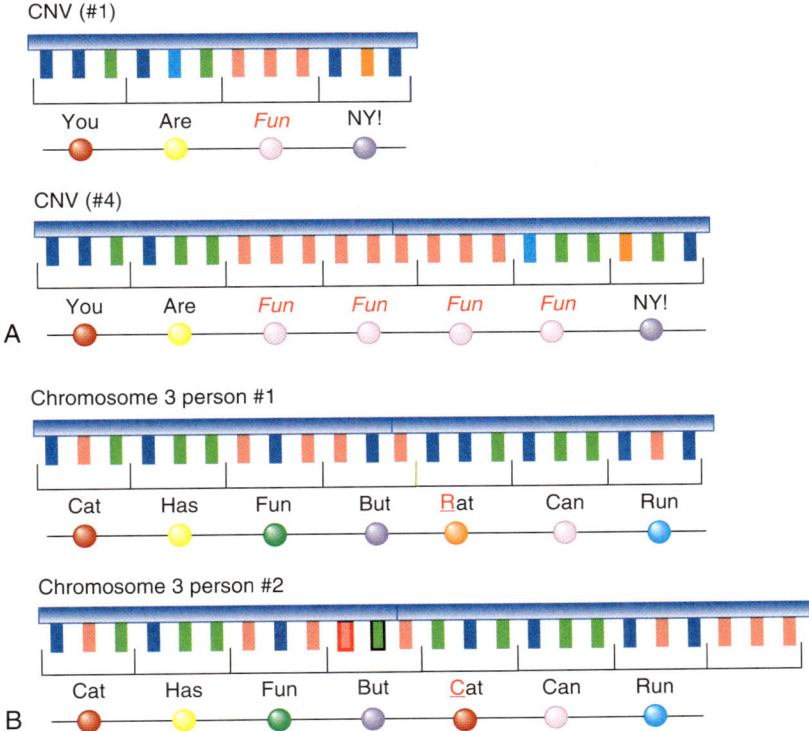

FIGURE 2-20 Genome DNA variation includes point mutations and copy number differences (A) This example shows copy number variation that alters copy number, changing word FUN from one copy to four copies. This copy number change is also represented in triplet bases. (B) The single base pair difference observed when the DNA from Person #1 is compared to the DNA from Person #2, is an example of a single nucleotide polymorphism (SNP).

mistakes during replication will probably not be experienced by both twins at the same time.

When CNVs increase or decrease the copy numbers of genes that are dosage-sensitive, the results can alter variable traits in humans, including the individual's susceptibility to diseases such as cancer. For example, the epidermal growth factor receptor gene (EGFR), which encodes a cell-surface receptor protein, is often present in higher copy numbers in people with non-small-cell lung cancer. An abnormally low copy number of the human CD16 immunoglobulin receptor protein gene (FCGR3B) is associated with increased susceptibility to the autoimmune disease lupus erythematosus. Research has also linked genome copy number variation to disorders such as autism, schizophrenia, and several learning disabilities.

Copy number variations can explain the range of diversity observed for humans, since CNV can make some individuals particularly susceptible to certain disorders and diseases. Recently, new techniques were developed to detect and quantify the DNA regions of the human genome that are duplicated in different individuals. Researchers from the 1000 Genomes Project, Agilent Technologies, and the University of Washington analyzed 159 different human genome sequences and counted the numbers of duplicated genes in each human genome. In other comparative studies they identified several genes that differ in copy number when human genomes are compared with those of apes such as gorillas, orangutans, and chimpanzees. Interestingly this includes duplications in genes that are involved in brain development, nerve cell growth, head size, neurotransmitter receptors, visual acuity, social deficits, epilepsy, spinal muscular atrophy, and intellectual ability.

Recently researchers have suggested that copy number variation has been implicated as a possible cause of autism spectrum disorder (ASD). People diagnosed with ASD exhibit traits that fall somewhere along a large range of developmental problems characterized by social impairment, repetitive behavior patterns, and difficulty with personal communication. ASD ranges from the most severe form, autism, to the much milder symptoms of Asperger syndrome. People of any age, ethnicity, or socioeconomic group can have ASD. Over the last decade an increase in the numbers of children diagnosed with autism has raised serious concerns about a possible autism epidemic in the United States.

The idea that infant and childhood vaccines cause ASD was based on one British clinical study that was later discredited. This incident, followed by debate and apparently contradictory information, raised serious public concerns about a putative connection between autism and childhood vaccinations. However, several carefully administered and controlled research studies including extensive research by the U.S. government have shown that there is no connection between childhood vaccines and autism. The original research claiming a link between vaccines and autism was withdrawn, and the medical license of the principal investigator was revoked in part as a result of a detailed investigation by a journalist who looked into the original research study that led to the international debate about autism and vaccines. He found clear evidence that the scientist who conducted the initial research on autism and vaccines had not just performed sloppy science but actually committed scientific fraud.

More than 60% of the human genes affected by CNV encode proteins that are involved in cognitive abilities in the brain. The genomes from people with autism contain 20% more sequences affected by CNV than the genomes from individuals who are not autistic. Several new genes involved in ASDs have been recently identified, including genes encoding neurotransmitter proteins that transmit nerve signals across the synapses between different nerves. In a search for genes that contribute directly to ASDs, scientists identified a region of human chromosome 5 that increases the risk of developing ASD by 30% when that region of chromosome 5 is mutated.

The causes of ASDs are not fully understood, but genetic inheritance does have a role in the development of autism. Genetic studies show that families with one ASD child have a one in 20 chance of having a second child with the disorder. In addition, if one genetically identical twin has autism, there is a 90% chance that the second genetically identical twin will also be affected.

Gene Families Encode Proteins with Similar Structure and Function

Genes that are members of a gene family encode proteins with similar amino acid sequences that often perform similar or related biological processes in the cells. Protein members of a gene family are identified by the amino acid sequence similarities among specific functional domains in the related protein family members. Even species as diverse as humans, worms, flies, and plants, have genes that code for similar proteins. However, only those gene families that express proteins involved in human cell and tissue development are expanded into large families in the human genome.

It makes sense that similar proteins found in different species can perform similar biochemical functions. All types of cells use similar proteins to conduct fundamental biochemical processes such as cellular metabolism and DNA replication. These genes are encoded by similar genes in very different organisms. Researchers found that ~60% of the genes in the fruit fly (*D. melanogaster*) and in humans have similar amino acid sequences and are predicted to have similar functions as well.

Genes coding for specialized proteins that perform in specialized processes such as photosynthesis, are found only in appropriate types of specialized tissues. Photosynthesis requires chlorophyll, a component of most plant cells. Different mammalian species such as mice and humans actually use similar master genes to control the stages of mammalian embryogenesis and regulate the development of tissues and organs. Mammals all have similar genes and proteins that specify the same overall mammalian body plan: one head, four limbs, and a tail (humans even have a hidden tailbone) (Unit 1).

Comparative Genomics

The exciting field of comparative genomics combines genome science, advances in DNA sequencing technologies and sophisticated computer programs to compare and analyze the genome DNA sequences from an ever increasing list of different organisms, including dog, rat, fruit fly, roundworms, budding and fission yeast, the mosquito and the parasite combination that transmit malaria, sea squirt, mustard weed, rice, and many microbial organisms. Other DNA sequencing projects are under way to analyze the genomes from the honeybee, chimpanzee, cow, and chicken. Comparing genome sequences allows researchers to identify similar DNA regions that could potentially have important functions in different organisms, most notably humans. DNA genome sequence information is a powerful scientific tool used to identify genes that confer unique, identifying characteristics to different species. For example, the dog genome (and the dogs!) helped scientists to identify the human genes involved in narcolepsy (sleeping sickness) and was used to develop treatments for a genetic cause of blindness in humans (Unit 4).

Chromosome number and gene number alone do not account for the wide range of different sizes, shapes, and functions exhibited by the large number of diverse biological organisms. Two popular research organisms, the mustard weed plant (A. thaliana) and the roundworm (C. elegans), have genomes of about 100 million bp each and contain approximately the same number of genes (25,000 and 19,000, respectively). The fruit fly (D. melanogaster) has a larger genome (137 million bp) but encodes only 13,000 genes. The single-celled eukaryotic yeast cells (S. cerevisiae) that make beer and bread, have a smaller genome (12.1 million bp) that encodes only about 6,000 genes. In comparison, the circular double-stranded chromosome in each E. coli bacterium contains 4.6 million bp and codes for about 3,200 genes.

Comparative genome research provides information used to develop novel treatments for human diseases. This research will also help scientists to learn more about the molecular mechanisms of species-specific diseases such as malaria and HIV disease (AIDS). Interestingly, species with genomes that are very similar in DNA sequence to the human genome, like the chimpanzee, are not susceptible to malaria or HIV disease (AIDS). Comparison of the genes implicated in disease susceptibility in these different species can help to explain the molecular differences between species and can sometimes suggest new approaches to prevent and treat human diseases.

Comparing Genes: Are You a Mouse or a Man?

For decades, researchers have used laboratory mice as a model system to study many human diseases because these animals are extremely well studied, second only to E. coli bacteria, and are also small mammals with affordable upkeep. Currently there are more than a hundred mouse models have been created to mimic different human diseases, which are based on studies of various mutations in mouse genes that cause symptoms in mice that resemble the symptoms of human diseases. These mutant strains of mice are developed for each disease by deleting a specific gene or set of genes from the mouse genome, which creates a "knockout" mouse strain for each disease. This type of approach is more difficult but still possible for some diseases that involve multiple genes, such as epilepsy, asthma, obesity, colon cancer, high blood pressure, and type II diabetes.

The DNA genome sequence of a female Black 6 mouse, which is a strain commonly used in biomedical research, was determined by a scientific team from 27 institutions in six countries funded by the NHGRI, NIH, and the

Wellcome Trust (Britain). Compared to the human genome of 3.2 billion bp, the Black 6 mouse genome is shorter with only 2.5 billion bp, but the mouse genome encodes about 30,000 genes, compared with only 18,000 to 22,000 human protein-coding genes. The human genome also contains more repeated noncoding DNA than does the Black 6 mouse genome (see below). Interestingly, in comparison to the human genome, the mouse genome is enriched in genes that are responsible for behaviors that are typical of rodents, such as large litter size, odor detection and a very robust immune system.

Despite the differences between mouse and man, comparison of the genome DNA sequences revealed that large regions of the human and mouse chromosomes have similar organization in chromosomes from the two species. For example, a region of mouse chromosome 16 contains DNA that is very similar in DNA sequence to several specific regions of DNA from human chromosomes 3, 21, 22, and 16.

Most of the Human Genome is Non-coding DNA

Scientists have known for some time that the human genome contains some fraction of DNA sequences that do not contain genes and do not code for proteins. However, the HGP discovered that humans have only 18,000 to 22,000 protein-coding genes and that these genes are encoded by only 1.5% of the total number of DNA sequences in the human genome! Another way to look at this information is that the remaining ~98.5% of the human genome, which is the vast majority of the human DNA, are non-coding DNA sequences that do not contain genes!

There is a huge amount of non-coding DNA sequences in the human genome and scientists do not yet completely understand the functions of much of the DNA sequenced by the HGP! A relatively small fraction of non-coding human genome sequences are control DNA elements that regulate the gene products involved in processes such as gene expression, chromosome packaging, and genome reprogramming (Unit 3 and Unit 6). This includes the non-coding enhancer and promoter regions associated with protein coding genes in the human genome (Unit 3). Large regions of DNA sequences are implicated in chromosome packaging and replication.

> **KEY CONCEPT**
> The human genome DNA contains a total of 3.2 billion bp, but by far the majority of the human genome DNA (3,152,000,000 bp) are non-coding sequences that do not contain genes.

For decades before the Human Genome Project, scientists studied the structures of genes in both prokaryotic and eukaryotic cells and were quite familiar with the enzymes that are required to express these genes. It is important to understand the functions of the RNA polymerase II enzyme to get an accurate picture of how gene expression is regulated in eukaryotic cells since one major point of gene control is when RNA polymerase initiates transcription, the process of copying the DNA gene into an RNA copy (Unit 3).

Eukaryotic cells use three major types of RNA polymerase enzymes to express different classes of genes (Unit 3). RNA polymerase II is the enzyme that transcribes all protein-coding genes in eukaryotic cells. RNA polymerase II copies genes into long pre-mRNA transcripts that are processed and transported to the cytoplasm for translation (Unit 3). The other two RNA polymerases I and III transcribe eukaryotic genes that are copied into RNA strands but these RNAs are never translated into proteins.

The transfer genes (tRNA genes) in the chromosome code for the transfer RNAs (tRNAs) that carry single amino acids to the ribosome to be converted into a long chain of amino acids. The tRNA strands fold into 'clover leaf' RNA structures that function in translation, but the tRNAs themselves are not translated (Unit 3). The ribosomal RNA genes (rRNA genes) are repeated many times in the chromosome DNA and code for highly abundant ribosomal RNAs (rRNAs) that fold into complex stem and loop structures in the cell. The rRNA molecules are essential for ribosome assembly and protein synthesis (translation), so an abundant supply of rRNA is necessary when cells are growing rapidly and must produce abundant amounts of all of the components involved in protein synthesis. Note that the noncoding rRNA genes are different from the genes that code for the many ribosomal proteins involved in ribosome assembly and function.

The third type of noncoding RNA genes code for a variety of short RNA molecules that bind to specific small proteins to make ribonuclear-protein (RNP) complexes, which perform diverse functions in the nucleus and the cytoplasm. This includes small nuclear RNAs (snRNAs) that bind directly to specialized proteins and form snRNPs that function in important cellular processes such as RNA splicing (Unit 3). The small nucleolar RNAs (snoRNAs) are involved in processing the precursor-rRNA strands into the mature rRNAs and are essential for translation. The nucleolus is a region inside the nucleus where the ribosomal protein and rRNA genes are expressed and come together to assemble the ribosomes. The nucleolus is not surrounded by a membrane, but the partially assembled ribosomal components must be transported through the pores in the nuclear membrane to travel to the cytoplasm and assemble mature ribosomes for translation.

Small Differences have a Big Impact: Single Nucleotide Polymorphisms

The success of the HGP allowed scientists to compare the genome sequences from different individuals, which demonstrated that the DNA sequences of genomes from *genetically unrelated* people are 99.9% identical. Among the 0.1% of the genome DNA sequences (3.2 million, 3,200,000 bp) that differ among individual human genomes are many single nucleotide polymorphisms (SNPs) that are distributed around the genome (Figure 2-20). SNPs are single base-pair differences that are detected when the linear DNA sequences from the genomes of two different individuals are compared, base by base.

Although SNPs change only single base pairs in the human genome sequence, these tiny differences can be detected and used to identify people. SNPs have become a key part of the DNA tests that are essential for human identification by forensic DNA fingerprinting (Unit 6). In addition, SNPs are also very important tools used to identify new genes involved in genetic diseases caused by more than one gene (Figure 2-20) (Unit 4). Most SNPs do not affect gene expression, but some SNP mutations are closely linked to human traits such as hair and eye color, while other SNPs have minor influences on the development of diseases, such as heart disease, diabetes, and stroke.

Noncoding, Ultraconserved DNA Sequences in Vertebrate Genomes

Computer studies comparing many genome sequences from different species have identified specific conserved DNA sequences that are present in most genomes and have not changed over millions of years of evolutionary time. Scientists discovered 481 ultraconserved DNA regions in the genomes that average about 200 bp each in length. These DNA sequences are 100% identical in vertebrate genomes (human, dog, and chicken) but are completely absent from the invertebrate sea squirt, fly, and worm genomes. Studies show that 3.5% of the human genome DNA sequences contain noncoding sequences that exist as highly conserved sequences in the mouse and rat genomes. These ultraconserved DNA sequences are intriguing and have been implicated in regulating gene expression but further research is needed to understand the functions of these regions of the genome. The ultraconserved DNA elements have 20-fold fewer naturally occurring SNPs than the surrounding genome sequences, emphasizing the functional importance of DNA sequences that exist unchanged through hundreds of millions of years of evolution.

Repeated DNA Sequences

The Human Genome Project found that at least half (50%) of the total human genome DNA sequences (3.2 billion) are made up of repeated non-coding DNA (1,600,000,000 bp; 1,600 million bp). A DNA base sequence that is repeated hundreds of times in the genome is often called "junk DNA" because scientists do not yet understand the role of many types of repeated DNA sequences in the structure, function and metabolism of human chromosomes. However, research shows that some types of repeated DNA sequences play very important roles in generating new genes by shuffling and reshuffling existing genes through a mechanism that relies on the presence of repeated sequences in the genome. These types of chromosome rearrangements over millions of years of evolutionary time, have imparted some of the genetic flexibility exhibited by the modern human genome sequences.

The two main types of repeated DNA sequences in the human genome are tandem repeats and interspersed repeats. In the case of tandem repeats, the repeat unit is arranged with one DNA repeat unit starting immediately after the end of the previous repeat unit, without unique sequences in between the tandem repeat units in the chromosome DNA. In the human genome the tandem DNA repeats are usually composed of short DNA sequences (4–20 bp) that are very abundant and can extend for many hundreds of base pairs in the human chromosome DNA.

Satellite DNA Repeats

About half of the noncoding regions of the human genome (800,000,000 bp; 800 million bp) contain highly repeated satellite sequences, a special type of short tandem DNA repeat that contains primarily A+T–rich DNA sequences, and contains very few GC base pairs. The different satellite DNA repeats in the genome have specific DNA sequences and repeat unit characteristics that depend on the location of the satellite repeat in the human chromosome.

Satellite DNA repeats were named because of the unusual properties of this DNA when the A+T-rich repeated DNA is separated on density gradients formed by centrifugation. The centrifuge is an instrument that is used routinely in the lab to separate a mixture of cellular molecules by taking advantage of their physical and chemical properties. The components are loaded into special tubes that already contain a liquid salt solution designed to change density during centrifugation. The tubes containing the samples are spun rapidly in a circle in the centrifuge, roughly like tying the tube to a string and spinning it above your head very, very rapidly. During centrifugation the molecules in the salt solution redistribute in the tube and form a gradient of different densities throughout the tube.

At the start of the experiment the chromosome DNA is sheared into smaller fragments, because the very long DNA molecules tend to get tangled in the tube during centrifugation. The sheared double-stranded DNA fragments are loaded on the top of the contents in the centrifuge tube. As the tube spins, a density gradient forms in the tube, with less dense material at the top and the more dense material at the bottom of the tube. The genome DNA fragments migrate in the centrifuge tube to the point where the gradient solution has the same density as the DNA fragments.

When this experiment is performed with prokaryotic chromosome DNA fragments, the result is always a single peak of DNA at a specific density. This is because bacterial chromosomes consist of mostly unique-sequence DNA occurring only once in the genome with very little repeated DNA. However, when the human chromosome DNA is sheared and the fragments are spun on a density gradient and analyzed, there are two peaks of human DNA. The very large dense peak of human DNA represents the majority of the unique single-copy human DNA sequences, which is bulk chromosome DNA. The much smaller, second 'satellite' peak contains less dense, highly A+T–rich satellite DNA.

Different types of satellite DNA repeat sequences and different numbers of satellite repeat units are located in distinct regions of human chromosomes and can be used to help identify different chromosomes. In addition some types of repeated DNA have direct functions in essential processes. The repeated DNA sequences at the ends of eukaryotic chromosomes have essential roles in duplicating the DNA at the ends of each chromosome. During cell division, the centromere region of each chromosome attaches to spindle fibers that guide the chromosomes when they move during cell division (mitosis and moiosis). The centromere is the region on each chromosome where the kinetochore is assembled, a large protein-DNA complex that functions during cell division to bind the centromere region of the chromosome to the microtubule proteins in the spindle fiber. The kinetochore also provides a molecular motor function needed to move the duplicated chromosomes to opposite poles during mitotic cell division.

In eukaryotic chromosomes, the functions of the centromere and kinetochore are essential for correct chromosome movement and segregation during cell division. These functions are essential to be sure that the new cells will inherit the correct chromosomes and the correct genes. Each human chromosome contains thousands of satellite DNA repeats located in the regions flanking the centromere. Some DNA repeats bind to specialized chromatin proteins that form various types of chromatin fibers that condense and decondense with the chromosome during the cell cycle.

Minisatellite DNA Repeats

Minisatellite repeats are short DNA sequences that are similar to the consensus sequence "GGGCAGGANG" (N is any base), with a strong bias toward purines (A and G) on one strand and pyrimidines (C and T) on the other strand. Minisatellite DNAs are located at hundreds of different positions in the human genome and are frequent at the sites in the chromosome DNA where double-strand DNA breaks occur during chromosome recombination events, also called crossing-over, in which DNA strands are physically exchanged between chromosomes during meiosis. A majority of minisatellite repeats having this DNA sequence are found adjacent to the telomere regions at the ends of chromosomes (Figure 2-21).

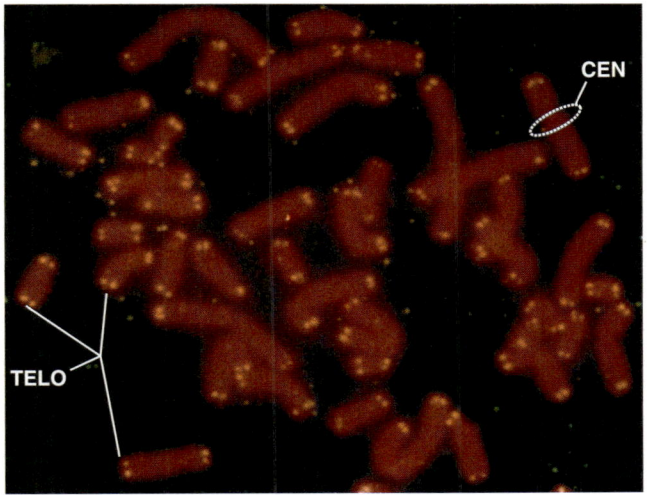

FIGURE 2-21 Human chromosomes have centromeres and telomeres Staining with fluorescent antibodies and DNA probes can visualize the locations of key regions, including telomeres and centromeres on eukaryotic chromosomes. The telomere regions at the ends of the chromosome contain repeated DNA sequences (yellow dots). These telomere DNA sequences are involved in replicating the ends of the chromosome DNA. The centromeres on the chromosomes are not stained but one centromere is circled (white dotted line).

Short tandem repeats (STRs) are an example of a type of short microsatellite repeat sequences (repeat unit of 1–6 bp) that are distributed around the human genome DNA (Unit 6). The number of repeats has been determined for 13 selected STRs located at different places in the human genome, which makes these STRs very valuable genetic tools used to identify humans by forensic DNA fingerprinting (Unit 6).

Jumping Genes move around Human Chromosomes

About half of the noncoding human genome sequences contain transposable DNA elements that physically move from one position to another position in the same or a different chromosome (Figure 2-22). Sometimes mobile DNA elements are called "selfish DNA" because these DNA elements encode the enzymes needed to allow the DNA elements to move from place to place in the genome. The chromosome DNA sequence that is targeted for insertion by the transposable element is cut by a special transposase enzyme that leaves 5 base sticky ends on the DNA. These sticky ends generate the flanking inverted repeat (IR) sequences that typically flank the transposable DNA element after it has inserted into the genome DNA (Figure 2-22).

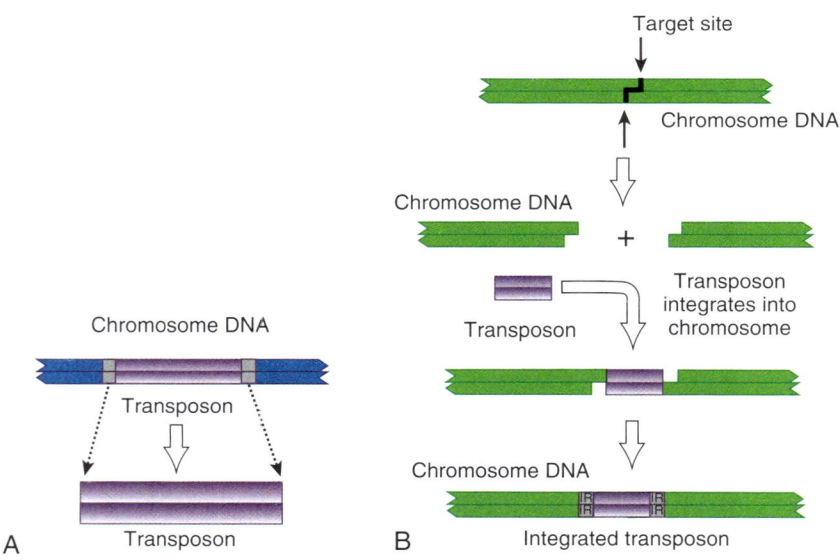

FIGURE 2-22 DNA transposons move around in the chromosomes (A) About half of the human genome DNA sequences do not code for proteins but do contain transposable elements that physically move from place to place in the chromosome DNA. (B) Some transposons code for enzymes that allow the transposon DNA elements to translocate (move) in the genome. A special enzyme cuts the chromosome DNA at the target site, leaving single-stranded extensions of 5 bases on the DNA ends. The transposable DNA element becomes integrated into the genome DNA in a process that copies the single-stranded DNA on the ends of the transposons. Integration of the transposon into the chromosome DNA creates inverted repeat sequences (IR) in the flanking genome DNA.

Retrotransposons Populate the Human Genome

To begin to understand the impact of mobile DNA elements on the human genome and to appreciate how they move around human chromosomes, it is important to review a few points about gene transcription in eukaryotic cells. First, eukaryotic cells normally use three different types of RNA polymerase enzymes to transcribe the different types of genes in the cell, including the protein-coding genes and the genes that are transcribed but the RNA copy is not translated into a protein (such as tRNA genes and rRNA genes). Second, in eukaryotic cells one type of RNA polymerase enzyme transcribes all of the protein-coding genes in the genome and a different type of RNA polymerase enzyme transcribes the tRNA genes and most rRNA genes. Third, researchers have studied how the RNA polymerase enzymes select the correct gene to transcribe (express) and found that the RNA polymerase enzymes select genes in part by recognizing the different DNA promoter sequences associated with the different types of genes. In addition different polymerase-specific transcription factors and other proteins bind to the correct promoter sequences of appropriate genes and help to recruit the appropriate RNA polymerase enzymes to transcribe a particular gene (Unit 3).

In contrast to tandem DNA repeats that are not interrupted by unique sequences, interspersed DNA repeats contain unique DNA sequences located in between the repeated units. Long interspersed repeated nuclear elements (LINEs) and short interspersed repeated elements (SINEs) both exhibit the ability to move around the human genome, although they use different mechanisms that depend on the locations of the gene promoters relative to the two types of genes. Protein-coding genes have RNA polymerase II promoters located in front of the coding region of the gene. As a result the RNA polymerase II promoters are not copied into RNA strands when the protein-coding genes are expressed. In contrast, the genes transcribed by RNA polymerase III have internal promoters located within the coding region of the gene and as a result the promoter sequence is included in the RNA transcripts copied from the gene (Figure 2-23).

SINEs are retro-transposons that move to different locations in human chromosomes using an RNA-mediated translocation mechanism. This mechanism explains not only how the SINEs move to different locations around the human genome, but also shows that these elements can duplicate in the process, increasing the number of SINEs in the genomes of human cells.

Each Human Genome has 1.5 Million Alu I DNA Repeats

The mechanism of retrotransposition begins when the internal promoter in the gene is used to make an RNA transcript encoding the internal promoter sequence. This is a key feature of the SINE retrotransposition amplification mechanism that is responsible for duplicating and distributing the Alu DNA sequences around the human genome (Figure 2-23). The RNA transcript encoding the internal promoter sequence is copied into DNA by the reverse transcriptase enzyme to produce a heteroduplex molecule, which is a double-stranded DNA-RNA molecule. Next the RNA bases in the heteroduplex

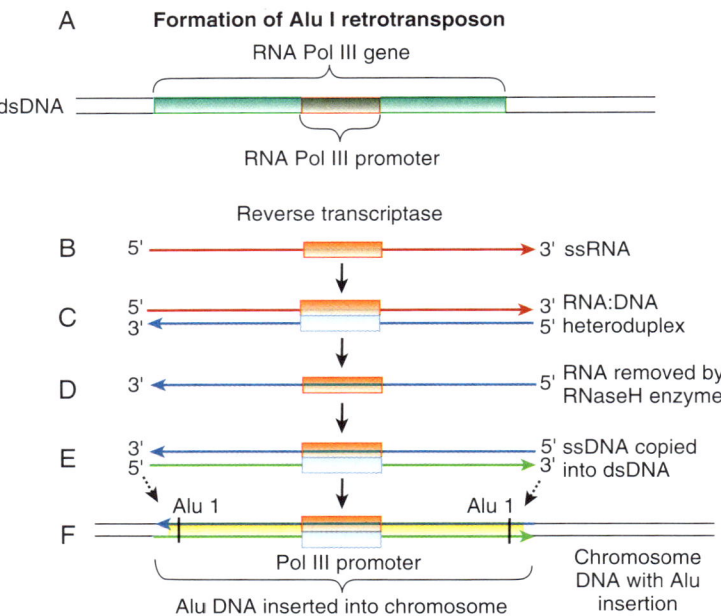

FIGURE 2-23 Alu repeats (SINEs) are retrotransposons (A) RNA polymerase III genes contain internal promoters within the coding region of the gene. (B) The RNA transcripts copied from the RNA polymerase III gene contain a promoter sequence within the RNA transcript. (C) The RNA transcripts are copied into complementary DNA strands by reverse transcriptase enzyme to produce a double-stranded DNA:RNA heteroduplex molecule. (D) The RNA strand in the heteroduplex is removed by the RNaseH enzyme, which leaves behind the single-stranded cDNA molecule. (E) This cDNA strand is copied into a double-stranded DNA molecule. (F) The dsDNA is inserted into the DNA at a new position in the chromosome.

molecule are removed by the RNaseH enzyme, and the remaining DNA strand is copied into a double-stranded DNA molecule. These retrotransposons are inserted into new positions in the human chromosomes, distributing the SINE elements in the genome. Over evolutionary time the retrotransposons amplified and distributed the SINE elements that are ubiquitous in modern eukaryotic genomes. The retrotransposons are particularly abundant in plant genomes, 49–78% of the maize (corn) genome contains retrotransposons and about 90% of the wheat genome consists of repeated DNA, of which 68% of the sequences are retrotransposon elements.

> **KEY CONCEPT**
> RNA polymerase III genes carry their own internal promoter and generate DNA elements that can transpose (move and leave a copy behind) to another chromosome location mediated by retrotransposition.

The Alu I DNA elements, which are named after the Alu I restriction enzyme, are the most common SINEs in the primate genomes, including the human genome. When human chromosome DNA is cut with the

Alu I restriction enzyme at its recognition sequence, 5'-AGCT-3', the reaction produces millions of DNA fragments of many different lengths. As expected, when resolved on an agarose gel and stained with ethidium bromide, the Alu I DNA fragments appear under UV light as a fluorescent smear. However, in addition to the DNA smear, a strong fluorescent DNA band is also visible above the fluorescent smear migrating in the gel at a length of about 280 base pairs, which corresponds to the length of the Alu I elements found in the human genome. This DNA band is composed of 1.5 million copies of the Alu I DNA repeat that were released from the human genome DNA by cleavage with the Alu I restriction enzyme. It is amazing that the Alu I DNA elements can be detected easily on the stained gel without amplification or the use of DNA probes for detection. The Alu I DNA sequences account for over 10% of the total sequences in the human DNA genome! The Alu elements are not only abundant in the DNA genome, the Alu elements are also widely distributed in human chromosome DNA.

Scientists think that the Alu I DNA elements probably originated in ancient primate genomes as derivatives of the 7SL RNA molecule, which is part of the signal recognition particle (SRP). In the cytoplasm of eukaryotic cells the SRP functions to target newly made proteins into the endoplasmic reticulum (ER), the membrane system that modifies proteins for secretion.

The Alu I repeat sequences are the most abundant mobile DNA elements in the modern human genome. Scientists have used the Alu I repeats for studies in human population genetics and for research on the evolution of primate genomes. However, SINEs, LINEs, and Alu I transposons have been implicated in causing human genetic diseases and cancers that are associated with the occasional insertion of a retrotransposon into an active protein-coding gene in the genome. Although such events are rare and are probably directly responsible for only a small number of mutations leading to human disease, there are still real people affected by this statistic. Interspersed repeats in the human genome have been linked to genetic diseases including hemophilia A (Factor VIII gene) and hemophilia B (Factor IX gene), X-linked severe combined immuno-deficiency (SCID), porphyria, predisposition to colon cancer (APC gene), and Duchenne muscular dystrophy (dystrophin gene).

Long Interspersed Nuclear Elements

LINEs are the second major type of repeated DNA elements that are found in eukaryotic chromosomes. RNA polymerase II transcribes protein-coding genes into pre-mRNAs which, like the LINE RNA transcripts, start just after the upstream promoter and do not contain a copy of the RNA polymerase II promoter. The RNA polymerase II transcripts contain introns, exons and the polyadenylation sequence (AAUAAA) in the RNA that signals a specialized polymerase to add a poly(A) tail to the 3' end of the RNA transcript (Unit 3).

The human genome contains a total of about 500,000 LINEs but most of these are only partial copies in the genome and only about 7,000 LINEs are full-length copies that have the ability to move around the genome DNA using a version of the retrotransposition mechanism described for SINEs. Some LINEs carry genes that code for reverse transcriptase and RNaseH, which are two enzymes involved in the retrotransposition of SINEs in the human genome. Apparently the LINEs sometimes provide support functions for the duplication and retrotransposition of the SINEs.

The DNA sequences of individual human genomes are currently determined by automated DNA sequencing machines and analyzed by very powerful computers. However, determining the entire base pair sequence of an individual's genome is not the most efficient way to identify a mutant gene, because the vast majority of the human genome DNA does not code for genes. However, recently a new approach to human genome analysis is called exome sequencing, which gets its name because only the protein coding exons in an individual's genome are sequenced, which in total represents less than 1.5% of the total human genome DNA sequences. Scientists have found that comparing the exome DNA sequences obtained from unrelated individuals suffering from the same genetic disease can quickly reveal the gene that causes the disease in question. This exome sequencing method is less expensive and more rapid than whole genome analysis to rapidly identify multiple genes responsible for genetic diseases, but it is limited to analyzing exon sequences only.

Unit 2 Questions

1. A karyotype analysis of human chromosomes shows:
   a. the eukaryotic nucleus with extended thread-like chromosome fibers
   b. the surface topology of the human sex chromosomes
   c. just the longest human chromosomes
   d. all of the condensed chromosomes in a single human cell nucleus

2. During the replication of eukaryotic chromosomes:
   a. DNA is copied into RNA
   b. RNA is copied into DNA
   c. DNA is copied into DNA
   d. RNA is copied into RNA

3. The ends of a single strand of DNA contain:
   a. A hydroxyl (OH) group on both ends (3′ OH and 5′ OH) of the DNA
   b. A phosphate (P) group on both ends (3′ P and 5′ P) of the DNA
   c. A hydroxyl (OH) group at one end (3′ OH) and a phosphate at the other end (5′ P)
   d. A hydroxyl (OH) group at one end (5′ OH) and a phosphate at the other end (3′ P)

4. Each DNA strand in a DNA helix contains:
   a. A hydroxyl (OH) group on both ends (3′ OH and 5′ OH) of each backbone strand
   b. A phosphate (P) group on both ends (3′ P and 5′ P) of each backbone strand
   c. A hydroxyl (OH) group at one end (5′ OH) and a phosphate at the other end (3′ P) of each backbone strand
   d. A hydroxyl (OH) group at one end (3′ OH) and a phosphate at the other end (5′ P) of each backbone strand

5. An enzyme that copies a single-stranded DNA template into a double-stranded DNA helix is called:
   a. A DNA polymerase
   b. A ribosome synthetase
   c. A restriction enzyme
   d. An RNA polymerase

6. An enzyme that expresses genes by copying the DNA template of a gene into RNA is called:
   a. An RNA polymerase
   b. A ribosome synthetase
   c. A restriction enzyme
   d. A DNA polymerase

7. DNA sequencing is best described as:
   a. A process where a DNA primer base-pairs to the target chromosome DNA spread on a glass microscope slide
   b. A method determining the linear order of the bases in a DNA strand, using enzymes and biochemical reactions
   c. A method using a DNA primer with a 5′ phosphate group (5′ P) to make a new RNA strand
   d. Any lab method using an RNA primer to generate the sequence of a DNA strand

8. When migrating through a gel under the influence of an electrical current, the DNA molecules move toward:
   a. The negative pole because DNA is positively charged
   b. The middle of the gel, because all the DNA strands are the same length
   c. The positive pole, because DNA molecules are negatively charged
   d. The top of the gel, because DNA can't enter the gel unless the DNA molecules are bound to proteins

9. The special enzyme that amplifies DNA in PCR reactions has which unusual characteristic?
   a. The special enzyme copies DNA much faster than other enzymes, so it can make a lot of DNA in a short time.
   b. The special enzyme is actually an RNA molecule and not a protein.
   c. The special enzyme is resistant to the high temperatures used in PCR reactions.
   d. The special enzyme is extraordinarily accurate and never makes a mistake.

10. Reverse transcriptase is an unusual enzyme because it can:
    a. Copy proteins into RNA even under high temperatures
    b. Copy a DNA strand into an RNA strand
    c. Copy an RNA strand into a DNA strand
    d. Bend the DNA helix while it copies the DNA into RNA

UNIT 3

Gene Expression Makes RNA and Proteins

The Flow of Genetic Information: DNA to RNA to Protein

The flow of genetic information required to make proteins in a eukaryotic cell begins with the genes encoded by the chromosome DNA in the nucleus (Figure 3-1). At appropriate times, selected protein-coding genes are expressed by copying the gene into RNA transcripts (pre-mRNAs), which are processed in the nucleus by special enzymes that produce mature messenger RNAs (mRNAs). These mRNAs are transported to the cytoplasm, where the ribosome binds to the mRNA and translates the coding region of the mRNA into a specific protein. Understanding the basic structures of the human protein-coding genes and how these genes are regulated in eukaryotic cells is important because RNA processing is one of the mechanisms that controls the transport of RNAs from the nucleus to the cytoplasm for translation into protein.

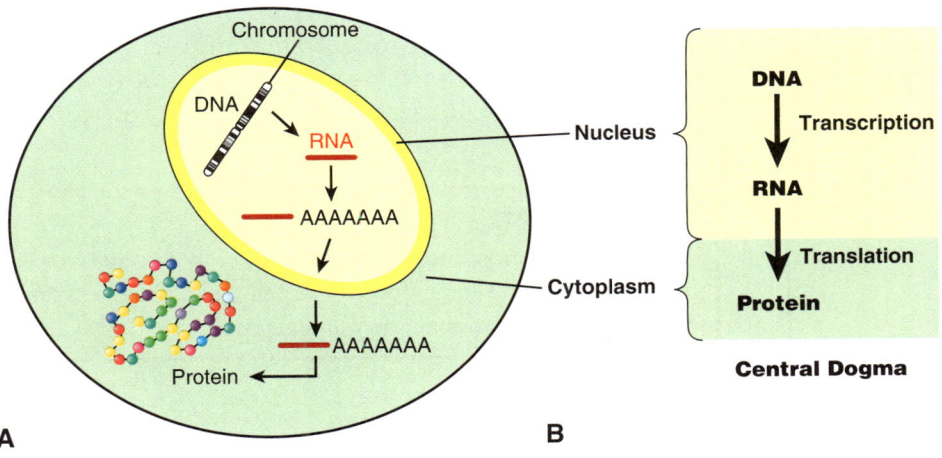

FIGURE 3-1 Overview of the flow of genetic information. (A) The flow of genetic information in a eukaryotic cell begins with the genes carried by the chromosomes in the nucleus. (B) DNA genes are copied into RNA by transcription in the nucleus. The RNAs are processed and transported to the cytoplasm where the RNAs are translated into proteins.

The human genome DNA sequence project has revealed valuable information about the genes and proteins that make us human and raised many new questions about gene expression. An adult human body has more than 600 trillion cells that perform a diverse range of biological functions in the body. The many different types of human cells are usually named by the tissue of origin. It is very important to remember that all of the cells in an individual human body contain the same DNA genome and the same genes. The human body produces many specialized types of cells with different functions that result from the expression of different genes that make specialized proteins. A very active area of research focuses on exploring how the cells in the human body, which all contain exactly the same genes, manage to make such a large number of different types of cells.

Understanding how different cells control gene expression requires information about the DNA control elements associated with genes and the regulatory proteins made in specific cells. Different types of cells produce specific regulatory proteins that control cell functions, regulate the development of tissues and organs, and even produce the proteins that influence personality traits and behavior.

Human genetic information is stored in the genome in chromosomes located in the nucleus (Figure 3-1). Because all of the cells in an individual's body have the same genes, the key to developing specialized cells is to control gene expression and produce only those proteins required for the specialized functions. Gene regulation at the level of transcription involves selecting the correct genes in the genome to be expressed in the correct cells at the correct time. Gene selection involves the actions of specialized transcription factors and other accessory proteins that interact with the DNA control elements in the genes.

Eukaryotic cells have a hierarchy of sophisticated mechanisms to control gene expression and generate the types of cells needed to perform specialized biological functions in the human body. This amazing process begins with fertilization and embryogenesis and continues after birth with the development of an adult body that relies on hundreds of types of cells that function in dozens of tissues and organs. To accomplish this feat the cells must flawlessly control the changes in the expression of many genes; some genes are turned on and other genes are turned off during growth and development.

> **KEY CONCEPT**
> In any one human body, all of the cells in that body have exactly the same DNA genome, and the same genes. The cells control gene expression to create different types of cells that have different functions in the body.

Expression of a Typical Human Gene

A clear picture of the general structure of a typical eukaryotic protein-coding gene is very important to better understand how human cells control gene expression. RNA polymerase II (also called RNAPII, Pol II) is the type of RNA polymerase enzyme that expresses the protein coding genes in eukaryotic cells, including human cells. The two other RNA polymerase I and

III enzymes express the genes in eukaryotic cells that code for RNAs that are *not* translated into proteins. RNA polymerase I transcribes ribosomal RNA genes (rRNA), and RNA polymerase III transcribes transfer RNA (tRNA) genes. The rRNAs and tRNAs are not translated into proteins but they do perform essential functions in the ribosome during protein synthesis.

The overall structure of the protein-coding genes expressed by the RNA polymerase II enzyme includes two major regions of DNA sequences (Figure 3-2):

(1) DNA control sequences that are not copied into RNA, and
(2) protein coding sequences that are copied into RNA.

The DNA control regions associated with different RNA polymerase II genes contain a promoter DNA region and a conserved TATAAAA sequence called a TATA box (Figure 3-2). The TATA box consensus sequence was identified by comparing the promoter DNA sequences of many different RNA polymerase II genes, showing that the majority of the protein coding regions are preceded by a promoter and a TATA box. The RNA polymerase II promoter region and TATA box are positioned in the RNA polymerase gene just before the site in the DNA (+1) where transcription starts. The RNA polymerase promoter and TATA box DNA are never copied into the RNA transcript (Figure 3-2).

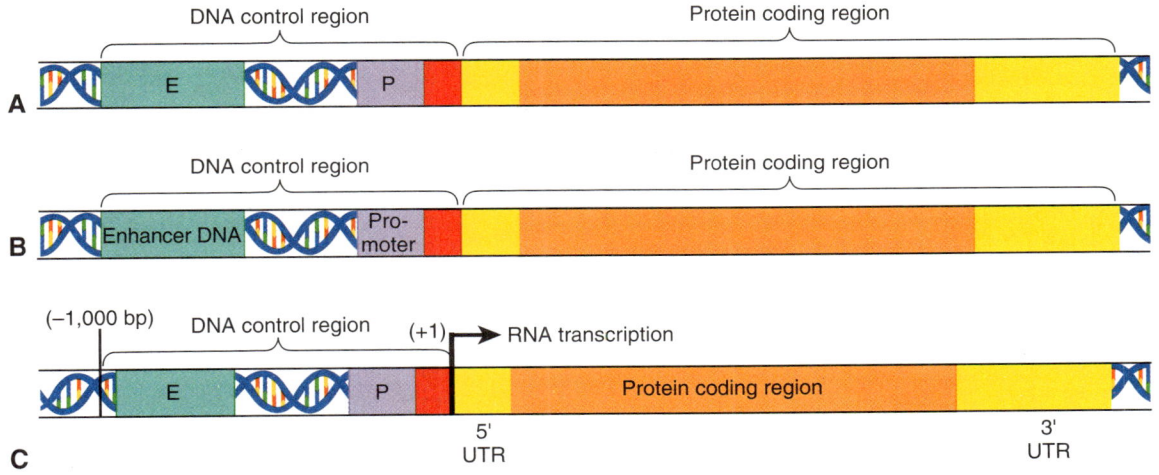

FIGURE 3-2 Structure of a eukaryotic protein-coding gene (A) Protein-coding genes are transcribed by the RNA polymerase II enzyme and contain two DNA regions with different functions.
- Control DNA sequences are located upstream of the protein coding region and interact with the RNA polymerase II enzyme. The control DNA sequences are not copied into the pre-mRNA strand.
- Protein coding sequences are located in the gene following the Control DNA region and are copied into the pre-mRNA strand.

(B) The Control DNA sequences in the protein coding genes include the Enhancer element (E) (green), the Promoter DNA element (P) (purple) and the TATA box DNA (red).

(C) The Enhancer (E) (green), the Promoter region (P) (purple) and the TATA box (red) are positioned in front of the transcription start site in the DNA, as indicated by (+1). The direction of RNA transcription is indicated by the arrow. The enhancer, promoter and TATA box sequences are not copied into RNA. The promoter and TATA box are always positioned near the transcription start site (+1) but the enhancer DNA elements can be located over 1,000 bp away from the gene on the same DNA helix.

The promoter regions of the RNA polymerase II genes play a key role in selecting which genes should be expressed, because cells produce only certain proteins at certain times. When conditions change, the cells must be able to turn on the expression of certain genes to produce the specific proteins needed to respond to the situation. Eukaryotic cells use different mechanisms to control gene expression at different points in the gene expression pathway, including promoter selection by RNA polymerase. A wide range of different specialized transcription factors (TF) play key roles in starting and controlling gene expression in different cells. Some specialized regulatory proteins bind to the conserved DNA control sequences in the promoter region of the protein-coding genes to exert control over gene expression.

RNA Polymerase II Enzyme Binds to Promoter DNA

For decades scientists have used many different approaches to study and characterize the protein and DNA components involved in RNA polymerase II transcription and to better understand how this enzyme executes the gene expression instructions in the cell. RNA polymerase II is a large, multi-subunit enzyme that can distinguish among the thousands of protein-coding genes in the cell with the help of specific transcription factors as well as specific DNA promoter sequences. The proteins associated with the RNA polymerase II enzyme bind to the promoters of the protein-coding genes are encoded by separate genes in the genome (Figure 3-3). Some accessory proteins interact with the polymerase enzyme only temporarily to perform a specific function needed for transcription initiation for example, while other proteins remain associated with the RNA polymerase II enzyme after the complex starts to copy the DNA template into RNA (Figure 3-3).

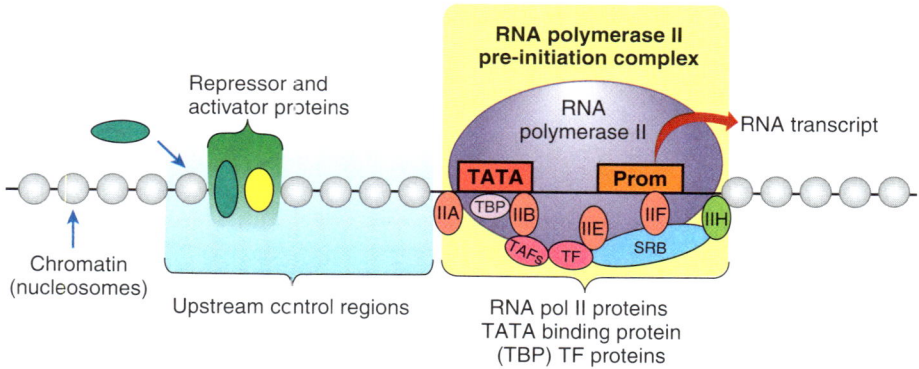

FIGURE 3-3 RNA polymerase II pre-initiation complex forms at the promoter The RNA polymerase II pre-initiation protein complex assembles on the promoter DNA to start transcription of a protein-coding gene. Promoter selection by RNA polymerase II is a key step in controlling gene expression because the RNA polymerase II enzyme must select the correct protein-coding gene to transcribe. Promoter selection involves many specialized proteins that help to determine which genes should be expressed by the RNA polymerase II enzyme in different types of cells at different times.

The specialized TF proteins play key roles in controlling gene expression in eukaryotic cells. The RNA polymerase II pre-initiation complex assembles on the promoter DNA region along with specialized transcription factor proteins, including transcription factor IIB (TFIIB) and a dimer of two TFIIE proteins (Figure 3-3, Figure 3-4). As the TATA box binding protein (TBP) binds to the TATA box DNA in the promoter, it bends the DNA helix, changing the direction of the promoter region of the DNA helix relative to the coding region of the gene (Figure 3-5). The TFIIE and TFIIF proteins interact with the carboxy-terminal domain (CTD) of the RNA polymerase II enzyme.

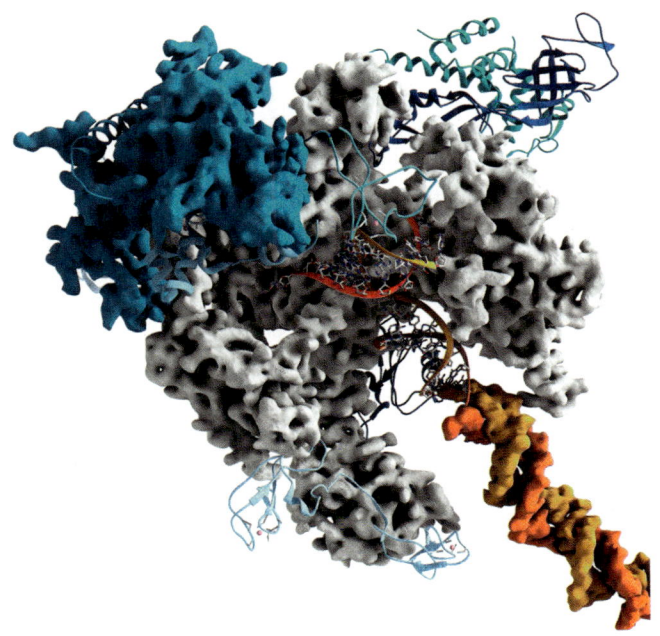

FIGURE 3-4 RNA polymerase II enzyme The RNA polymerase II enzyme (blue and white) begins transcription of the template DNA (yellow) and makes a single-stranded RNA (pre-mRNA).

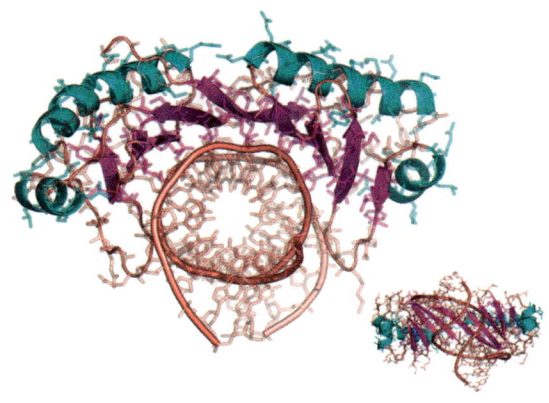

FIGURE 3-5 DNA helix is bent by the RNA polymerase II pre-initiation complex The TATA box binding protein (blue and purple alpha helices) binds to the TATA box DNA sequence (DNA helix (pink) is shown on-end) and creates a bend in the DNA helix that is characteristic feature of the RNA polymerase II promoter DNA region.

The TFIIH protein adds a phosphate to the CTD, which triggers the RNA polymerase II enzyme to initiate transcription of the gene.

Research has clearly established that the promoter region and TATA box located upstream of the protein coding region are sufficient to express RNA polymerase II genes outside of cells in the lab (*in vitro*). However, additional research showed that authentic expression and regulation of the RNA polymerase II genes in living cells (*in vivo*) requires the function of additional DNA sequences located at some distance from the gene being expressed (Figure 3-6). Scientists were intrigued to find that important regulatory sequences called enhancer and silencer DNA elements are located hundreds or thousands of base pairs away from the gene on the same chromosome DNA. Enhancer and silencer DNA elements are usually found upstream of the gene they control, but enhancer elements have also been identified far downstream of genes and are occasionally found in the intron DNA associated with a neighboring gene.

Enhancer proteins bind to the enhancer DNA element and can also interact with the TF proteins that bind to the promoter DNA at the transcription start site of the gene. These DNA-protein interactions create a physical loop in the chromosome DNA that brings together two distant regions of the same chromosome DNA helix to promote interactions between the enhancer elements and the core promoter sequences (Figure 3-6). In this way the proteins bound to the enhancer DNA element are brought in contact with the transcription factors and other specialized proteins associated with the pre-initiation complex bound to the promoter region of the gene (Figure 3-6). Interactions between an enhancer DNA element and the promoter region of a RNA polymerase II gene typically increase the transcription rate of the gene (the number of new RNAs initiated in an interval of time) and increases the total number of RNA copies made from the gene. Similar interactions involving a silencer DNA element will typically suppress transcription of the gene, decreasing the number of RNA copies made in the cell from that gene.

> **KEY CONCEPT**
> The RNA polymerase II control regions include enhancer and silencer DNA elements, the core promoter, and the TATA box sequences. None of these DNA control regions are copied into RNA during transcription.

Gene Expression is often Controlled at the Start of Transcription

The promoter DNA sequences of different protein-coding genes are partly conserved but also have unique sequences that help RNA polymerase II enzymes to select the correct genes to transcribe. The promoter sequences in the control region of each gene are bound by promoter-specific TF proteins that help the RNA polymerase II enzyme to recognize specific genes to express.

The upstream enhancer and silencer elements, the promoter DNA sequences and the polymerase-specific TF proteins all have key roles in the

UNIT 3 Gene Expression Makes RNA and Proteins

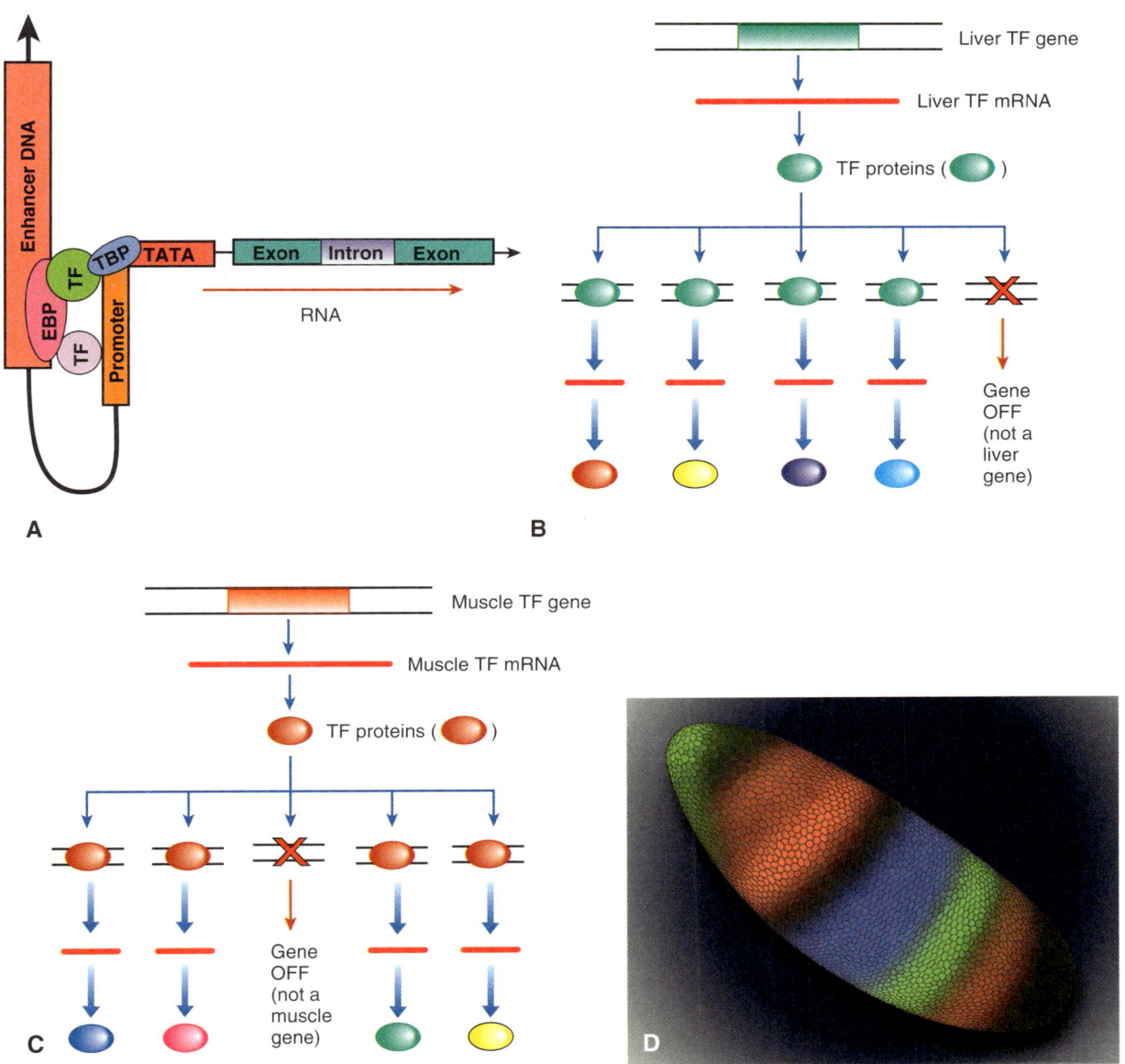

FIGURE 3-6 Enhancer elements control tissue-specific gene transcription (A) The promoter DNA alone is sufficient to initiate RNA polymerase II transcription in the lab (in vitro), but authentic regulation of the RNA polymerase II genes in living cells (in vivo) also requires the function of the upstream DNA sequences such as the enhancer (and silencer) DNA elements, which form a looped region of double-stranded chromosome DNA. The red arrow indicates the direction of RNA transcription. (B) In liver cells, the liver-specific TF proteins act to turn on liver-specific genes to produce proteins that function only in liver cells. (C) Other cell types such as muscle cells, express tissue-specific TF proteins to turn on muscle-specific genes to make proteins that function only in muscle cells. (D) All the cells in an organism contain a complete copy of the same DNA genome. In the fruit fly embryo shown here, some genes are expressed while other genes are switched off and are not copied into RNA. The expression of the regulatory proteins in the embryo are detected using fluorescent antibodies including Knirps (green), Kruppel (blue) and Giant (red), which are all proteins that bind to upstream control DNA elements of the genes. The black regions do not express these regulatory proteins.

regulatory mechanisms that activate and inhibit gene expression. In eukaryotic cells the promoters contain specific DNA sequence variations that influence the binding of the regulatory proteins to the promoter region to control expression of a particular gene. This regulatory system can start or block gene transcription based on interactions among the RNA polymerase II enzyme and specialized proteins, including TFs. Certain TF proteins are expressed in certain cell types, allowing the TF proteins to influence the expression of tissue-specific genes. For example, liver-specific TFs are expressed only in liver cells and bind to the promoter regions of certain liver-specific genes that should be expressed in liver cells (Figure 3-6). Specialized proteins such as TFs function to selectively express appropriate genes in different cell types such as muscle (Figure 3-6).

The genetics of *Drosophila melanogaster* are extremely well known, including the genes involved in specific cell and tissue development during embryogenesis. Like other eukaryotic organisms, all the cells in a fruit fly embryo contain a complete copy of the DNA genome. As expected, gene expression in the fruit fly embryo is coordinated by regulatory proteins that are made in the different types of cells and control gene transcription by binding to upstream control DNA elements. This gene expression mechanism can be visualized in the fruit fly embryo using different colored fluorescent antibody proteins that bind to specific proteins made in the embryo. The antibodies fluoresce different colors by detecting the production of different regulatory proteins during fruit fly development. The regulatory proteins observed in the embryo included Knirps, Kruppel and Giant (Figure 3-6). Fruit fly scientists are well known for the innovative names given to *Drosophila* genes and mutations.

The "simple" step of turning on the transcription of a gene to make an RNA copy to translate into a protein actually involves many steps and a host of specialized proteins and enzymes. The proteins must perform specific functions at specific times to recruit RNA polymerase II to the correct gene promoter. The RNA polymerase I and RNA polymerase III enzymes must recognize and bind to very different promoter DNA sequences in the ribosomal RNA (rRNA) genes and transfer RNA (tRNA) genes. All three eukaryotic RNA polymerase enzymes (I, II, and III) rely on specific transcription factors to express the appropriate selected genes.

Processing and Splicing RNA Polymerase II Transcripts

In addition to transcription initiation, the expression of the RNAs transcribed by the three different RNA polymerase enzymes can be regulated at many steps along the gene expression pathway from the nucleus to the cytoplasm (DNA to RNA to protein) (Figure 3-1). The RNA polymerase II enzyme transcribes the DNA encoding each protein into a long precursor RNA (pre-mRNA) molecule containing a 5' UTR, a coding region that includes exon and intron sequences and a 3' UTR (Figure 3-7). The untranslated regions of the transcript (5' UTR and 3' UTR) are copied into the precursor RNA molecule but are not translated into protein. RNA processing represents another step in the gene expression pathway that can be regulated to control the amount and timing of gene products.

UNIT 3 Gene Expression Makes RNA and Proteins

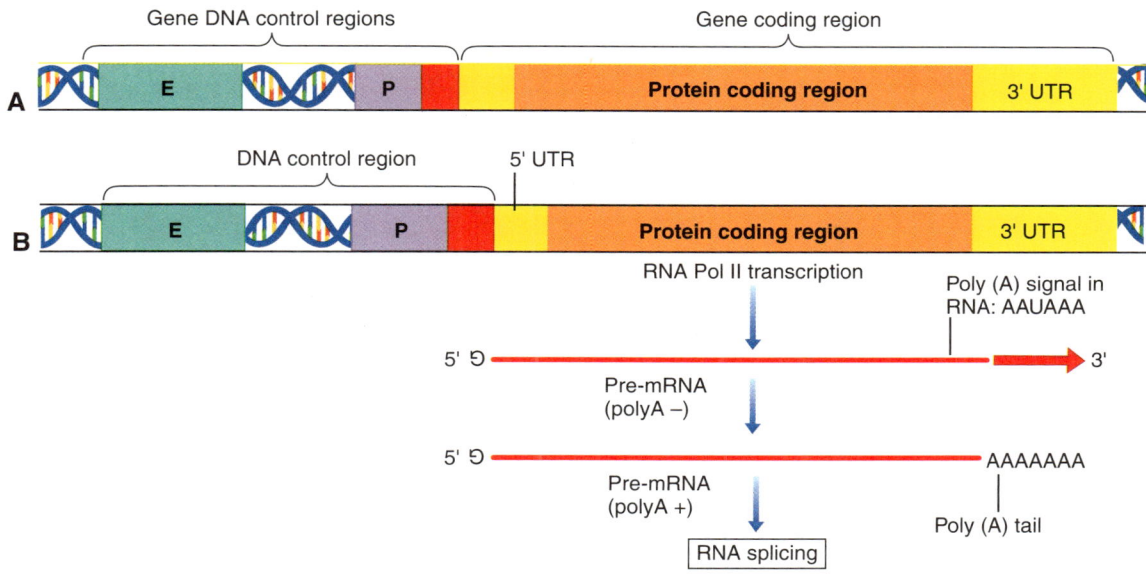

FIGURE 3-7 RNA polymerase II genes are transcribed into pre-mRNAs (A) The RNA polymerase II enzyme copies a protein coding gene into a long precursor RNA (pre-mRNA) molecule that contains a 5′ UTR (5′ untranslated region), a coding region with both exons and introns and a 3′ UTR (3′ untranslated region). The untranslated regions in the RNA transcript are included in the precursor RNA molecule but are not translated into protein. The Control Regions include the promoter (P) (purple), TATA Box (red), and enhancer (E) (green). (B) Polyadenylation is the process of adding a long poly (A) tail onto the 3′ end of the pre-mRNA strand. The pre-mRNA is cut at the AAUAAA signal by one enzyme and another enzyme adds the poly(A) tail to the 3′ end of the mRNA.

The RNA sequences encoded by the UTRs in the pre-mRNA have important functions in the processing and splicing events, which convert the pre-mRNAs into the mature messenger RNAs (mRNAs) that contain the molecular features required for translation. To be exported from the nucleus to the cytoplasm for translation, the pre-mRNA molecules must be processed into mRNAs in the nucleus. As soon as the 5′ end of the newly synthesized pre-mRNA molecule is available, a "cap" is added to the 5′ end and a poly(A) tail is added to the 3′ end of the pre-mRNA (Figure 3-7). The cap is actually an upside down G base added to the 5′ end to protect the end of the RNA strand during processing.

Poly(A) "Tail" is used to Isolate mRNAs

Gene expression research often involves isolating and purifying specific protein coding mRNA molecules, which are typically made at low copy levels in the cells. In contrast massive numbers of tRNAs and rRNAs are needed to accommodate the high rates of protein synthesis required in growing cells (Unit 2). Scientists have developed staining methods to distinguish cells that are expressing one protein from other cells in the population that are not expressing that protein but a different approach was needed to physically isolate the relatively rare mRNA molecules. To solve this problem, scientists exploited the structure of mRNAs and devised an affinity column strategy to isolate mRNAs for many types of analysis, including Northern blotting (Unit 6) and *in vitro* translation methods to make protein products for study.

To be exported from the nucleus to the cytoplasm, a eukaryotic mRNA must contain a poly(A) tail, which is a string of about 200 "A" bases that is added to the 3' end of the pre-mRNA by the process of polyadenylation (Figure 3-8). The pre-mRNA strand is cleaved by a special enzyme at a site in the RNA strand located several nucleotides past the "AAUAAA" consensus sequence in the 3' UTR of the pre-mRNA. The "AAUAAA" sequence also signals the action of a second enzyme that adds a poly(A) tail to the 3' end of each pre-mRNA, one A base at a time. The A bases in the poly(A) tail are added after transcription by a post-transcriptional processing event and are not encoded in the DNA sequence of the gene.

The fact that almost all the mRNAs in the cell contain poly(A) tails offered scientists an excellent way to efficiently separate the relatively rare mRNAs from the large number of rRNAs and tRNAs normally made in living cells. Scientists made an affinity column containing a matrix of beads with oligo(dT) strands (5'-TTTTTTTTTTTT-3') attached. The total RNA preparation is applied to the column and poly(A) tails on the ends of the mRNAs form hydrogen bonds with the T bases in the oligo(dT) on the beads, causing the mRNAs to "stick" to the column (Figure 3-8). The poly(A-) RNA molecules that do not bind to the oligo(dT) beads pass through the column and are collected in the Poly(A-) fraction, which includes rRNAs and tRNAs. The mRNAs remain bound to the oligo(dT) beads on the column. The column is washed with a special buffer that breaks the H-bonds between the poly(A) tail and the oligo(dT) releasing the mRNAs from the beads. The Poly(A+) fraction is collected from the column and contains partially purified mRNA molecules that are used in many different applications (Figure 3-8). The mRNA samples are often converted into double-stranded DNA copies of the mRNA (Unit 2). The mRNA from the affinity column can be copied into single-stranded DNA by the reverse transcriptase enzyme and then converted into double-stranded complementary DNA (cDNA) by the DNA polymerase enzyme. The double-stranded cDNA molecule is much more stable than the single-stranded mRNA molecules and DNA can be manipulated in the lab in ways that are not possible with mRNA. For example, a double-stranded DNA fragment can be inserted into a vector in a recombinant DNA cloning experiment and used to express a foreign protein in a host cell. This is possible because the cDNA represents only the exon sequences in the gene that were expressed as mRNA, the cDNAs were copied from a spliced mRNA (Unit 2).

Specific mRNAs are Analyzed by Northern Blot Transfer

The Southern blot transfer is a very powerful method used to identify specific coding and non-coding DNA sequences among the entire genome of 3.2 billion DNA base pairs (Unit 6). This is feasible because the Southern blot method uses highly specific DNA probes that are complementary to a

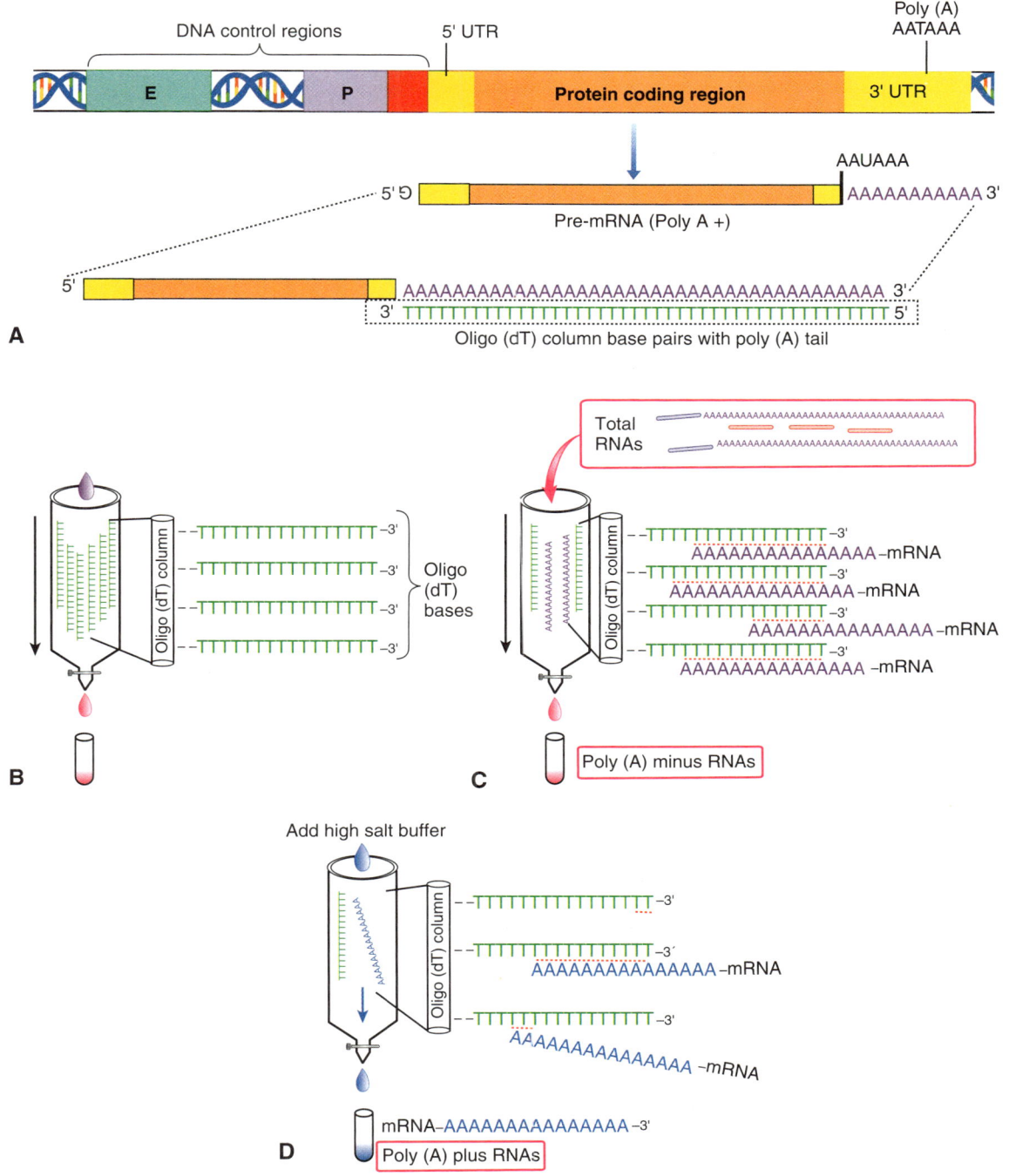

FIGURE 3-8 RNA polymerase II mRNAs are 'hooked' by their poly(A) tails (A) The long string of A bases on the 3' end of the mRNA, the poly(A) tail, can be used to purify mRNAs using an affinity column. (B) The affinity column contains beads attached to strings of oligo(dT) bases that are used to isolate the mRNAs. (C) Total RNAs harvested from cells are added to the column and the A bases in the mRNA poly(A) tails base pair to the oligo(dT) bases on the column. RNAs without poly(A) tails will pass through the column and are collected in the Poly(A−) fraction. The Poly(A+) RNAs (mRNAs) remain on the column. (D) The addition of a high salt buffer to the column disrupts the H-bonds between the oligo(dT) and the poly(A) tails on the mRNAs. The mRNAs are released from the oligo(dT) column and are collected as the Poly(A+) RNA fraction.

DNA (or RNA) target strand, and are used to detect the target DNA (or RNA) among a large number of other DNA (or RNA) strands. The Southern blot method was modified for the forensic DNA fingerprinting technology used to identify humans, as well as other applications in medicine and biotechnology (Unit 6).

The Southern blot method was originally developed to analyze DNA molecules, but soon after a modified version called the Northern blot was developed to analyze RNAs, which allowed scientists to easily analyze the specific mRNA transcripts made in living cells, even mRNAs that are present at low copy number in the cell. In a Northern blot experiment the RNA strands are separated on a gel much as described for DNA and then the RNAs are transferred from the gel onto a special membrane sheet that represents a replica of the RNA strands in the gel (Unit 6). The DNA probes used in Northern blot hybridizations are complementary to the RNA target strands, and the results are visualized by exposing the X-ray film to the membrane, as described for a Southern blot.

The ability to determine which mRNAs are copied from which genes in the genome is essential to our ability to study and understand cell functions. All human genomes carry the same genes, but different cells express different genes in the genome. The differential expression of genes in specialized cells and tissues can be monitored by Northern blot analysis using specific DNA probes designed to detect only the mRNAs expressed from selected genes. For years Northern blots have been used in countless experiments to examine the expression of many human genes in different cell types and under different growth conditions. Northern blot analysis has been used to compare the expression of a specific gene in a healthy tissue such as kidney, with the RNAs expressed from the genes in the cells from a kidney tumor. In other applications, Northern blots are used to compare the mRNAs made in normal cells with the RNA expression patterns in the same cells with a genetic disease or disorder.

Protein-Coding Genes Contain Intron and Exon Sequences in the Genome

Human protein-coding genes are transcribed into very long pre-mRNAs that contain intron and exon sequences (Figure 3-9). The pre-mRNAs must be spliced to remove the introns and connect the RNA exons together with base-pair accuracy. A look at a very short segment of genome DNA sequence encompassing just three exon/intron splice junctions makes it easy to appreciate the difficulties encountered in detecting genes just by visual inspection of the primary DNA sequence (Figure 3-10). The ability to produce an mRNA transcript that codes for a specific amino acid sequence depends on the accuracy of the RNA splicing mechanism to remove the intron sequences and create the final mRNA.

The complex RNA splicing mechanism in eukaryotic cells involves molecular components that contribute to the accuracy of the RNA splicing process, including small nuclear RNA (snRNA) molecules (U1, U2, U4, U5, and U6) that bind to small specialized proteins and make small ribonuclear protein particles (snRNPs). These molecular components assemble

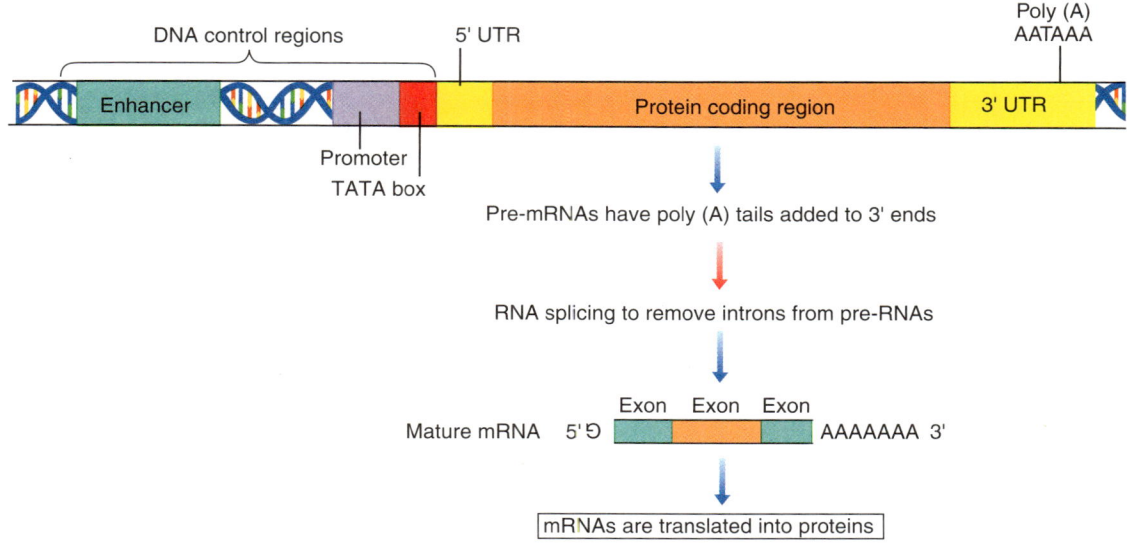

FIGURE 3-9 Pre-mRNAs are spliced to create the protein coding region in the mRNAs The pre-mRNAs transcribed in the nucleus must be processed into mature mRNAs before they can be transported to the cytoplasm for translation into proteins. The pre-mRNAs are processed so that the final mature mRNAs contain upside down G caps on the 5′ ends, the exon sequences that are spliced together to create the protein coding region and a poly(A) tail is added to each 3′ end. After RNA splicing is completed, the mature mRNAs are transported to the cytoplasm and translated into specific proteins as dictated by each mRNA coding region.

into a large ribonucleoprotein (RNP) complex called a spliceosome. This RNA-protein complex removes the intron sequences from the RNA transcript by catalyzing a lariat mechanism that removes the intron and joins the two remaining exon sequences together into the final mRNA. The RNA in the intron contains an "A" base at the lariat branch point in the RNA sequence, which plays an essential role in the molecular mechanism of the enzyme splicing function (Figure 3-11).

The accuracy of the splicing mechanism depends on the accuracy of the spliceosome components, which produce an accurate coding region with an open reading frame in the mRNA that codes for the protein. The spliceosome loops the intron RNA strand as described in the lariat mechanism, bringing together the ends of the exons, and sealing the bond with each RNA junction accurate to a single base (Figure 3-11). Even an error of a single base has the potential to destroy the protein coding region of the spliced, mature mRNA, which must contain a specific open reading frame to code for the correct amino acid sequence of the protein.

Alternative RNA Splicing Increases Human Genome Flexibility

Scientists studying eukaryotic gene expression at the molecular level have found that cells use alternative RNA splicing mechanisms to permit cells to produce more than one type of protein from the same gene. Alternative RNA splicing takes place when the pre-mRNAs copied from the same gene are

Close up of a protein coding gene in the genome DNA:

Red arrow in DNA sequence indicates the RNA splice site where two exons are spliced together in the final mRNA.

Intron	GTCCATATCCTCGATATTAAAAA
	TTTATATAGGCCTATACACACAGGGG
	ATTTTTAGTATATCTACCACATACACACA
	TATATATAGAGAGATACCCCCTTTAGT
	AAGTTACGGTTATATATAGAGAGAGG
Exon	GATTATATATAGCGCGCGAGAG
Intron	TGGAAACCTAACCCATCAGCATC
	AATTGGTCCATATCCTCGATATTAAAAA
	ATTGGTCCATATCCTCGATATTAAAAAT
	TTATATAGGCCTATACACACAGGGGA
Exon	ATATATATACCACACACACAT
	ATATATAGAGAGAGACCCCCTTTAGTAA
	GTTACGGTTATAGATATAGAGAGAGGAA
	GGGATTATATATAGCGCGCGAGAGAT
	AGATGATAACCTACCCATCATCATCATC

FIGURE 3-10 **RNA splicing is amazingly accurate!** The DNA sequence (one strand only) of a fictitious protein coding gene is shown as the DNA sequences would appear in the human genome DNA, including the intron (highlighted in grey) and exon (highlighted in yellow) sequences as indicated. The red arrow indicates the location of an RNA splice site junction (highlighted in red) as it appears in the DNA sequence at the point where the two RNA exons on either side are spliced together to make the final mRNA. The RNA splicing mechanism must be accurate to the base-level because even a single base mutation could alter the final protein product of the spliced mRNA.

spliced differently, producing more than one type of protein. For example, an exon could be removed from one of the alternatively spliced pre-mRNAs but is not removed from the other pre-mRNA. As a result, two (or more) different proteins can be produced from a single gene in human cells that use an alternative splicing mechanism.

Figure 3-12 shows an example of two proteins made from one gene by alternative splicing. The gene is copied into a pre-mRNA containing three exons (exons 1, 2, 3) that are spliced together to make mRNA #1, which contains an open reading frame coding for protein #1. The gene is then transcribed into another copy of the same pre-mRNA, but this time the pre-mRNA undergoes alternative splicing to remove exon 3 and splice together exons 1, 2, and 4 to produce protein #2. In this case the two different proteins made by alternative splicing are structurally related and potentially perform similar or functions (Figure 3-12).

> **KEY CONCEPT**
> The pre-mRNAs transcribed from the same gene can be alternatively spliced to generate different mRNAs that code for different proteins.

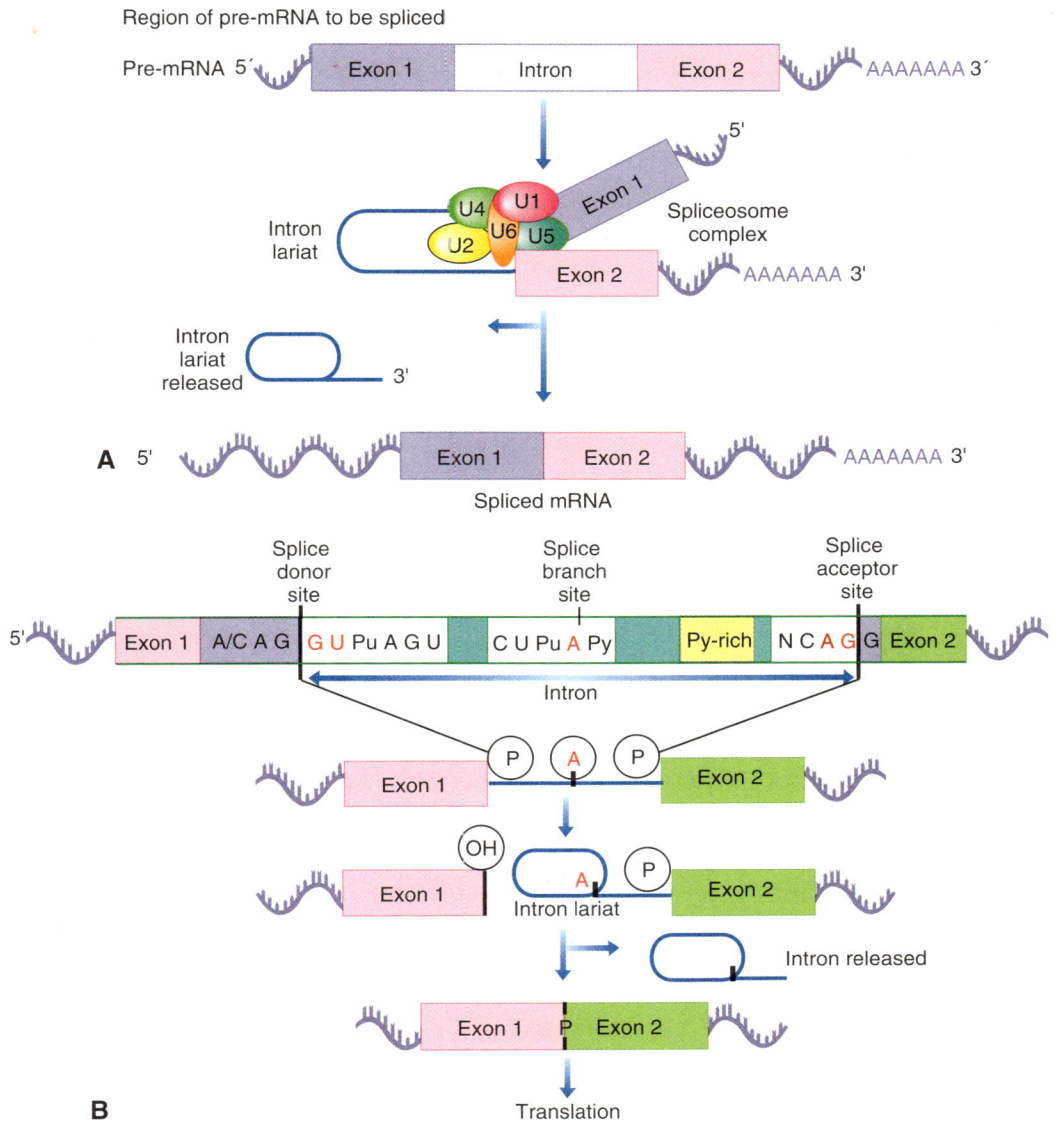

FIGURE 3-11 Pre-mRNA splicing involves the spliceosome (A) The process of RNA splicing requires the spliceosome complex, which contains many small RNAs and small proteins with specialized functions that work together to bring together the appropriate exon sequences (purple and pink). This intricate lariat mechanism of RNA splicing must be extremely accurate to avoid introducing mutations into the spliced mRNA strands, possibly altering the encoded proteins as well. (B) The spliceosome catalyzes the splicing of the RNA polymerase II pre-mRNAs using a mechanism that generates a circular lariat RNA intermediate structure. When RNA splicing is complete, the lariat RNA is discarded and the mature, processed mRNA is transported to the cytoplasm for translation into protein.

The human alpha tropomyosin gene is an example of a gene that is normally expressed by alternative RNA splicing pathways in different cell types. The same human alpha tropomyosin gene is used to code for different but related proteins in nerve, brain, fibroblasts, striated and smooth muscle cells. Alternative splicing of the alpha tropomyosin pre-mRNA transcripts produces different types of alpha tropomyosin proteins with specific functions in the different

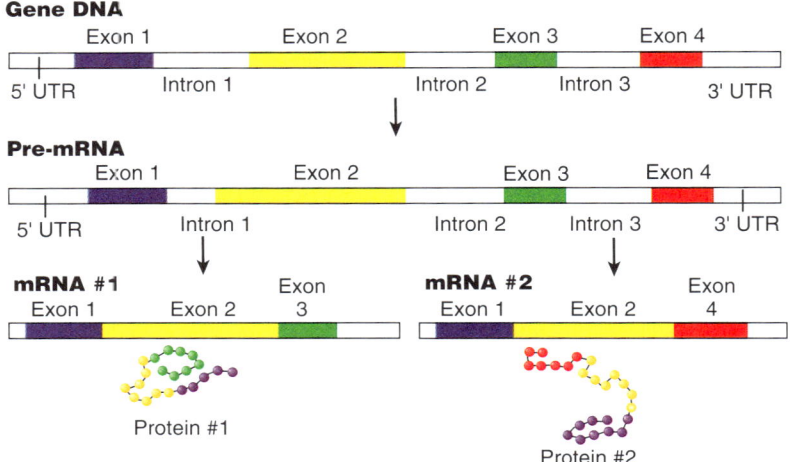

FIGURE 3-12 Alternate RNA splicing produces more than one protein from just one gene
The pre-mRNA splicing mechanism allows eukaryotic cells to make more than one protein from just one gene DNA sequence. This feat is accomplished by an alternate RNA splicing mechanism. The pre-mRNAs copied from the same gene are spliced differently to make different mRNAs that are translated into different proteins. In the example shown here a pre-mRNA is spliced to make mRNA #1, which contains the amino acids encoded by exons 1 (purple), 2 (yellow) and 3 (green) in protein #1. A different pre-mRNA transcript is spliced to make mRNA #2, which includes the amino acids encoded by exons 1 (purple), 2 (yellow) and 4 (red), making protein #2. Protein #1 and protein #2 contain different amino acids depicted in different colors that indicate the exons included in the fully processed, mature mRNAs #1 and #2.

types of cells. For example, the type of tropomyosin protein expressed in smooth muscle cells and in fibroblasts is encoded by an exon located near the 3′ end of the mRNA that is removed from the alternatively spliced mRNA expressed in striated muscle cells.

Translating the mRNA Genetic Code into Protein

Protein synthesis in prokaryotic and eukaryotic cells takes place on ribosomes (Figure 3-13). The order of the amino acids added to the growing amino acid chain, is dictated by the order of the 3-letter codons in the mRNA sequence. A series of intricate molecular maneuvers allow the ribosome to read the genetic information encoded in the mRNA sequences. At the start of translation, the large and small ribosome subunits assemble with the mRNA strand positioned so that the ribosome can read the individual 3-letter codons written in tandem in the mRNA molecule, in order, starting at the 5′ end (Figure 3-14). Each 3-letter RNA codon in the mRNA specifies one of the 20 amino acids encoded in the mRNA. The cell translates the order of the mRNA codons into the specific order of the amino acids in the encoded protein, using "adaptor" transfer RNA (tRNA) molecules to make the molecular connections between the identity of the mRNA codon and the identity of the amino acid added to the protein.

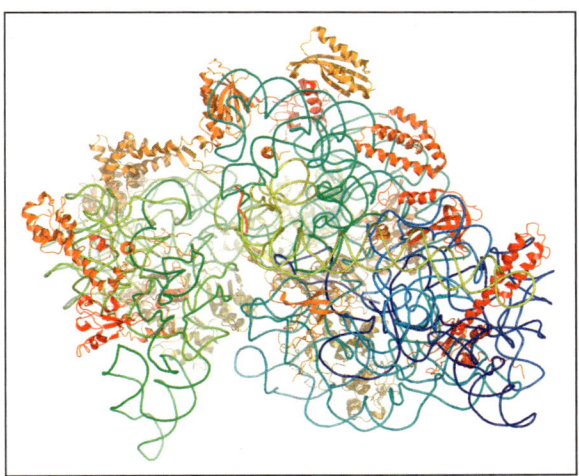

FIGURE 3-13 Macromolecular ribosomes synthesize proteins Eukaryotic and prokaryotic ribosomes are enormous protein-RNA complexes composed of dozens of ribosome proteins and ribosomal RNAs (rRNA). The atomic-level structure of the bacterial ribosome was published in 2001, a major scientific achievement. This computer model shows the structure of the small ribosomal subunit from bacteria (large subunit is not shown), which contains dozens of ribosome proteins (alpha-helices in red) and rRNAs (blue lines).

In 1973, Alexander Rich (Massachusetts Institute of Technology) published the first image of a tRNA molecule folded into a "clover leaf-like" RNA secondary structure, which indicated the position of each atom and chemical bond in the short tRNA molecule. Rich showed that the tRNAs that carry different amino acids all fold into similar "clover leaf" RNA structures because the tRNAs form similar intra-strand base pairs in the tRNA molecules (Figure 3-14). Having an accurate 3D picture of how tRNAs fold in the cell allowed scientists to perform experiments designed to delineate the functions of the tRNAs in protein synthesis.

In preparation for translation, each tRNA molecule is "charged" with one of the 20 amino acids available for protein synthesis in the cell, as dictated by the specific 3-letter anti-codon in each tRNA molecule in the universal Genetic Code Table (Figure 3-15). The base-pairing interactions between the 3-base codon in the mRNA and the 3-base anti-codon in the tRNA are very specific. The formation of hydrogen bonds between bases decodes the genetic information specifying the protein sequence carried in the mRNA codons (Figure 3-15). The anti-codon is a very important functional feature of the tRNA structure because during translation the bases in the anti-codon base-pair with the complementary bases in the mRNA codon, which effectively decodes the genetic code.

Many decades of genetic experiments performed by amazing scientists have revealed the coding relationship between the mRNA codons and the amino acids in the encoded protein. This research established the universal genetic code, which makes it possible for scientists to translate the coding region of an mRNA sequence and predict the encoded chain of amino

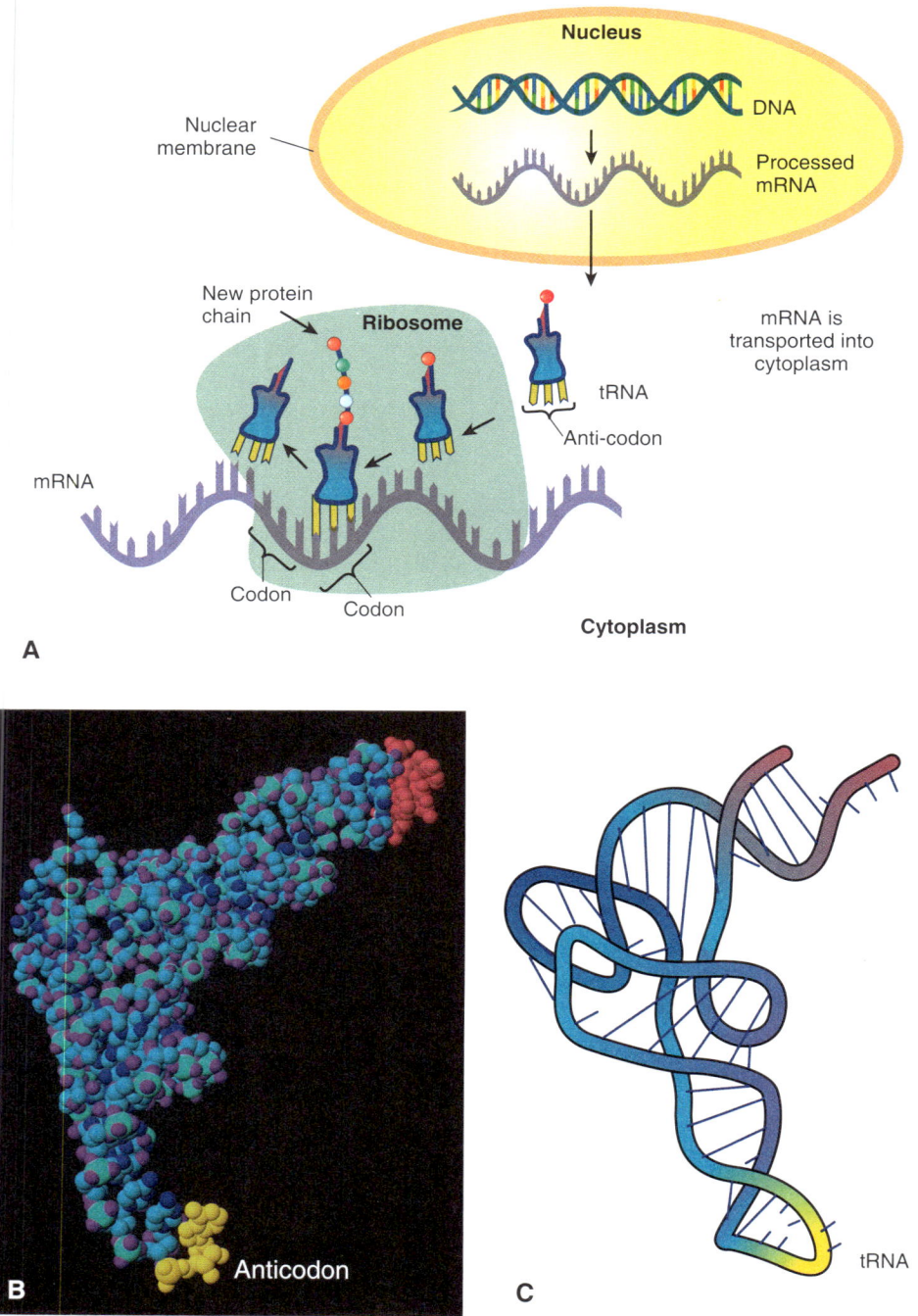

FIGURE 3-14 Transfer RNAs bring an amino acid to the ribosome (A) The mRNA codons are read by the ribosome in tandem starting with AUG and are translated into a chain of amino acids. Each transfer RNA (tRNA) molecule carries one specific amino acid to the ribosome. (B) The serine transfer RNA, tRNA(ser), carries the serine amino acid to the ribosome. The seryl tRNA synthetase enzyme adds the amino acid serine onto a specific tRNA, tRNA(ser), which contains a serine-binding site (red) and an anticodon (yellow). In the ribosome, the anticodon base pairs to a complementary codon in the mRNA sequence, and the serine amino acid is transferred onto the end of the growing chain of amino acids. (C) After the tRNA(ser) has released the amino acid, it exits the ribosome to pick up another serine amino acid.

UNIT 3 Gene Expression Makes RNA and Proteins

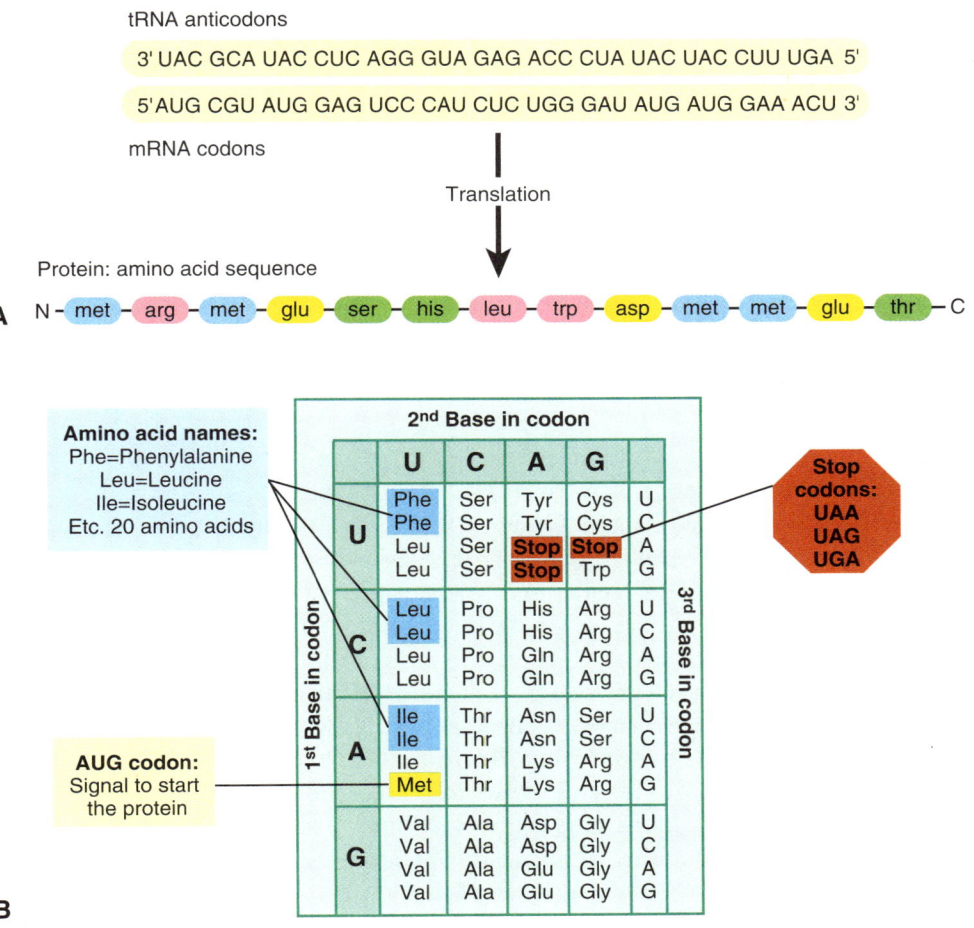

FIGURE 3-15 Link between the tRNA anti-codon, the mRNA codon and an amino acid
(A) There is a direct relationship between the in-frame mRNA codons (after RNA splicing) and the tRNA anti-codons. The base-pairing hydrogen bonding interactions between the codons and anti-codons specify the order that the amino acids should be added to the growing protein. (B) The Genetic Code Table identifies the mRNA codons that specify the amino acids to be incorporated into the growing protein chain during translation in the ribosome. The amino acids are listed as three letter abbreviations in the center of the table. To identify the amino acid that is specified by a particular three letter codon, find the first base in the codon among the choices presented down the left side. Then find the second base in the codon among the choices across the top, and find the third base in the codon among the choices presented on the right side of the table and read the corresponding amino acid.

acids (Figure 3-15). The genetic code is degenerate so there are more codons than there are amino acids so some amino acids are specified by more than one codon.

The AUG Codon in mRNA Initiates Protein Synthesis

The ribosome initiates protein synthesis by reading the base sequence of the mRNA molecule, starting with the AUG codon located closest to the 5′ end of the mRNA strand (Figure 3-15). The ribosome continues to read the

3-letter codons in the mRNA starting with the universal AUG, in tandem, in-frame.

An AUG codon in the mRNA will potentially send one of two signals, depending on the position of the AUG codon in the mRNA strand:

(1) The AUG codon located closest to the 5′ end of the mRNA strand signals the start of a new protein and always dictates that the protein will start with the amino acid methionine (Met) (Figure 3-15).
(2) In frame AUG codons that are located downstream of the AUG start codon signal the incorporation of methionine residues into the new protein chain. These internal AUG codons do not signal the start of another new protein (Figure 3-15).

The codons and the corresponding amino acids listed in the Genetic Code Table show that the amino acid methionine is specified by the AUG codon (Figure 3-15). In the Genetic Code Table, the amino acids are listed using three letter abbreviations in the center of the table and the choices for the first base in the codon are presented down the left side; the choices for the second base in the codon are across the top, and choices for the third base in the codon are presented on the right side.

There are 64 different codons available to make a 3-letter genetic code, but there are only 20 amino acids, indicating that the genetic code is degenerate because one amino acid can be encoded by more than one 3-letter codon. In the genetic code, the third base of the 3-letter codon for an amino acid can vary without changing the amino acid specified by the 3-letter codon (wobble hypothesis). For example, the amino acid threonine (Thr or T) is specified by four 3-letter codons that differ only in the third base position in the codons (read 5′ to 3′): (A C U; A C C; A C A; A C G).

> **KEY CONCEPT**
> The translation mechanism matches the 3-letter codon in the mRNA sequence with the complementary 3-letter anti-codon in the tRNA, indicating which of the 20 different amino acids to add to the growing protein chain.

As the ribosome continues to read each mRNA codon in-frame, it continues to add the appropriate amino acid to the growing protein, as indicated by the Genetic Code Table (Figure 3-15). The ribosome catalyzes the formation of a covalent peptide bond between the incoming amino acid and the amino acid previously incorporated into the growing protein chain (Figure 3-16). As a result the protein is made up of a chain of amino acids connected to each other by covalent peptide bonds. Protein synthesis terminates when the ribosome encounters one or more in-frame translation stop codons (UAA, UGA, UAG) in the mRNA sequence. At this point a translation termination factor inserts into the growing protein chain in place of an amino acid, and the ribosome subunits separate, releasing the new amino acid chain into the cytoplasm.

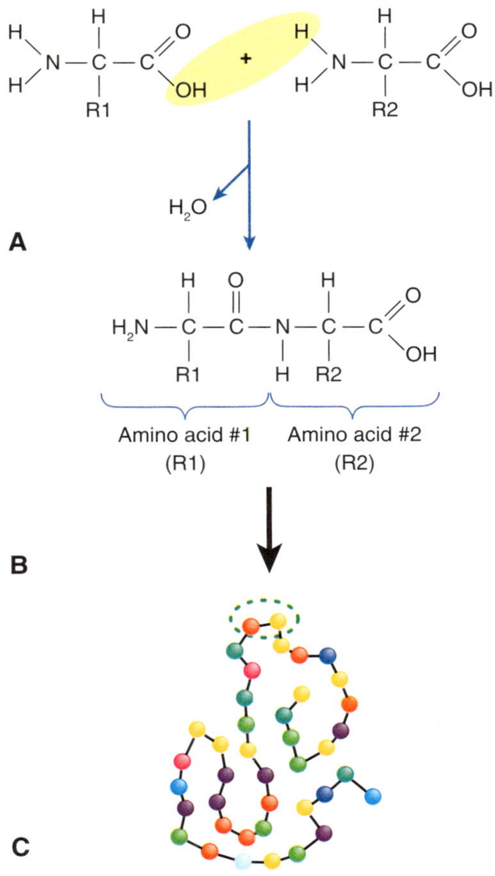

FIGURE 3-16 Amino acids are connected together by peptide bonds in a protein (A) The protein synthesis reaction in the ribosome joins two amino acids together by making a covalent peptide bond connecting the amino acid backbones, which releases one molecule of water (H_2O). (B) The product of this reaction is a short protein containing two amino acids linked together by their backbones called a dipeptide. The different amino acids contain peptide backbones with the same chemical structure. However, the different side chains on the amino acids (R1 and R2) confer specific chemical properties to each amino acid. (C) The typical protein product of translation contains many amino acids connected together by peptide bonds in the linear order dictated by the mRNA codons. The peptide bond between two amino acids is indicated by a dotted circle.

> **KEY CONCEPT**
> The tandem 3-letter codons in the mRNA sequence can be translated with the help of the Genetic Code Table to identify the specific amino acid to be added to the protein chain.

Protein Folding is Essential for Protein Function

Ribosomes synthesize proteins as linear chains of amino acids that are linked together with peptide bonds. However, for a protein molecule to function properly in the cell, it must be able to fold into specific 3D shapes, as needed for the protein to perform various functions. The study of how proteins fold is a very active area of modern research that integrates approaches and technologies from many scientific fields, ranging from genetics and molecular biology to biochemistry and biophysics. Some biological proteins fold spontaneously into functional 3D shapes that are based on information in the amino acid sequence and the surrounding environment. However, some proteins are unable to fold properly without the assistance of proteins called chaperones, which catalyze the folding of newly synthesized amino acid chains.

The primary structure of a protein is the linear chain of amino acids created by protein synthesis on a ribosome. The linear amino acid strand folds into a secondary structure that is represented by the path followed by the peptide bonds that make up the (polypeptide) backbone of the protein. The amino acid strands fold into two common types of secondary structures, the alpha helix and the beta sheet (Figure 3-17). The alpha helix forms when the chemical bonds between the side groups on the amino acids in the same chain, cause the protein backbone to twist into a spiral alpha helix. The beta sheet structure forms between two amino acid strands that belong to different protein molecules or can link together different regions of the same amino acid strand. The alpha helix and beta sheet structures are connected to each other through beta amino acid strands.

The 3D shape made up by all of the atoms in one protein molecule reflects the final tertiary protein structure. Protein complexes that contain multiple peptides are called quaternary structures and the individual polypeptide chains that make up a protein complex are referred to as the subunits. These levels of hierarchal protein structure build on each other as the newly synthesized protein chain folds. Most proteins made in living cells are classified as either fibrous or globular proteins based on their tertiary structures. In general, fibrous proteins are composed of amino acid chains that are organized into long fibers that function as fibrous structural components of the cytoskeleton and provide a framework for the overall cell morphology. The globular proteins are typically more compact than the fibrous proteins and sometimes include glycolytic enzymes, DNA and RNA polymerases, and regulatory factors.

Proteins have regions of amino acids that fold somewhat independently of the rest of the protein and are involved in a specific function. These amino acid motifs occur in many different protein molecules, and sometimes form domains that represent regions of the protein that have particular molecular functions, such as a DNA binding motif. Different molecular machines are assembled in the cell from combinations of protein parts that fold to fit together like the 3D pieces of a complex puzzle. The different molecular machines in the cell carry out diverse biochemical processes in the cells such as gene expression, protein synthesis, RNA splicing, and transport by molecular motors.

The leptin hormone, which helps regulate human appetite, contains five alpha helices connected by single amino acid strands (Figure 3-17). The chaperone heat shock protein 90, which increases dramatically in cells that are exposed to physical stress such as high temperatures, contains both alpha helices and beta sheets. The high-density lipoprotein (HDL) has an amazing structure made up of extremely long alpha helices that contain

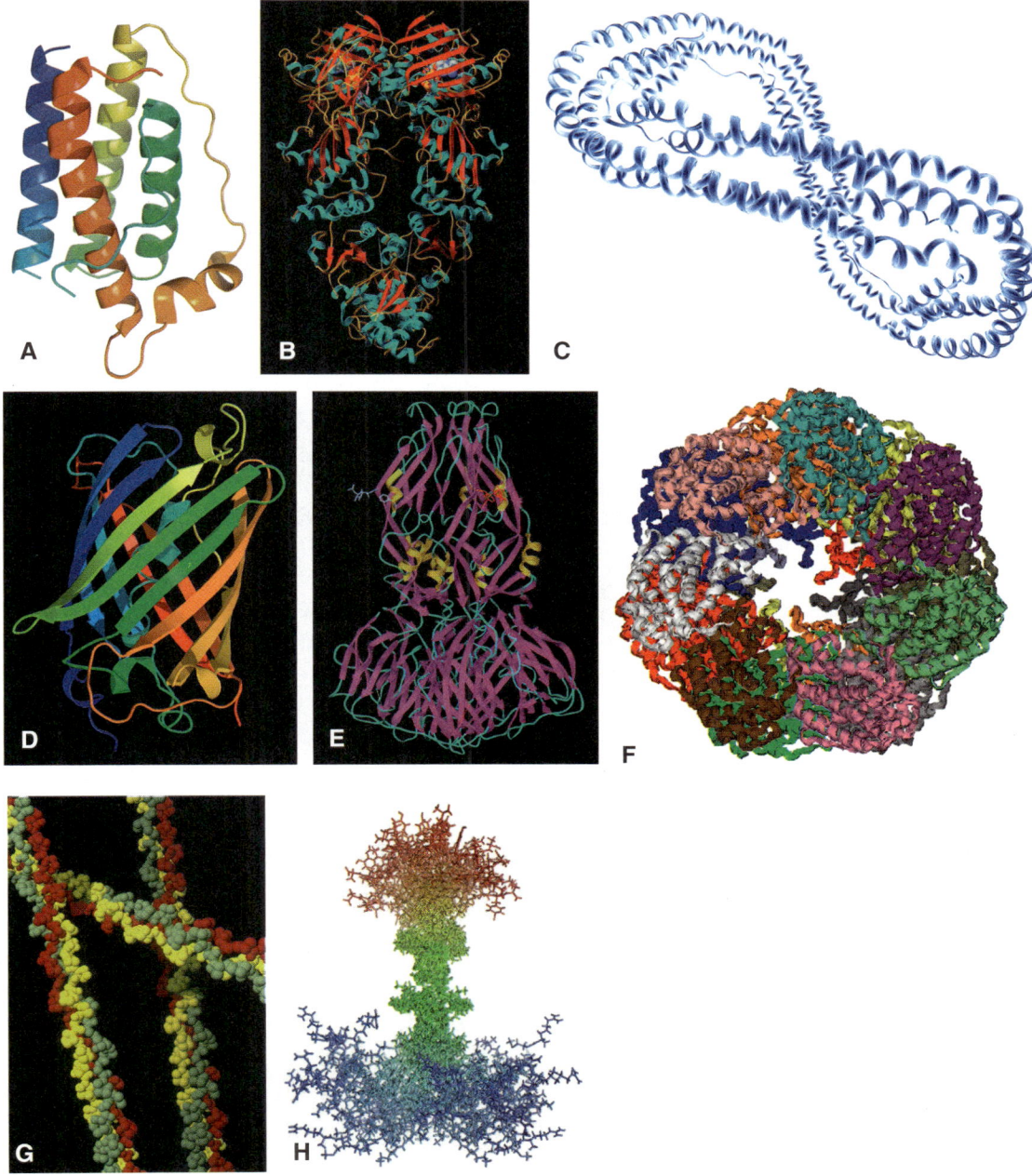

FIGURE 3-17 Amino acid chains fold into different 3D protein shapes Newly made proteins fold into alpha helix and beta sheet secondary structures, which become integrated into the stable tertiary shape of the completely folded protein. (A) The leptin hormone regulates appetite and contains primarily alpha helices connected together by short strands. (B) Heat shock protein 90 is made in cells exposed to stress and contains both alpha helices and beta sheets. (C) This high-density lipoprotein (HDL) requires extremely long hydrophobic alpha helices to function as a 'good cholesterol' that carries lipids (fat) in the bloodstream. (D) The green fluorescent protein (GFP) is used to tag proteins under study and contains mostly beta sheets. (E) This glycoprotein resides on the surface of the dengue hemorrhagic fever virus and binds to a target cell to allow the viral DNA to enter the cell. (F) The barrel-like GROEL chaperone assists in folding new proteins. (G) Collagen proteins secreted by fibroblast skin cells make up the collagen fibers in connective tissue. (H) The syndecan-4 protein works with integrin proteins to attach the cell to a substrate and allow cell movement.

hydrophobic amino acids. The HDL is called the "good" cholesterol in the bloodstream because it carries lipids (fat) around the body and does not contribute to blood clots (Figure 3-17).

Some proteins are made predominantly of beta sheet structures connected by strands including the famous green fluorescent protein (GFP) (Figure 3-17). The first green fluorescent protein (GFP) gene was cloned from the jellyfish *Aequorea victoria*. GFP is a very popular and effective fluorescent protein "tag" widely used in research. The GFP gene is fused to the gene coding for the protein under study and the cells and tissues expressing the resulting GFP fusion protein will fluoresce. Another protein containing many beta strand structures is located on the outside surface of the envelope surrounding the dengue hemorrhagic fever (DHF) viral capsid. This protein binds the virus to the target cell allowing the viral genome to enter and infect the cell (Figure 3-17).

The proteins in cells often adopt 3D shapes that are directly related to the biological functions of the proteins. GROEL is an enormous chaperone complex shaped like a barrel (Figure 3-17). GROEL helps proteins fold; new proteins enter one end of the barrel and exit the other end as a folded protein. Collagen fibers contain three linear protein chains that are wrapped around each other (Figure 3-17). Collagen fibers are made by the skin cells and become part of the connective tissues that support the layers of human skin. The wrinkles that appear in the skin result from depletion of the collagen fibers underlying human skin cells as we age. Integrin proteins are sticky molecules located on the surfaces of cells that are involved in cell adhesion and transmitting signals between cells (Figure 3-17). The integrin proteins change shape when transmitting cell signals. The integrin proteins work with the syndecan-4 protein to attach the cell to a substrate, allowing the cell to move. A large region of the syndecan-4 protein extends outside the cell, interacting with components in the surrounding extracellular matrix.

> **KEY CONCEPT**
> The synthesis of a protein begins with a methionine (Met) amino acid at the amino-terminus of the protein, incorporated in response to the AUG codon located near the 5' end of the mRNA. The ribosome reads the mRNA codons in tandem, adding the correct amino acids as directed by the coding region.

Water-hating and Water-loving Amino acids affect Protein Folding

Although the proteins in cells are all made of a linear string of amino acids, proteins actually represent a very diverse group of molecules that can adopt myriad 3D shapes as needed to perform thousands of different biological functions in the cells. The functional shape adopted by a folded protein is influenced by many factors, including the chemical properties of the amino acid side chains, the local cellular environment, and the activities of any chaperone proteins that assist with protein folding (Figure 3-18).

UNIT 3 Gene Expression Makes RNA and Proteins

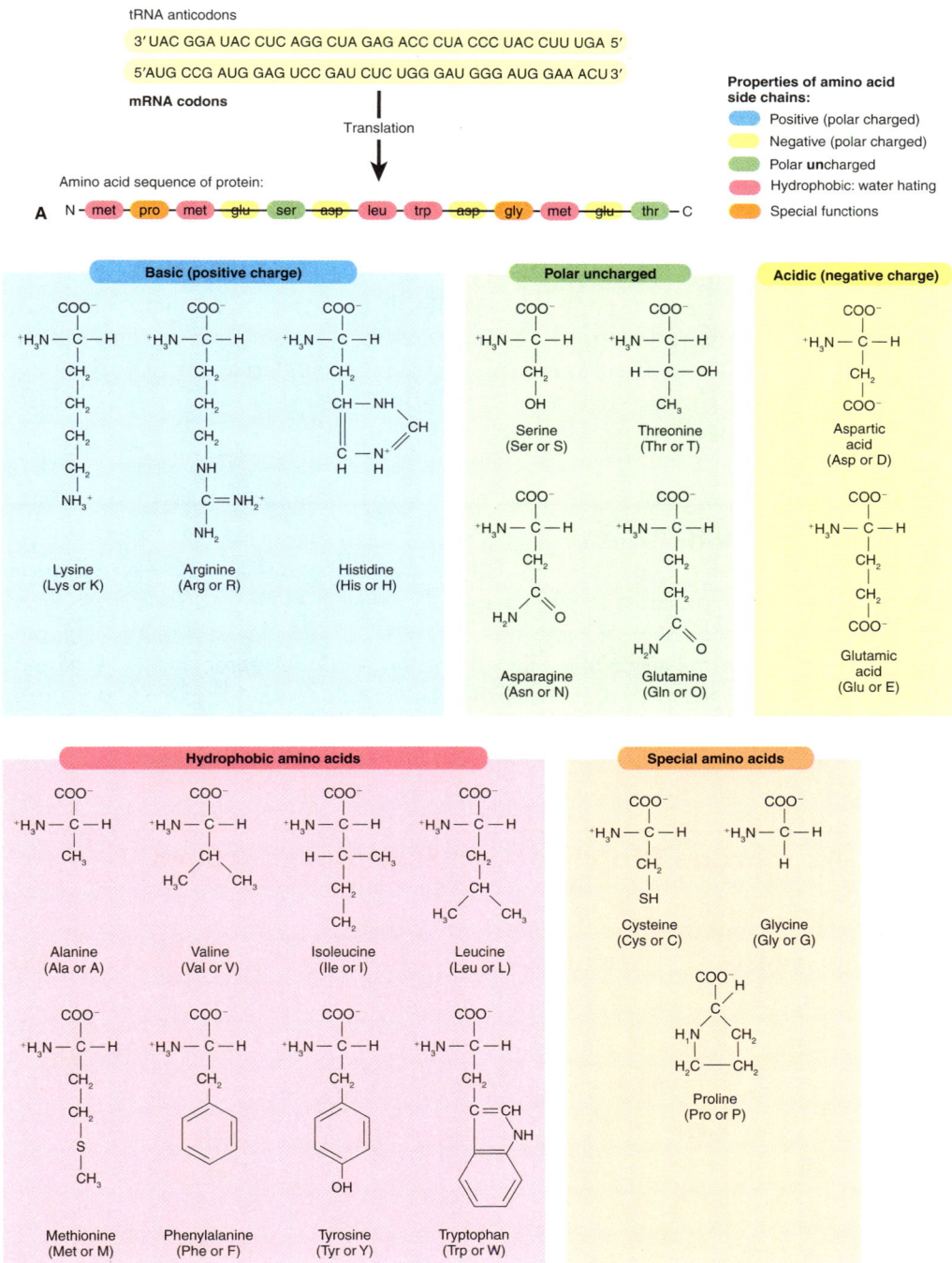

FIGURE 3-18 Amino acid side chains influence protein folding and function (A) The linear relationship between the tRNA anticodons, the mRNA codons and the amino acids in the protein chain is shown. (B) The amino acids have the same chemical backbone structure and are grouped according to the different chemical properties of the side chain group (R). Basic amino acids have positively charged R groups (blue), and acidic amino acids have negatively charged R groups (yellow). The amino acids that associate with membranes have hydrophobic hydrocarbon chains (R groups) with water-hating properties (pink) that cling to each other to escape contact with water. Hydrophilic (water loving) amino acids have charged polar side chains (green). Amino acids with special chemical properties include cysteine, glycine, and proline (orange).

Each of the 20 naturally occurring amino acid contains a differenttype of chemical side chain (R) that confers distinct chemical properties to the amino acid and affects the folding of the overall protein (Figure 3-18). The water-loving (hydrophilic) amino acids have polar side chains containing overall positive or negative charges. In contrast, the water-hating (hydrophobic) amino acids have long hydrocarbon side chains. The amino acids with the longer hydrocarbon side chains have the most hydrophobic (water hating) characteristics (Figures 3-18). The amino acids classified as 'special' include cysteine, glycine, and proline. Cysteine has an SH group that can form covalent disulfide bonds with other cysteines, either in the same protein or in different proteins. Glycine, the smallest amino acid, fits well into the alpha helical structure, but the incorporation of a proline amino acid will typically cause a "kink" or bend to form in the protein structure.

> **KEY CONCEPT**
> The chemical side chain on each amino acid contributes to the overall chemical and physical characteristics of the folded protein structure. The newly synthesized protein folds spontaneously into a structure with the hydrophobic (water-hating) amino acids on the inside of the folded structure, and the hydrophilic (water-loving) amino acids in the protein are positioned on the outside surface of the folded protein (Figure 3-19).

Membrane Proteins have Essential Functions

The cytoplasm of eukaryotic cells contains a large membrane network, the endoplasmic reticulum (ER), which is connected with the double envelope

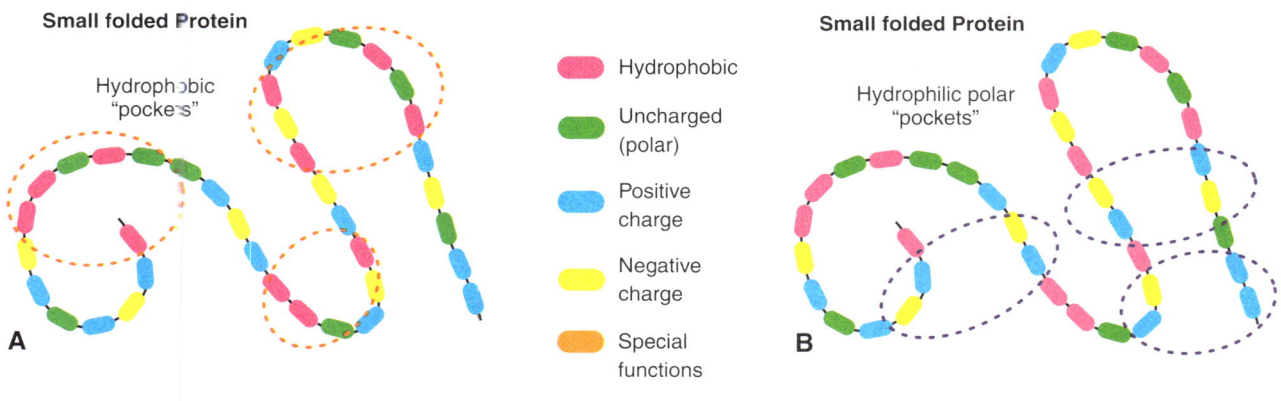

FIGURE 3-19 Properties of amino acid side chains influence protein folding (A) As the amino acid chain folds, the hydrophobic amino acids (pink) move to the inside of the folding protein structure away from the water (aqueous) surroundings. (B) At the same time the hydrophilic amino acids (blue, yellow) form pockets of oppositely charged residues on the inside of the folded protein or move to the outer surface of the structure to maintain contact with the aqueous surroundings.

that surrounds the nucleus. Sometimes the newly synthesized proteins remain in the cytoplasm to function while other proteins made in the cytoplasm are transported into the nucleus to function. Other proteins such as hormones are destined to exit the cell and travel to distant locations in the body to function.

Proteins that require post-translational modification including phosphorylation (the addition of phosphates) and glycosylation (the addition of sugar groups) are translated on ribosomes that are attached to the rough endoplasmic reticulum (RER). The newly translated proteins are transported through the lumen of the RER and are enclosed in membrane vesicles that bud off of the RER membranes and carry the proteins to the Golgi apparatus, which sorts and directs the protein traffic in the cell. The vesicles in the stack of Golgi membranes carry selected protein molecules to the plasma membrane, where they are secreted out of the cell (Figure 3-20).

The proportion and distribution of hydrophobic amino acids in a folded protein structure influence how the protein interacts with the membranes in the cell. The biological membranes are primarily phospholipid bilayers that surround the contents of the cells and control the movement of molecules and ions in and out of the cell. The exact composition of a particular plasma membrane depends on the type of cell it surrounds and the membrane proteins, including receptors, which are expressed in the cell.

The plasma membrane bilayers surrounding cells, nuclei, and organelles are made up of phospholipid (PL) molecules, which have a charged phosphate head region and long, hydrophobic fatty acid tails (Figure 3-21). The water-loving and water-hating domains and characteristics of the individual

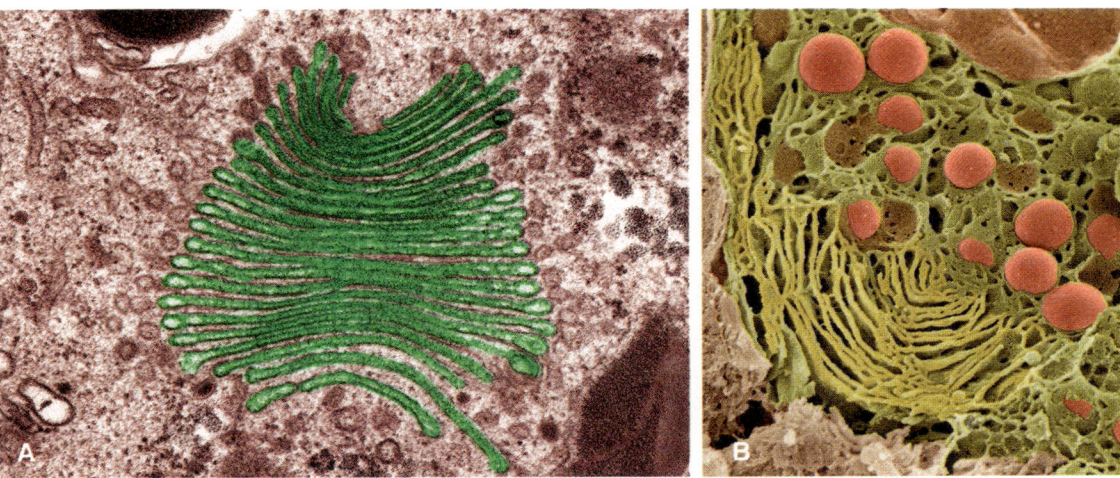

FIGURE 3-20 Newly translated proteins can be modified (A) Proteins translated on ribosomes on the rough endoplasmic reticulum travel through the ER to the disk-like membranes of the Golgi apparatus (green) where a specialized delivery system modifies, stores and transports proteins and lipids to various destinations in the cell. (B) In addition to insulin, pancreatic cells make digestive enzymes that are packaged into zymogen granules (red) in the Golgi (yellow) and are later activated in the small intestine to help digest carbohydrates, fats and proteins.

PL molecules strongly influence how PLs interact with other PLs and with other cellular components. These principles are illustrated in a wonderful simulation by Concord Consortium. In this simple animation, the PL molecules are placed in a container with two separated liquid phases, oil on the top and water on the bottom. The monolayer of PL molecules spontaneously forms at the interface between the oil and water phases, with the charged head region of the PLs in the water phase and the long fatty acid tails of the PLs inserted into the hydrophobic oil phase. The PLs demonstrate that they do not need instructions from an external source to dictate their behavior. In an aqueous environment, PLs will spontaneously form bilayer structures that resemble the plasma membranes found in living cells (Figure 3-21). The charged PL head domains are located on the external surfaces of the bilayer, but the long hydrocarbon tails are located inside the hydrophobic inner layer of the cell membrane (Figure 3-21).

Proteins that function in and around cellular membranes contain regions of hydrophobic amino acids that allow some protein domains to interact with the interior of the plasma membrane bilayer (Figure 3-22). Proteins containing hydrophobic amino acids can function entirely within the bilayer or span the membrane and can contain a single membrane-spanning domain or have several domains that span the membrane. A good example is the cystic fibrosis transmembrane receptor (CFTR) protein, which spans the membrane several times. Mutations in CFTR

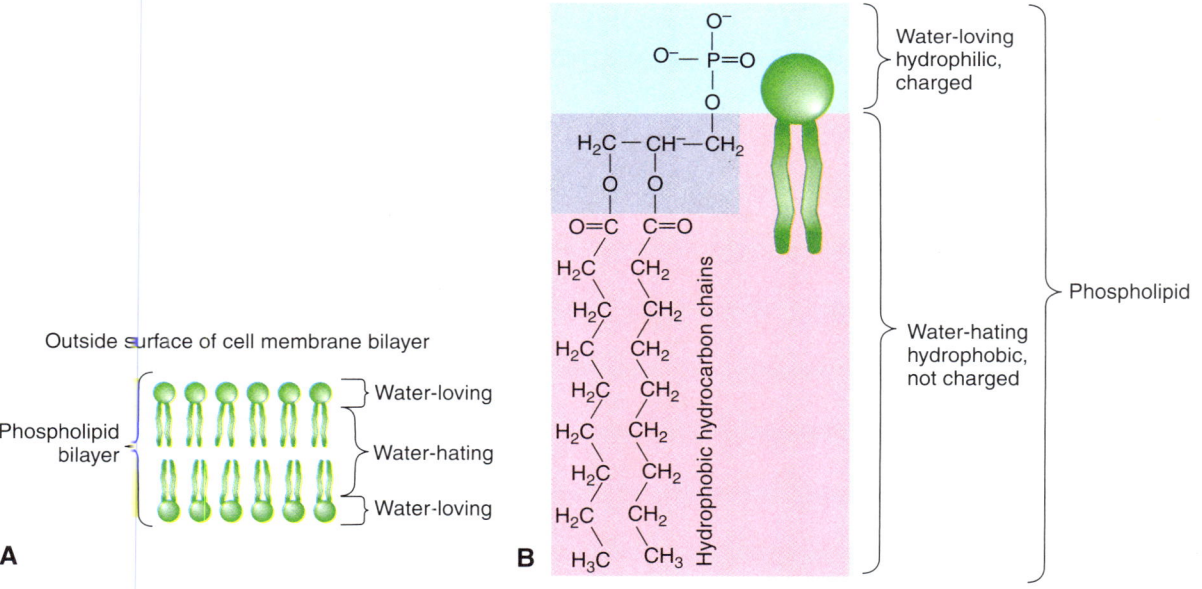

FIGURE 3-21 Cell membranes contain phospholipids (A) Cell membranes are typically made up of two layers (bilayer) of phospholipid molecules (PL). The inside and outside surfaces of the cell membrane contain the PL head groups which are charged and interact with the watery environment. The inner region of the membrane is composed of the long hydrophobic tails of the PL molecules. (B) PLs have a charged head domain and hydrophobic fatty acid tails with the chemical structures as shown.

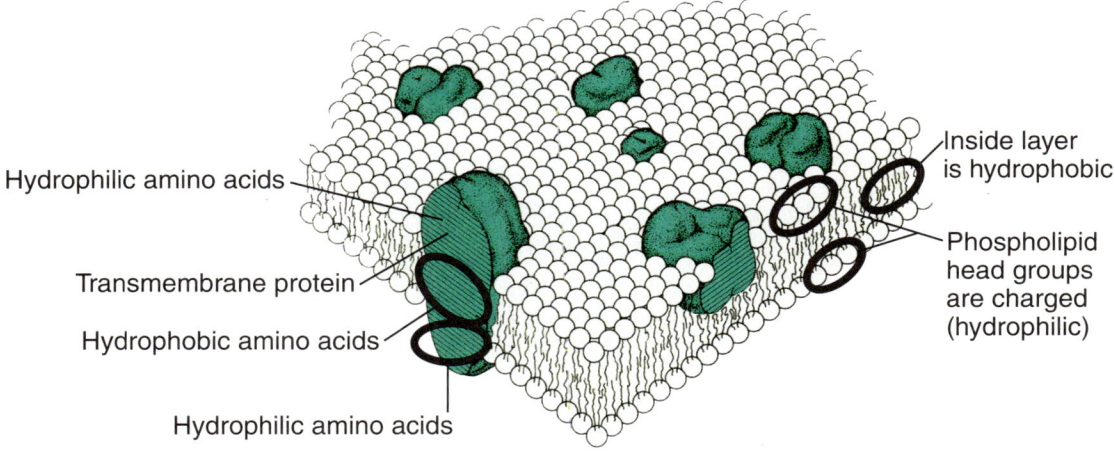

FIGURE 3-22 Cell membranes contain proteins Cell membranes contain many different types of proteins with different functions. Some membrane proteins 'float' in the PL layer while other proteins span the membrane completely (transmembrane protein). The hydrophobic amino acids make up the parts of the proteins that interact with the inner region of the membrane, while the hydrophilic protein domains interact with the charged PL head groups on the surface of the membrane. Some transmembrane proteins have hydrophobic domains that span the membrane several times. Membrane proteins can also interact with non-membrane proteins.

cause the genetic disease, cystic fibrosis (Unit 4). Proteins that are associated with the cell membrane can interact with proteins either on the inner or outer faces of the membrane or within the membrane bilayer. Receptor proteins on the outside of the cell are anchored to the outer layer of the plasma membrane through hydrophobic amino acids and they also interact with molecules outside the cell through hydrophilic domains extending away from the cell surface (Figure 3-22).

Enzyme Proteins Drive Essential Biochemical Reactions in the Cell

The major purpose of the gene expression pathway is to convert the genetic information in the gene DNA sequence into a functional protein that performs a specific function in the cell. Proteins are amazing biological molecules. Each protein is a chain of amino acids made up of a selection from only 20 different amino acids, yet proteins are extremely diverse in size, shape, and physical and chemical properties. The proteins made from only 20 different amino acids can perform the biochemical reactions required for life to exist.

Two proteins with the same amino acid sequence are in fact two copies of the same protein and under similar conditions in the cell they both fold into the same 3D shape and perform the same biological function. However, in some rare prion diseases, a harmless protein in the body can adopt a different 3D shape that causes a serious, even fatal illness (Unit 4). Two proteins coded from different genes and with different amino acid sequences usually fold into two different shapes and perform unrelated functions. Sometimes the 3D shapes of the folded proteins can best accommodate the function of the protein. Recall the example of the PCNA clamp proteins that form a ring around the DNA helix function to hold the DNA polymerase enzyme onto the DNA helix during replication (Unit 2). Another example is a hypodermic syringe–like protein structure that is used by bacteriophage to inject the phage DNA genome into the bacterial cell to establish an infection (Unit 1).

Living cells require the successful completion of thousands of different biochemical reactions for the cells to continue to grow and survive. However, under normal physiological conditions in the human body, without the support of proteins most biochemical reactions would not proceed fast enough to support life. Enzymes are proteins that solve this problem by interacting temporarily with the reacting molecules to make and break covalent bonds, rapidly catalyzing the conversion of the reactants into the products of the reaction. Once the biochemical reaction is complete, the enzyme reverts to its original state, unchanged and ready to catalyze the next reaction. Enzymes play very important roles in cells because they can break and make the strong covalent bonds that connect the atoms in essential biological molecules, including proteins, carbohydrates, fats, and nucleic acids.

Enzymes can catalyze simple biochemical reactions such as combining two molecules (A + B) to generate a new product (C): [A + B = C]. Although ribosomes are extremely large complexes, the biochemical reaction catalyzed by the ribosome is fairly simple; the two amino acids (A + B) are linked together with a peptide bond to make the product (C), a dipeptide. As protein synthesis continues, this reaction is repeated with new amino acids added to the growing protein as specified by the codons (mRNA), anti-codons (tRNA), and the genetic code.

In general, each enzyme catalyzes one specific biochemical reaction by interacting with a specific substrate and breaking and making the appropriate chemical bond(s) to rapidly create the finished product. The enzymes that degrade proteins break the covalent peptide bonds and release the amino acids. Other enzymes are required for the synthesis of amino acids and other building blocks needed for the assembly of biological polymers in the cell such as DNA and RNA.

Enzymes also play a key role in cellular metabolism. Glycolysis is a major metabolic pathway in cells that consists of a series of biochemical reactions catalyzed by different glycolytic enzymes that convert glucose (sugar) into pyruvate. The glycolytic enzymes include glyceraldehyde-3-dehydrogenase (GAPDH), which catalyzes the biochemical conversion of glyceraldehyde-3-phosphate into 1,3-bisphosphoglycerate, an early step in glycolysis in human cells. The purpose of the glycolytic cycle is to produce the high-energy compounds ATP (adenosine triphosphate) and NADH (reduced nicotinamide adenine dinucleotide), which store energy in the form of high-energy chemical bonds that are used later to drive energy-requiring reactions in the cell.

The seryl tRNA synthetase enzyme represents a group of special enzymes that add a specific amino acid onto the end of a specific tRNA (transfer

RNA) molecule in preparation for translation. The seryl tRNA synthetase enzyme ensures that the correct amino acid, serine, is attached to a tRNA containing the correct anti-codon sequence (UCU, UCA, UCG, UCC, AGU, AGC) (Figure 3-15). This matches the correct amino acid with the correct anti-codon during protein synthesis (Figure 3-23). Probably not surprising to anyone who mows lawns and fights weeds, the most abundant protein complex found in nature is an enzyme called ribulose bisphosphate carboxylase/oxygenase, or Rubisco, which catalyzes biochemical reactions involved in fixing carbon dioxide during photosynthesis in plants.

> **KEY CONCEPT**
> Biochemical reactions catalyzed by enzymes involve breaking and making covalent chemical bonds. Enzymes often make temporary chemical intermediates that function during the reaction, but after the reaction is complete, the enzyme returns to its pre-reaction state, ready to catalyze the next reaction.

RNA Ribozymes

Without the assistance of proteins, the DNA molecule can store information, but DNA cannot replicate or be copied into RNA without the help of protein enzymes. Although DNA has no known enzymatic properties, some special RNA molecules have been identified as ribozymes because these RNA molecules possess enzymatic activity, even in the absence of proteins (Figure 3-24). In cells, the RNA component of the RNA/protein ribozyme acts as a very efficient enzyme that rapidly catalyzes specific biochemical reactions. Most ribozymes catalyze self-cleavage, or cleave other RNAs. The discovery of ribozymes supports the idea that RNA was the first form

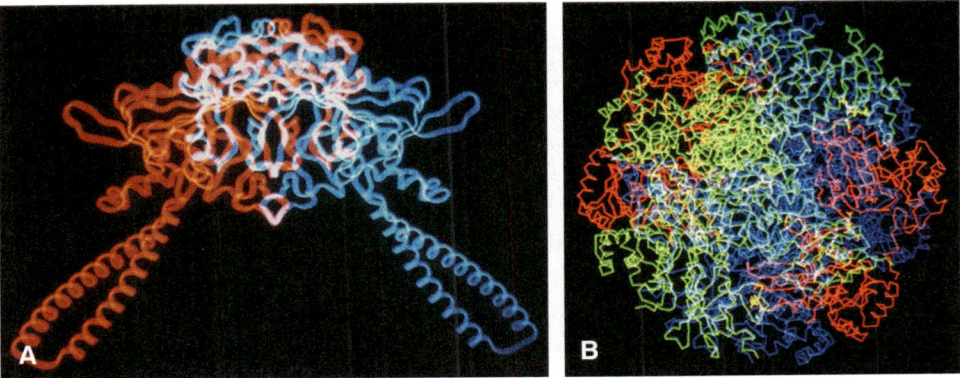

FIGURE 3-23 Enzyme proteins speed things up (A) The seryl tRNA synthetase enzyme adds a specific amino acid (serine) onto a specific tRNA (transfer RNA) molecule that functions in protein synthesis. (B) The rubisco complex has a total of 8 small subunits (red) and 8 large subunits (green, blue), and is the most abundant naturally occurring protein on Earth. Rubisco is an enzyme needed for photosynthesis in plants!

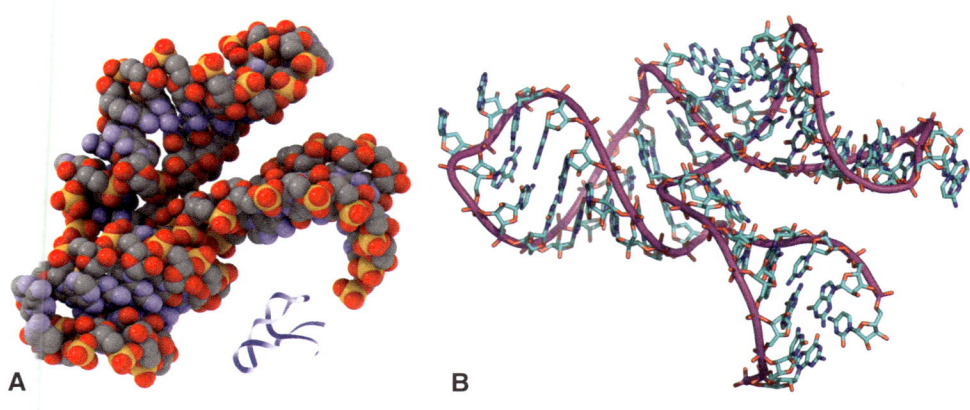

FIGURE 3-24 Hammerhead ribozyme self-cleaves RNA (A) Most hammerhead RNAs are part of the genomes of viral plant pathogens and are called ribozymes because they can exhibit enzyme-like properties. The hammerhead RNA gets its name from the shape of a small conserved tertiary RNA structure that self-cleaves RNAs during viral replication. The atoms in the space filled model of the hammerhead RNA are color coded as carbon and hydrogen (grey), oxygen (red), phosphorous (yellow) and nitrogen (blue). The blue ribbon structure shown below the model shows the overall secondary structure of the hammerhead ribozyme. (B) Some ribozymes catalyze their own cleavage, or cut other RNAs, but some function in protein synthesis in the ribosome.

of life on Earth because RNA can perform the functions that are carried out by DNA and proteins in modern cells.

Motor Proteins Transport Components Inside the Cell

Proteins perform many functions in cells in addition to catalyzing biochemical reactions and expressing genes. An intricate system of cytoskeletal fibers and motor proteins exist in eukaryotic cells (Figure 3-25). The cytoskeleton defines overall cell shape, is essential for vesicle transport, chromosome segregation, and cell mobility. The major components of the cytoskeleton are microtubules, composed of tubulin proteins and actin filaments. Actin, a contractile protein found in muscle, functions in the cytoskeleton to anchor cells to surrounding tissues. Cytoskeletal microtubules that provide structural functions are often cross-linked together by microtubule-associated proteins (MAPs). Another cytoskeletal protein component, tau, is implicated in degenerative neurological diseases such as Parkinson's and Alzheimer's disease. Tau proteins are abundant in normal nerve cells, but the large aggregates of tau proteins seen in people with brain disorders damage the nerve cells.

The cell cytoskeleton also has important functions in the process of apoptosis. Cells to self-destruct because the cell is damaged, old, diseased or to remove certain tissues during development such as the cells between the fingers of a developing fetus (Figure 3-26). This process is also called programmed cell death because death occurs under the direction of specific genes that are expressed in the cell slated to die. Inside the apoptotic cell

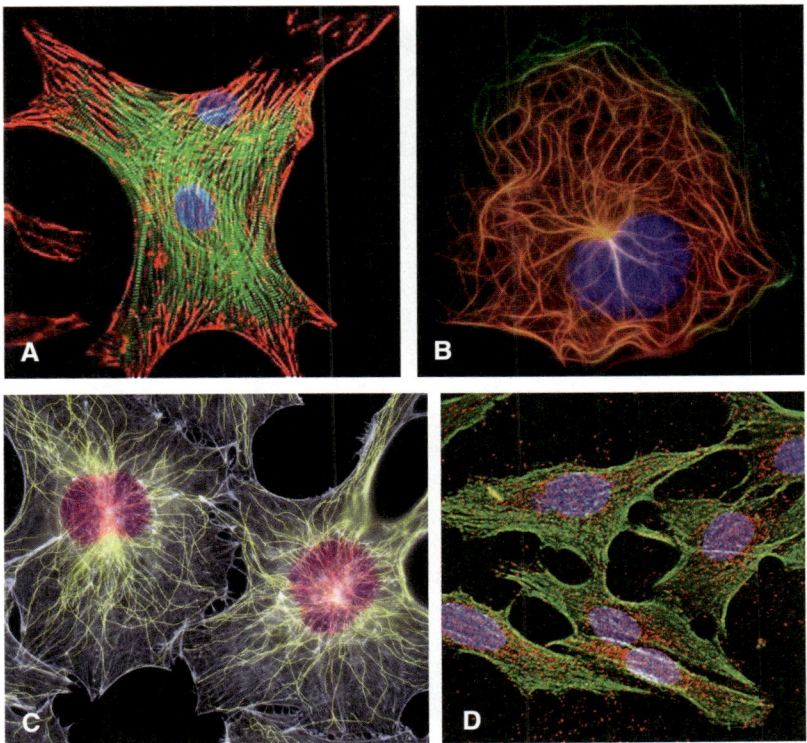

FIGURE 3-25 Cytoskeletal proteins determine cell shape and size (A) These smooth muscle cells (red) were produced by the heart and contain actin filaments (green). Actin is a contractile protein that can anchor the cell to surrounding tissues. (B) This shows a typical cell growing on the surface of a dish in the lab with a large number of cytoskeletal microtubules (red) and a prominent nucleus (blue). (C) The nuclei (purple) in these fibroblast cells are surrounded by a cytoskeleton made up of microtubules containing tubulin proteins (yellow) and actin filaments (blue). (D) These brain cells with nuclei (blue) were obtained from a Parkinson's patient and stained for actin (green). Tau proteins (red and pink) are abundant in normal nerve cells, but tau proteins also form large aggregates in the nerves of people with Alzheimer's disease and possibly Parkinson's disease as also well.

nucleus the chromosome DNA is broken into fragments of DNA. Apoptotic cells also undergo cytoskeletal changes that cause the cells to become round, the cell surface membrane forms large blebs, and the cell is divided into apoptotic bodies that are engulfed and discarded by special cells.

Cells have many types of motor proteins that typically convert chemical energy into mechanical movement for many different biochemical processes in cells. Kinesin and dynein motors are examples of molecular machines that are involved in essential cellular processes such as chromosome segregation during cell division and moving vesicles and organelles along microtubule fibers inside the long axons of nerve cells (Figure 3-27). The cell makes elaborate preparations for mitosis that begin after the chromosome DNA has completely replicated (Unit 2). Before mitosis the components in the cytoplasm of the cell must duplicate in order to have a supply of mitochondria, ER, RER, membrane components, Golgi, ribosomes and chloroplasts in plant cells available to distribute to the two progeny cells. After the chromosomes have

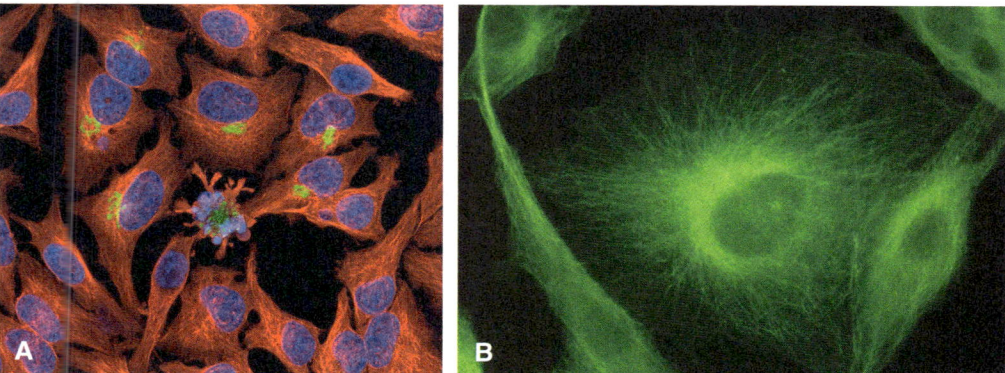

FIGURE 3-26 Cytoskeletal proteins and apoptosis (A) Selected cells that are damaged or are no longer needed by the body perform apoptosis (programmed cell death). The apoptotic cell shown here (center of field) has undergone cytoskeletal changes causing the cell to round up, then the cell surface formed blebs, the genome DNA fragmented and the cell was divided into apoptotic bodies. The healthy cells surrounding the apoptotic cell contain DNA (blue) and tubulin (red). (B) These healthy baby hamster kidney cells growing in the lab have been labeled with anti-tubulin antibodies to show the microtubule structures in the cell. The dark oval region in the center of each cell is the space occupied by the nucleus.

moved to the poles, the spindle apparatus rapidly disassembles, a nuclear envelop appears around the chromosomes at each pole, and the cytoskeletons assemble inside the two new cells.

From the standpoint of genes, DNA and inheritance, the process of chromosome segregation is the most critical stage of cell division because this is the time when the chromosomes are inherited by the offspring.

The kinesin family of motor proteins has many important functions in the mitotic spindle apparatus that are required for chromosome segregation at each cell division. As mitosis approaches much of the cytoskeleton disassembles into tubulin proteins and the nuclear envelop disappears. The tubulin proteins reassemble into the spindle fibers apparatus that spans the previous location of the nucleus.

The spindle fibers attach to the centromere region of each chromosome pair lined up at the midline of the spindle apparatus. It is essential that the centromeres of the two chromosomes in each pair attach to fibers from opposite poles so that the same numbers of chromosomes are equally distributed to each of the two progeny cells. Any molecular defect that interferes with the spindle fibers binding to the centromere can cause the affected chromosome to be left behind when the chromosomes move, resulting in cells that inherit the incorrect number of chromosomes (aneuploidy). When this type of mistake occurs during meiotic cell division when sperm and egg cells are made, the consequences can be a genetic disorder such as Down syndrome (trisomy 21) where the affected person inherits a third copy of chromosome 21.

Myosin V is a member of a large family of motor proteins that are involved in the contraction of individual muscle cells and in the contraction exhibited by muscle tissues (Figure 3-27). Inside live cells the 'muscle' needed to move cargo is provided by motor proteins that "walk" along microtubule fibers while "carrying" the cargo. The kinesin motor is a large, multi-subunit machine with heavy chain motor proteins and two light chain proteins attached to long protein stalks that bind to the cargo. ATP hydrolysis (ATP to ADP + Pi) generates the energy necessary for the motor machine to move the cargo one step forward for every molecule of ATP that is used.

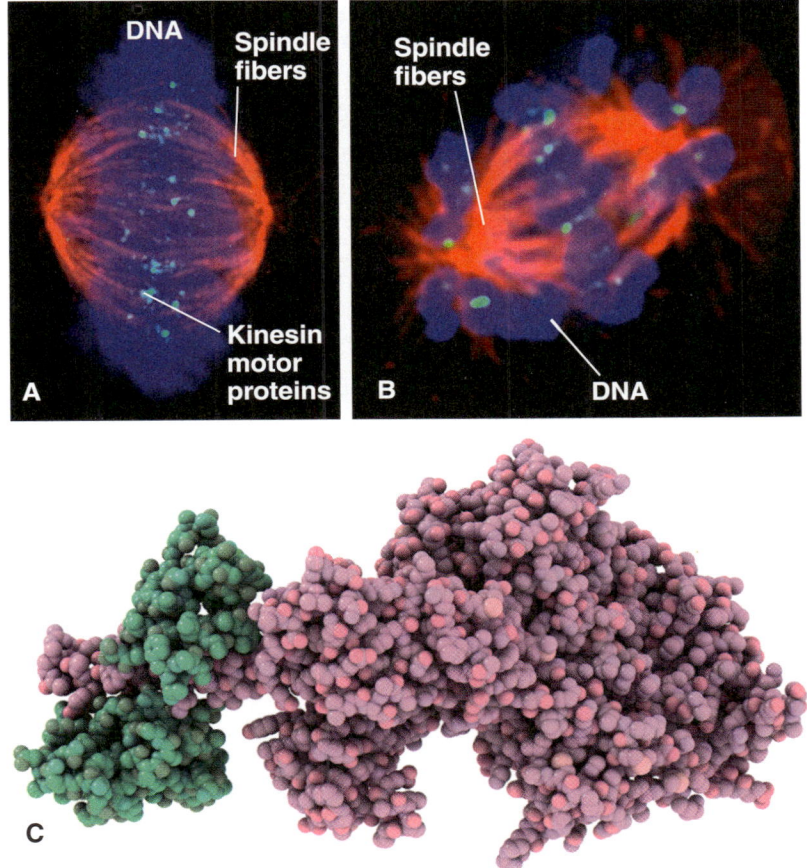

FIGURE 3-27 Motor proteins move chromosomes and contract muscles (A) The two daughter cell nuclei that form from one parent nucleus, contain an equal number of chromosomes. To make this happen at every cell division, the spindle apparatus assembles from microtubules (red) and the spindle fibers (red) attach to the chromosomes (dark blue cloud) lined up along the center of the cell. The kinesin motor proteins (light blue dots), help move the chromosomes to opposite poles during mitosis. (B) The chromosomes in this mutant cell should be moving to opposite poles to ensure that each daughter cell receives the correct number of chromosomes, but the chromosomes (dark blue) and the kinesin motors (light blue dots) appear to be located at the wrong places among the spindle fibers (red). This cell has a mutation that causes the chromosome DNA and the kinesin motor proteins to be miss-located, resulting in defective mitosis and abnormal chromosome segregation. (C) Myosin V is a member of a large family of motor proteins that convert chemical energy into mechanical movement in cells such as muscle contraction.

> **KEY CONCEPT**
> To ensure that the two daughter cell nuclei inherit an equal number of chromosomes at the end of cell division, the spindle fibers attach to the centromere on each chromosome and move the chromosomes to opposite poles when the cells divide. Mutations that alter the motor proteins, the spindle fiber or the centromere can cause defects in mitosis that result in abnormal chromosome segregation.

DNA Binding Proteins

Many different types of proteins function by binding to the chromosome DNA in the cell. These DNA binding proteins have essential roles to play in key processes such as controlling gene expression, packaging DNA into chromosome structures, replicating DNA before cell division, and repairing damaged DNA. Proteins that bind specifically to DNA use different regions of amino acids called DNA binding motifs. These protein motifs comprise different sequences of amino acids that fold into various 3D shapes in the protein. These 3D protein shapes are capable of binding to the DNA helix by means of a variety of binding mechanisms.

The helix-turn-helix (HTH) DNA binding motif is commonly used by regulatory proteins to bind to genes in eukaryotic and prokaryotic cells. The HTH motif contains two α-helical regions that are connected to each other by the "turn" amino acid strand. The two α-helical regions in the HTH motif are positioned to make interactions with specific bases in the DNA helix (Figure 3-28). The recognition helix is one of two α-helices in the HTH motif that interacts with the major groove of the DNA helix. The amino acid side chains in the recognition helix form hydrogen bonds with the oxygen and nitrogen atoms in the DNA backbone and make highly specific interactions with the DNA base pairs.

Scientists have studied many prokaryotic genomes, which has provided an extremely important foundation for the development of modern genome sciences applied to a wide range of studies on eukaryotic genomes. Scientists probably know more about *E. coli* than any other microbe because *E. coli* has been studied as a model genetic system for decades.

E. coli bacteria use glucose as a primary source of energy for the cell when the bacteria are growing in an environment where glucose is plentiful (Figure 3-29). In *E. coli* cells the *lac* repressor protein regulates the transcription of genes involved in lactose metabolism. When glucose levels are low, the cells turn on genes that allow them to metabolize other carbon sources such as lactose. This mechanism involves the messenger molecule, cyclic AMP (cAMP). Low glucose levels in the cells cause an increase in cAMP, which binds to the catabolite activator protein (CAP) (Figure 3-29). The activated CAP protein complex bends the DNA helix and activates the gene, permitting RNA polymerase to copy the gene into mRNA. The *lac* repressor protein turns off the expression of genes activated by the CAP complex. When lactose levels are low in the cell the *lac* repressor protein binds to the genes and forms a DNA loop that traps the

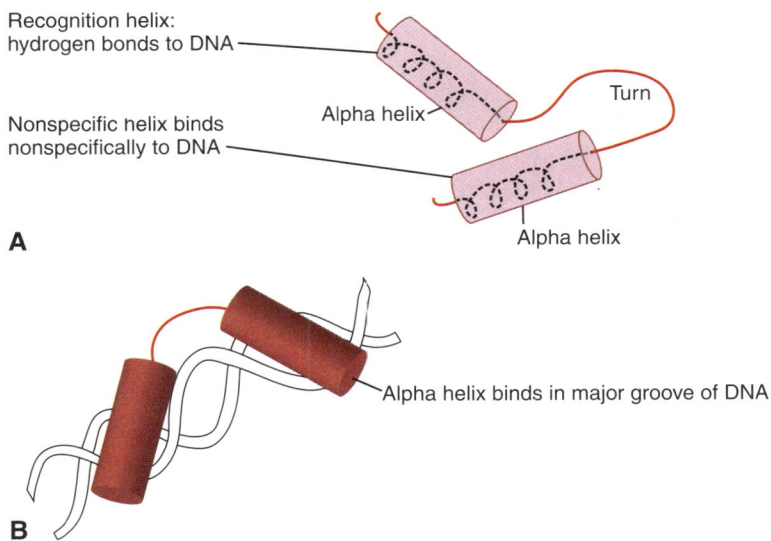

FIGURE 3-28 The HTH protein domain binds to the DNA helix (A) The helix turn helix (HTH) domain in a DNA binding protein contains two alpha-helices (cylinders) connected by a short strand of amino acids. The alpha-helical path of the protein backbones are shown by the dotted lines. (B) The HTH alpha helices bind to two locations in the major groove of the same DNA helix. The amino acids in the HTH recognition helix form specific H-bonds with the bases in the DNA molecule, making this a sequence-specific interaction. The amino acids in the HTH nonspecific alpha helix make nonspecific interactions with the DNA. (The base pairs were omitted from the DNA for clarity.)

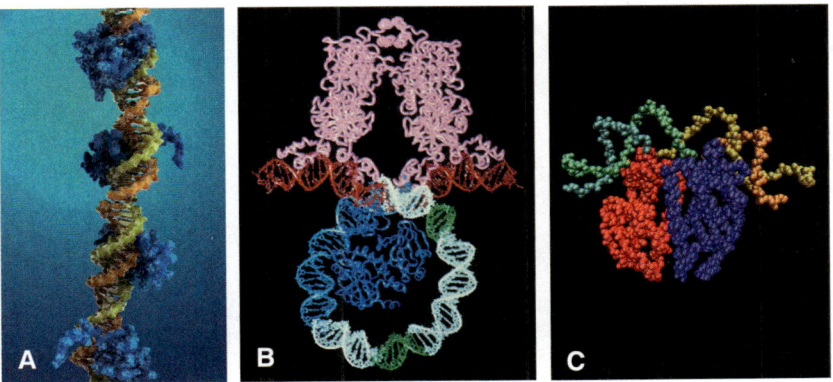

FIGURE 3-29 Sequence-specific DNA binding proteins (A) The TRF2 protein (blue) specifically binds to the telomere DNA (yellow, orange) at the ends of eukaryotic chromosomes. (B) The bacterial lactose (lac) repressor protein (pink) binds to the DNA genes in E. coli that control lactose metabolism. When lactose levels are low in the cell the lac repressor protein binds to the DNA helix (light blue) and forms a repression loop in the DNA that traps and silences the CAP activator protein (blue), turning off expression of an enzyme needed to metabolize lactose. (C) Bacteria metabolize glucose but when glucose is low, a dimer of the catabolite gene activator protein (CAP) (blue, red) binds to and activates the genes encoding enzymes that metabolize lactose. cAMP binds to CAP and the cAMP-CAP complex binds to the gene and bends the DNA helix, signaling the RNA polymerase enzyme to copy the gene into RNA.

CAP activator protein and represses the genes involved in lactose metabolism. The establishment, maintenance, and replication of eukaryotic chromosomes require many different proteins that recognize and bind to specific DNA sequences in the chromosome DNA. An example is the telomere binding protein TRF2, which is one of four proteins that bind to the short repeated DNA sequences at the ends of eukaryotic chromosomes. The TRF2 protein protects the chromosome ends from degradation in between cycles of chromosome DNA replication. The telomere DNA also adopts specific secondary structures that play important roles in replicating the ends of the chromosome DNA (Figure 3-29).

Many leucine zipper (LZ) DNA binding proteins are involved in regulating transcription in cells. The LZ DNA binding domain is composed of two identical protein subunit that bind to each other through long α-helical dimerization regions (Figure 3-30). Each LZ dimerization protein subunit contains leucine amino acids located at every seventh position in the amino acid chain. The leucine amino acid residues have long hydrocarbon side chains that zip together with the leucine side chains on the other protein subunit like the teeth on a zipper, binding the two proteins together into a very tight dimer (Figure 3-30). The regions of the LZ proteins that bind to DNA contain basic amino acids such as arginine and lysine, which are positively charged and bind tightly to the negatively charged DNA helix.

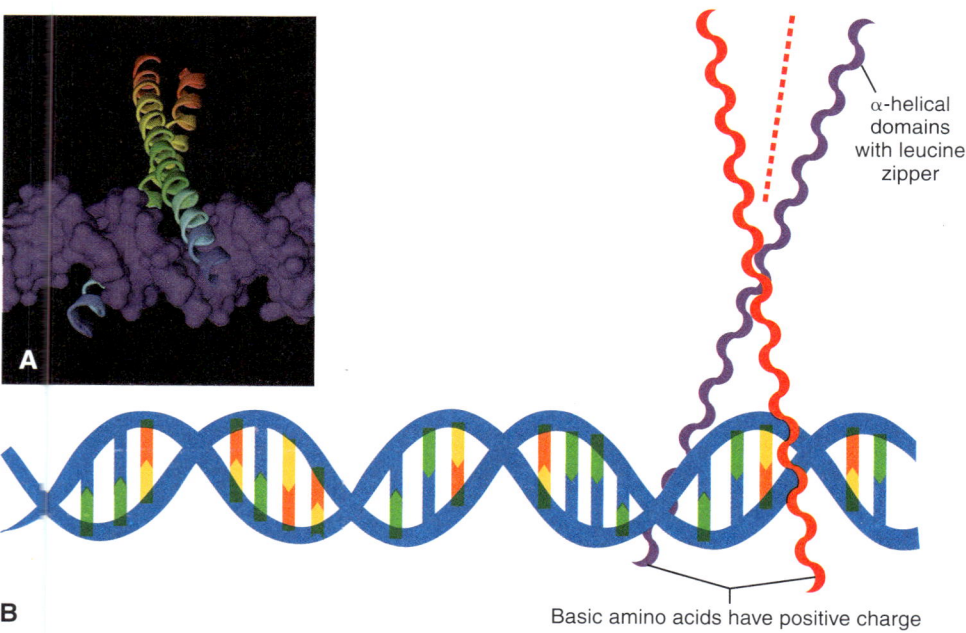

FIGURE 3-30 Leucine zipper DNA binding proteins control gene expression (A) The leucine zipper (LZ) DNA binding protein is a dimer of two proteins that bind to each other and also bind to the DNA helix through two long -helical "tong-like" regions. The LZ proteins dimerize through interactions between the side chains on the leucine amino acids in the protein-protein binding domain. The side chains on the leucines in the LZ regions in between the two proteins actually zip together like the teeth on a zipper (dotted line), holding the proteins together in the dimer. The other ends of the LZ proteins contain basic amino acids such as arginine and lysine that are positively charged and bind tightly to the negatively charged DNA helix.

UNIT 3 Gene Expression Makes RNA and Proteins

Many transcription factor proteins bind to DNA using zinc finger DNA binding motifs that bind to the promoter sequences of a gene (Figure 3-31). The transcription factor protein TFIIIA contains nine zinc finger motifs and functions during transcription by the RNA polymerase III enzyme. The amino acids in the α-helix and β-strand regions of each zinc finger domain adopt a finger-like configuration to interact with the DNA helix. In the bound protein the zinc fingers are stabilized by chemical bonds between the zinc metal ion and the conserved cysteine and histidine amino acids in the protein (Finger 3-31).

Histone Proteins Bind Nonspecifically to the Negatively charged DNA Helix

Histones are small DNA binding proteins that differ from the leucine zipper and zinc finger proteins because histone proteins bind to DNA nonspecifically, regardless of the DNA sequence. All eukaryotic organisms code for the four universal histone proteins (H3, H4, H2A, and H2B), which are the most highly conserved proteins known. In other words, the amino acid sequence of the H3 protein from a pea is almost identical in amino acid sequence to the H3 protein from humans. This suggests that histone proteins perform fundamental, essential functions that depend on the amino acid sequences of the histone proteins, and cannot vary and still function properly.

Histones are small, globular proteins containing charged amino-terminal (N−) tails, and short carboxy-terminal (C−) tails. Histones are key proteins

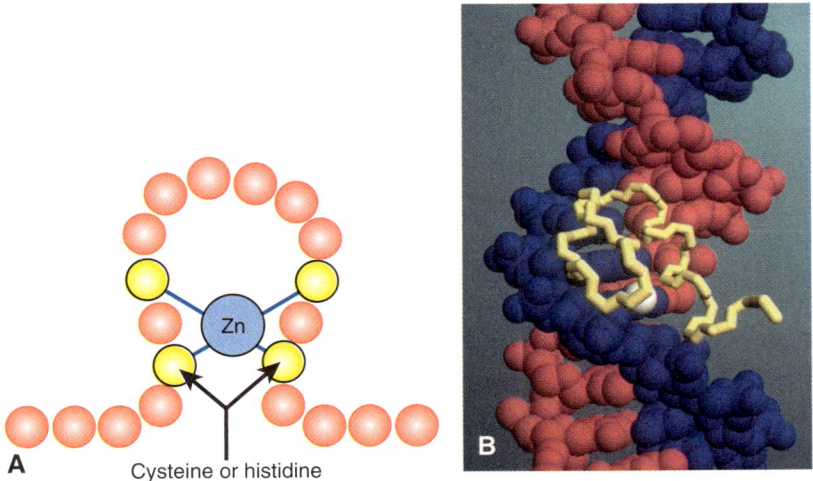

FIGURE 3-31 Zinc finger proteins bind to the DNA helix (A) TFIIIA is a zinc finger (Zn finger) transcription factor protein that works with the RNA polymerase III enzyme to control gene expression in eukaryotic cells. The nine zinc finger domains in the TFIIIA protein each contain conserved cysteine and histidine amino acids (yellow) that form bonds with a zinc metal ion (blue) and stabilize the amino acid structure of the zinc finger protein as it binds to the major groove of the DNA helix. (B) The zinc finger domain shown here (yellow backbone ribbon) is bound to the major groove of a DNA helix (red and blue strands) and is stabilized by the zinc ion (white).

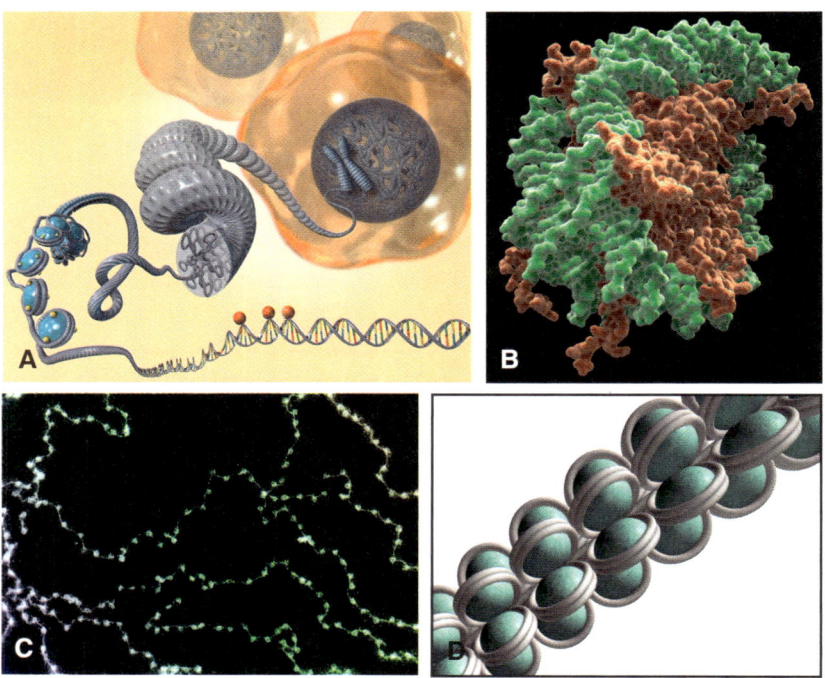

FIGURE 3-32 Histone proteins package DNA into a chromatin fiber (A) Over 6 feet (2 meters) of DNA helix is condensed to fit inside a human cell nucleus by packaging the DNA into nucleosomes in chromatin fibers that are further assembled into larger coils and then into condensed chromosomes. (B) The universal nucleosome contains an octamer core of histone proteins (orange-brown) with the DNA helix (green) wrapped twice around the outside of the core. (C) Chromatin fibers contain nucleosomes with linker DNA in between that resemble "beads on a string". The chromatin fibers shown are released from chicken erythrocyte (RBCs) nuclei that are lysed in a low salt buffer. Unlike human RBCs, chicken RBCs retain the nucleus throughout development. (D) The nucleosome fibers are thought to coil further into a 30-nm wide solenoid fiber, which in turn forms loops that further package the chromatin into a mature condensed chromosome.

that are essential for the assembly of nucleosomes to package DNA into chromosomes in cells. The structure of the nucleosome has been studied extensively through the use of many approaches, including X-ray crystallography to determine the 3D structure of the DNA-protein nucleosome (Figure 3-32). Nucleosomes have the same structure and composition regardless of the source of the chromatin under study. All the evidence indicates that nucleosomes are the universal subunits of eukaryotic chromatin and are the fundamental building blocks of eukaryotic chromosomes.

Nucleosome assembly begins with interactions among the histone proteins to make a histone octamer core structure containing two copies each of H3, H4, H2A, and H2B proteins. The DNA helix wraps twice around the outside of a histone core before exiting on either side of the nucleosome. Most eukaryotic organisms also make a fifth histone, called H1 or H5, which are not as conserved in amino acid sequence as the core histones. H5/H1 is sometimes called the linker histone because it associates with the linker DNA and is not part of the histone core proteins.

In the chromatin fibers the nucleosomes are positioned along the DNA helix such that the fiber can be extended or condensed. When the fiber is extended, the linker DNA between adjacent nucleosomes is visible and the chromatin fiber resembles beads on a string (Figure 3-32). The globular middle regions of the histone proteins assemble together in the nucleosome core, while the amino terminal ends of the histone proteins extend a significant distance away from the core, where they can bind to other proteins. The N-termini of the core histone proteins in adjacent nucleosomes along the chromatin fiber can interact with each other as well as with nonhistone proteins in the cell. Interactions between the N-termini of the core histones in adjacent nucleosomes can cause the entire chromatin fiber to contract and condense into a much shorter, much thicker fiber. Other changes altering the histone N-termini cause the nucleosomes to repel each other, causing the overall chromatin fiber to extend and become less condensed.

This introduction to the dynamic world of chromatin is also an introduction to the newest area of human genetics, called epigenetics. Epigenetic changes do not alter the DNA sequence but they are transmitted to relatives. Epigenetic changes depend on the distribution of tiny chemical groups on the chromosome DNA (Unit 6). Histone proteins also have important functions in genome and nuclear reprogramming and are involved in the epigenetic regulation of eukaryotic gene expression that can be transmitted over many generations (Unit 6).

Unit 3 Questions

1. In eukaryotic cells the protein-coding genes are expressed by an enzyme or enzymes called:
   a. RNA polymerase I
   b. RNA polymerase II
   c. RNA polymerase III
   d. RNA polymerases I, II and III

2. A eukaryotic protein-coding gene typically contains at least one DNA control region called:
   a. An RNA polyadenylation site
   b. A protein-coding region
   c. A 3′ untranslated region
   d. A promoter region

3. An enhancer DNA element is unusual because:
   a. The DNA helix can bend at the enhancer element
   b. The enhancer is often located a long distance from the start of the gene
   c. The enhancer can function in either orientation relative to the chromosome arms
   d. All of the above
   e. None of the above

4. An in-frame codon in an mRNA coding region specifies:
   a. A base pair
   b. A protein
   c. An amino acid
   d. A hydrogen bond

5. Ribosomes make proteins by synthesizing:
   a. Linear strands of DNA bases linked together with phosphodiester bonds
   b. Linear chains of amino acids linked together with phosphodiester bonds to make proteins
   c. Linear strands of RNA bases linked together with phosphodiester bonds
   d. Linear chains of amino acids linked together with peptide bonds to make proteins

6. The amino acids in a protein that interacts with the cell membrane exhibit what types of chemical characteristics?
   a. Water-hating properties
   b. Water-loving properties
   c. Alpha helix structure
   d. Beta sheet structure

7. Enzymes are specialized proteins that:
   a. Can speed up biochemical reactions in the cell
   b. Can make and break chemical bonds
   c. Can perform a function without permanently changing the molecular structure of the enzyme
   d. All of the above
   e. None of the above

8. Which of these proteins can bind nonspecifically to a DNA helix (without regard to base sequence)?
   a. Restriction enzymes
   b. TATA DNA binding protein
   c. Histone proteins
   d. All of the above
   e. None of the above

9. The universal DNA-protein structure called the building block of eukaryotic chromosomes is a:
   a. Ribosome
   b. Nucleosome
   c. Chromatin fiber
   d. Microtubule fiber

10. Which of these different types of RNA molecules contains a protein-coding region?
    a. rRNA
    b. tRNA
    c. mRNA
    d. All of the above
    e. None of the above

UNIT 4

Genetic Testing to Diagnose Genetic Diseases

DNA Mutations Change Gene Sequences

Some people think of a genetic disease as being similar to a "time bomb" in their chromosome DNA. The DNA mutation that causes a genetic disease is present in the chromosome DNA at birth, but the impact of the mutation might occur at any time in life. The mutation remains silent until the product of the mutant gene is needed and the mutant gene is expressed. If the mutant protein made in the cell is defective and does not function then any tissue or organ made up of the mutant cell type will not work properly either. Genetic diseases and disorders are caused by mutant genes transmitted from parent to biological child by mutant genes inherited through the egg and the sperm cells (Figure 4-1).

Normally a biological child receives one copy of each chromosome (and each gene) from Mom and one copy of each chromosome (and each gene) from Dad (Figure 4-1). In this way a child inherits two copies of most human genes, either:

- two normal (wild-type) genes, or
- two mutant (altered) genes, or
- two different copies of the same gene (one wild-type copy and one mutant copy or two mutant copies).

DNA mutations change the order of the base pairs (DNA sequence) along the DNA template strand, which in turn alters the sequence of the RNA copied from the mutant gene and the amino acid sequence of the encoded protein (Figure 4-2). When a normal or wild-type gene is expressed, the genetic information in the DNA sequence of the gene represents the RNA words or codons that specify the amino acid sequence of the encoded protein. The expression of a mutant gene follows the same pathway in the cell. The altered genetic information encoded in the mutant DNA sequence is transmitted through altered RNA codons that specify a mutant protein product (Figure 4-2).

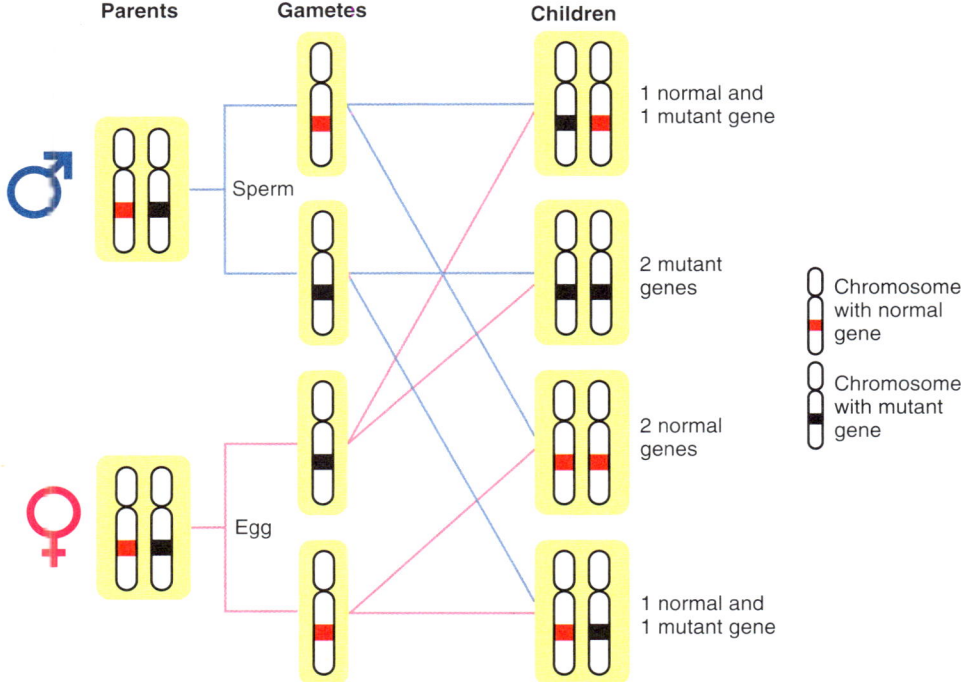

FIGURE 4-1 Human traits and genetic diseases are transmitted by inherited genes Human chromosomes carry DNA that codes for genes. The male and female parents donate chromosomes to produce the haploid sperm and egg cells. Fertilization produces a diploid zygote cell. The zygote divides and gives rise to the embryo who inherits parental genes: normal (wild-type) allele (red), mutant allele (black).

Human chromosomes can have different types of DNA mutations, ranging from single base-pair changes (point mutations) (Figure 4-2) to mutations involving the insertion (addition), deletion (removal), or rearrangement of large chromosome regions containing thousands of DNA base pairs (Figure 4-2). Genetic diseases that are caused by very small DNA changes such as single base-pair mutations (point mutations) can be detected by DNA sequence analysis or genotyping (see below). Genetic diseases that are associated with large chromosome mutations such as translocations, inversions, deletions, and insertions can be visualized by karyotype analysis of human chromosomes.

DNA mutations that pre-exist in the chromosomes of the biological parents are passed to children by the inheritance of gametes (Figure 4-1). In contrast, somatic DNA mutations are base changes that occur spontaneously in the chromosome DNA during a person's life span. These mutations are caused by environmental agents such as ultraviolet radiation (sunlight), ionizing radiation (medical and dental X-rays), and environmental chemicals (pesticides, etc.) that damage DNA. Flat, planar chemicals (dyes, benzenes, etc.) can insert between the base pairs of the DNA helix, which causes mutations when the DNA replicates or is copied into RNA.

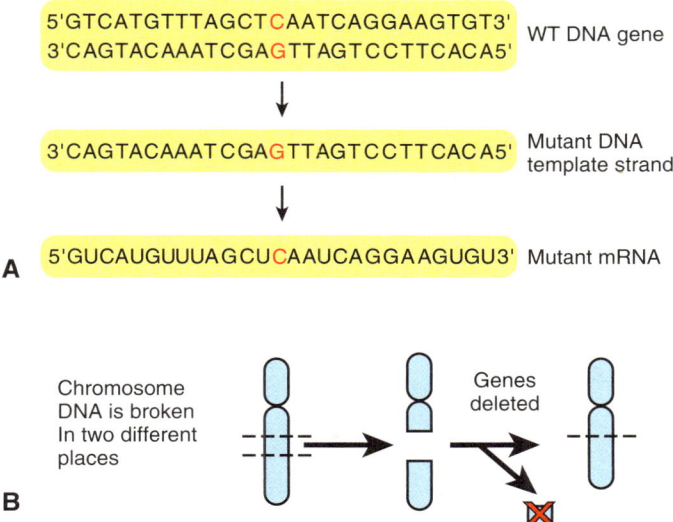

FIGURE 4-2 Large and small chromosome DNA mutations (A) A chromosome DNA mutation can be as small as a single base pair change (red), in this case in a gene that is copied into RNA, potentially altering the protein product. (B) A chromosome deletion can remove a large region of thousands of DNA base pairs that potentially encodes many genes.

Random mutations are most likely to affect the vast regions of noncoding sequences in the human genome DNA, which will not alter the protein coding sequences. However, damage to the chromosome DNA can have very serious consequences, even in cases where gene expression is not directly affected. Ionizing radiation such as medical X-rays can physically break double-stranded chromosome DNA even though the DNA is protected by the chromosome proteins (Figure 4-3).

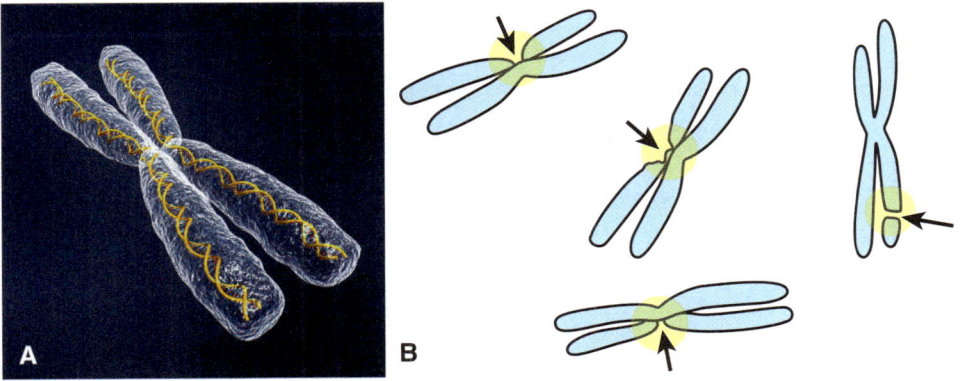

FIGURE 4-3 Radiation exposure can damage human chromosome DNA (A) Each chromosome contains a DNA helix that extends from one end of the chromosome to the other. This chromosome has replicated and contains two DNA helices, one in each copy. (B) These broken chromosomes show how exposure to radiation can break one of the two DNA helices as indicated by red arrows and yellow circles. Broken chromosome DNA that is not repaired can often cause cancer to develop.

Exposure to environmental agents poses a significant risk for damage to human chromosome DNA, which must be fixed by the DNA repair enzymes in the cell. The DNA repair enzymes are sent to the site of DNA damage to repair the broken chromosome DNA. A cell that fails to repair its broken chromosome DNA before beginning the next cell division cycle will accumulate abnormal chromosomes with potentially extensive DNA rearrangements. This is one reason that unstable, rearranged chromosomes are a characteristic feature of cancer cells.

DNA Mutations Alter Control Regions and Protein Coding Regions of Genes

The impact of a given DNA mutation on gene expression depends on the location of the altered base pairs relative to the genetic information encoded in the gene (Figure 4-4). Typical eukaryotic protein-coding genes contain two main regions, the DNA control promoter region and the protein-coding region (Unit 3). A DNA mutation in the control region of a protein-coding gene does not change the amino acid sequence of the protein product of the gene. This is because the promoter DNA in the gene is not copied into RNA or translated into protein, so the promoter mutations will not affect the amino acid sequence of the encoded protein (Figure 4-4) (Unit 3). However, a mutation that alters the promoter DNA sequence can completely block gene expression by preventing the RNA polymerase II enzyme from binding to the gene promoter DNA. However, a different mutation in the promoter DNA might reduce but not abolish transcription by decreasing the affinity of the transcription factor proteins or the RNA polymerase II enzyme for the promoter DNA sequences.

Mutations that alter the DNA sequences in the protein-coding region of a gene will probably change the amino acid sequence and possibly alter the

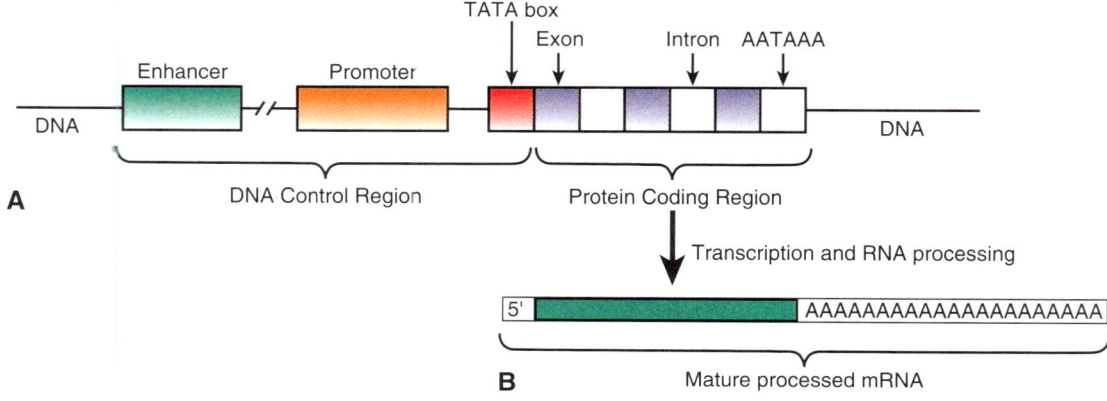

FIGURE 4-4 DNA mutations alter eukaryotic gene expression The impact of a DNA mutation on gene expression and protein function depends on the location of the mutation in the gene. (A) A mutation in the DNA control region (enhancer, promoter, TATA) can abolish gene expression by preventing interaction with the RNA polymerase II enzyme and TF proteins. (B) Any mutation in the protein coding region is copied into mutant RNA transcripts. The mutation might alter RNA processing and prevent protein expression. A mRNA can be translated into a mutant protein that might not function at all or might be partially defective.

function of the encoded protein. However, some mutations in the coding region do not affect the protein product (Figure 4-4). The RNA polymerase II enzyme copies most eukaryotic genes into pre-mRNAs containing introns and exons (Unit 3). The accuracy of the RNA splicing process is essential to ensure that the spliced mRNA strands contain the correct exon sequences to encode the protein of interest. DNA mutations that change the exon sequences will usually (but not always) produce a mutant protein with altered amino acids (Figure 4-4). In contrast, a DNA mutation in an intron sequence will be removed by RNA splicing before the mRNA is translated and therefore will not alter the final protein. Other mutations that alter the conserved RNA sequences involved in splicing the pre-mRNAs (the splice site sequences and the branch site sequences) can cause mutations that interfere with RNA splicing.

> **KEY CONCEPT**
> There are many points along the pathway of genetic information in the cell where a mutation can block or change the level of gene expression.

DNA mutations that alter the exon sequences in the protein-coding region of a gene can have a significant impact on the protein made in the cell or will have no effect, depending on the sequence altered by the mutation in the coding sequence (Figure 4-5). For example, a DNA base inserted into a gene that changes AGC to AAG alters the mRNA sequence and causes a lysine amino acid to be inserted into the protein in place of a serine in the wild-type protein (Figure 4-5). A missense mutation that alters a UUU codon to a UUG codon causes a phenylalanine amino acid in the wild-type protein to be replaced with a leucine amino acid in the mutant protein (Figure 4-5).

Sometimes a small DNA mutation can have a large impact on the cell, while other mutations can have little or no effect. For example, a point mutation that changes a UUU codon to UUC does not alter the amino acid sequence of the protein product because both the UUU and UUC codons specify the same amino acid, phenylalanine (Figure 4-5). However, a nonsense point mutation that changes UGG to UAG in the mRNA introduces a new translation stop codon (UAG) that prematurely terminates protein synthesis and makes a short protein product (Figure 4-5) (Unit 3). The genetic code in the Genetic Code Table contains three special codons that function in the genetic code as punctuation signals in the mRNA that indicate the end of a protein. A DNA mutation that changes the UGG codon (tryptophan) into a translation stop codon (UAG) causes the synthesis of a prematurely shortened mutant protein (Figure 4-5).

The Importance of Inheriting two Copies of each Gene (Alleles)

The mutant genes located on human autosomal chromosomes are normally inherited as autosomal recessive or autosomal dominant mutations (Figure 4-6). The term *autosomal* refers to the fact that the mutant gene is carried on one

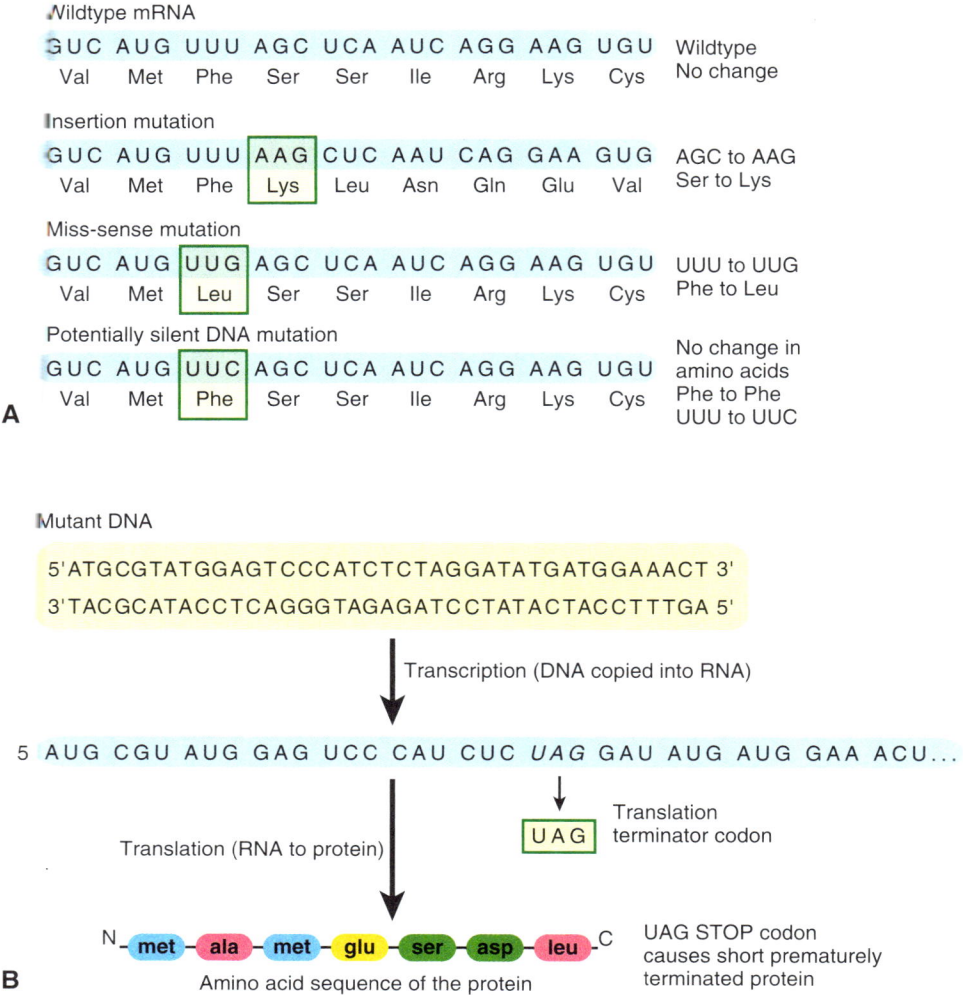

FIGURE 4-5 DNA mutations can change amino acids in the protein products (A) These gene mutations alter the protein coding region of the gene in different ways. The insertion mutation shown alters AGC to AAG and changes a Serine (Ser) to a lysine (Lys). A miss-sense mutation that changes UUU to UUG causes a phenylalanine (Phe) to be replaced with a leucine (Leu) in the protein. Some mutations can change the DNA sequence and the mRNA but do not change the amino acids. For example, the UUU to UUC mutation replaces a Phe with a Phe. (B) This point mutation produces premature termination of protein synthesis by creating a translation stop codon (UAG) in-frame in the coding region.

of the 22 pairs of autosomal chromosomes and not on a sex chromosome (X or Y chromosome).

To evaluate how a given DNA mutation might affect gene expression, it is important to investigate the function of both inherited copies of the gene in question and to identify the wild-type (normal) and mutant alleles. The copies of each chromosome inherited from Mom and from Dad have very similar DNA sequences, although each human chromosome contains a specific subset of the total number of human genes. The copy of chromosome 6 inherited from Mom has the same genes as the copy of chromosome 6 inherited from Dad. This is how people normally inherit two different versions (alleles) of every gene,

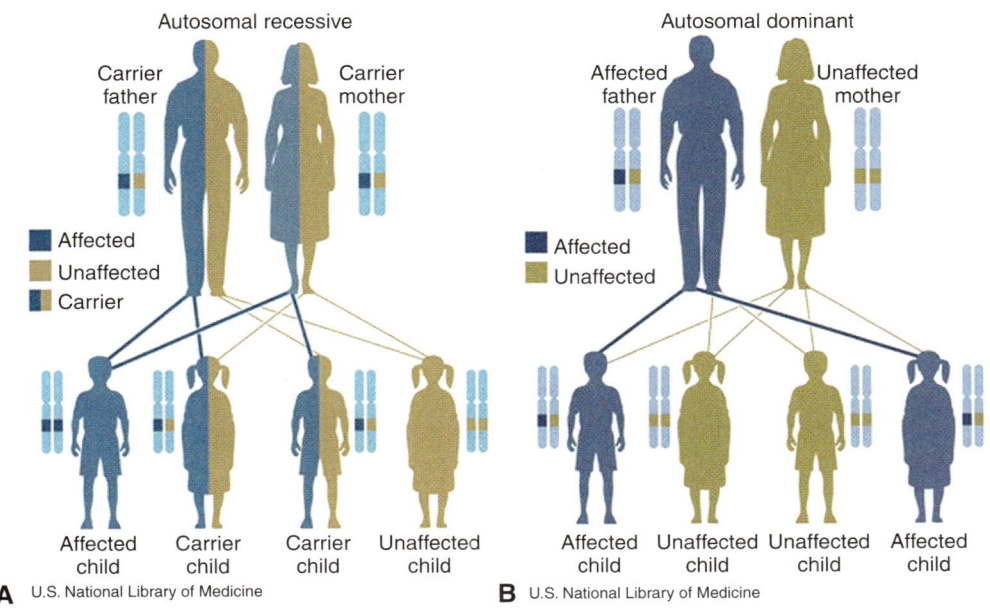

FIGURE 4-6 Inheritance of human autosomal chromosome mutations An inherited gene mutation can be a recessive or a dominant allele. A human autosomal mutation is carried on one of the 22 pairs of autosomal chromosomes, and not on the sex chromosomes (X or Y). (A) Someone who inherits a wild-type gene and a recessive mutant copy (allele) of the same gene does not usually experience disease symptoms because the mutation is recessive. This is the case for people who are carriers of a genetic disease. (B) A person who inherits a wild-type gene and a dominant mutant allele of the same gene usually experiences symptoms that are caused by the expression of the dominant mutant allele.

with rare exceptions for copy number variation (Unit 2). The consequences of inheriting different combinations of mutant and wild-type alleles (or two mutant alleles) depend on the functions of the genes and alleles in question. For example, if a person inherits a wild-type allele and a mutant allele of the same gene and does not experience negative effects or disease symptoms, it is likely that the inherited mutation is recessive (Figure 4-6).

Cells that carry both a wild-type allele and a mutant allele of a particular gene usually produce both the wild-type and mutant proteins at the same time in the same cells. If the mutant protein does not function, the mutant cells must rely for survival on the function of the normal proteins expressed from the wild-type gene. The expression of the normal gene and protein are usually unaffected by the production of the mutant proteins in the same cell. However, in cases where the mutant protein interferes with an essential function performed by the normal proteins, the cells will die because they lack an essential protein function.

Some People are Carriers of a Genetic Disease

People who inherit one normal (wild-type) allele and one mutant allele of a disease gene and who do not experience disease symptoms are called genetic disease "carriers." Unfortunately, most people who are genetic carriers do not know that they carry a dangerous genetic allele and are not aware that they can pass a disease gene to their biological children (e.g., cystic fibrosis and sickle cell disease). However, people who inherit a wild-type gene along with a mutant allele of the gene do not automatically avoid the negative effects of the mutant proteins. Genetically dominant mutant alleles can cause cells to be defective even when the cells are making functional wild-type proteins in addition to the mutant proteins. Other types of dominant mutations produce mutant proteins that in some way interfere with the functions of the wild-type proteins and block essential biochemical reactions, causing an overall negative impact on the mutant cells and organism.

> **KEY CONCEPT**
> If one copy of the gene is altered by a mutation and the other copy of the same gene is a wild-type gene, the wild-type gene can provide the missing protein function when the mutation is recessive.

The copy number of some genes is set during fertilization, when one of the two copies of a small number of human genes is permanently inactivated by genome imprinting, a process used by the cell to inactivate one copy (allele) of a gene by DNA methylation (Unit 6). Some gene copies are inactivated only when inherited from the father, while other gene copies are inactivated only when inherited from the mother. Although people normally inherit two functional copies of each gene, some genes are inherited in more than one or two copies per cell by copy number variation (Unit 2).

This type of epigenetic gene regulation involves special methylation enzymes that attach small methyl ($-CH_3$) groups to the promoter DNA sequences of the genes to be silenced. The methyl groups prevent the expression of genes without changing the sequence of the promoter DNA (Unit 6).

Human Genetic Diseases and the Involvement of Gene Mutations

Research shows that almost all noncontagious human diseases have a genetic component, but human genetic diseases are rarely caused by a mutation in a single gene. Most genetic diseases typically result from interactions among multiple mutant genes and proteins and include the influential contributions of lifestyle, personal experience and environmental factors.

Human genetic diseases are grouped by the number of mutant genes involved in causing each disease:

(1) Single gene diseases and disorders (monogenic diseases) are rare and are caused by a mutation in a single gene that is transmitted by inheriting one copy or two copies of the mutant gene (depending on the disease). Examples of single gene diseases include cystic fibrosis (CF), Huntington's disease, hemophilia, sickle cell disease, and Duchenne muscular dystrophy.

(2) <u>Complex multigenic diseases</u> (polygenic diseases) are much more common than single gene genetic diseases and are caused by interactions among a number of different mutant genes and proteins. The fact that environmental factors are also involved makes these diseases much more difficult to diagnose and treat compared to diseases that are caused by a single mutant gene. Scientists are now using modern genetic approaches like SNP mapping to identify the multiple genes involved in these complex polygenic diseases. The highest healthcare costs in the United States include the treatment of complex polygenic diseases such as heart disease, Alzheimer's disease, type II diabetes, colon cancer, autism, type II diabetes, and most forms of cancer.

(3) <u>Chromosome abnormalities</u> (polygenic) result from deletions, duplications, and rearrangements that exchange large segments of chromosome DNA, and can potentially involve changes in many genes. In some cases the disorder occurs at conception when a person inherits an abnormal number of intact human chromosomes. Examples include Down syndrome or trisomy 21, where the affected person inherits an extra copy of chromosome 21 in all of his or her cells. Other inherited chromosome abnormalities include the Klinefelter syndrome and Turner syndrome.

Cancer cells often contain abnormal numbers of chromosomes (aneuploidy) caused by changes in somatic cells that occur during cancer development. These chromosome changes are not passed to the next generation because they usually do not affect the gametes (egg and sperm cells). In some rare cases, chromosome rearrangements are directly responsible for the development of a specific cancer. An example is the Philadelphia chromosome translocation, which creates a unique oncogene (cancer gene) that causes a form of blood cancer.

Inheritance Patterns of Genetic Diseases

(1) <u>Autosomal dominant</u>. Inheriting only one copy of a mutant gene is sufficient for an individual to be affected by an autosomal dominant disease (Figure 4-6). In the case of autosomal dominant disorders, each affected person must have one affected parent who carries the mutant gene. Examples include Huntington disease and neurofibromatosis type 1.

(2) <u>Autosomal recessive</u>. In this case a person must inherit two copies of the mutant gene in order to be affected by the disease (Figure 4-6). The affected individual usually has parents who each carry a single mutant gene (and are genetic carriers) but do not have symptoms. Examples include cystic fibrosis and sickle cell anemia diseases.

(3) <u>X-linked recessive</u>. X-linked recessive diseases are caused by recessive mutant genes carried on the X chromosome (Figure 4-7). The pattern of transmission of an X-linked dominant mutation differs between males and females, but males are more frequently affected than females. Only mothers carry X chromosomes, so fathers cannot pass X-linked genetic traits to their sons (no male-to-male transmission), and all of Dad's X chromosomes are used to conceive daughters (X, X). Examples of X-linked recessive diseases include hemophilia and Fabry disease.

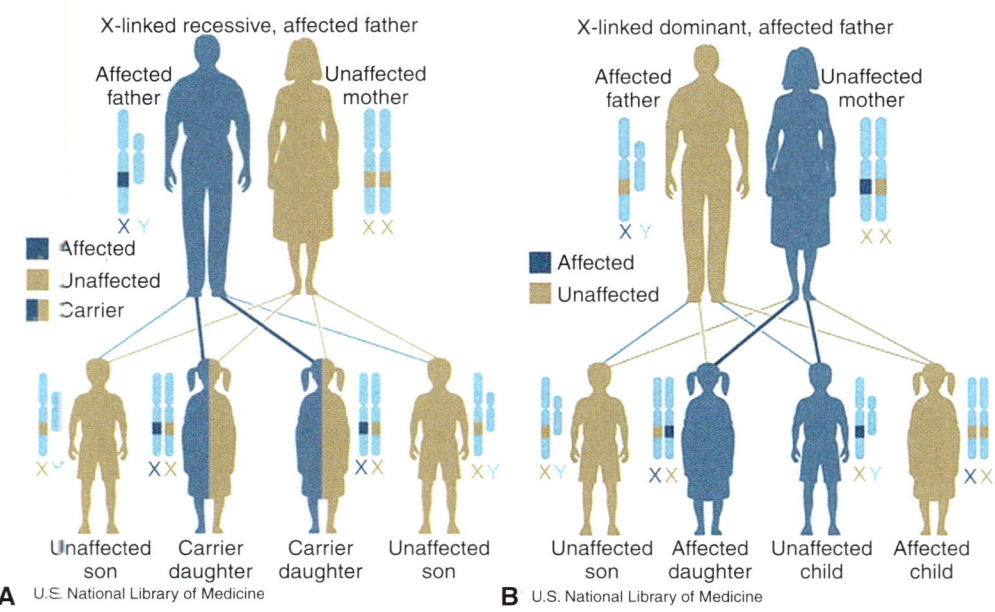
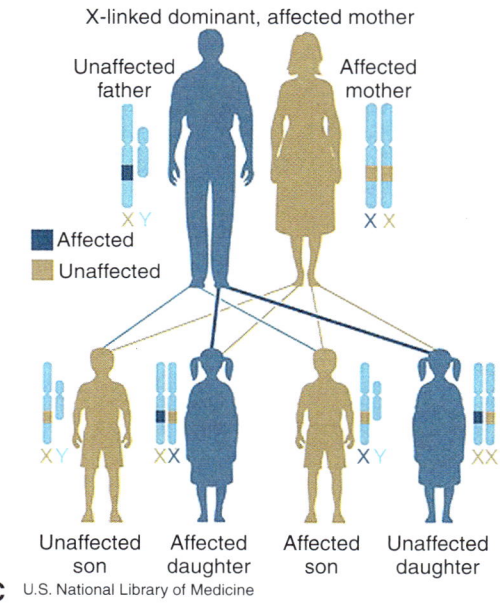

FIGURE 4-7 X-linked chromosome mutations have distinctive inheritance patterns X-linked gene mutations are located on the X chromosome (X, X female; X, Y male). (A) The X-linked recessive gene mutation is inherited from the father on the X chromosome (father is X,Y). The sons are not affected because they inherit an unaltered X chromosome from Mom and a normal Y chromosome from Dad. However, the daughters are carriers because they inherit one mutant X chromosome from Dad and one normal X chromosome from Mom. (B) An X-linked dominant gene mutation is inherited on the X chromosome from the father but does not affect his sons, who inherited their X chromosome from Mom (and their Y chromosome from Dad). However, both daughters inherit an X-linked dominant mutation and are affected by the disease symptoms. (C) An X-linked dominant mutation is carried on the X-chromosome by the mother. The affected sons and daughters who inherit and express the disease gene mutation are affected by disease symptoms.

(4) X-linked dominant. The diseases and disorders that are X-linked dominant are caused by mutations in genes carried on the X chromosome (Figure 4-7). Females, who naturally inherit two X chromosomes, are more frequently affected than males, and the pattern of transmission of an X-linked dominant mutation differs between men and women. Fathers cannot pass X-linked genetic traits to their sons (no male-to-male transmission). An example is fragile X syndrome.

(5) Codominant inheritance. This type of inheritance pattern occurs when two different mutant alleles of the same gene are expressed at the same time in the same cells. Each allele produces a slightly different version of the mutant protein and both mutant proteins can partially function in a way that influences the disease symptoms. Examples include the ABO blood group genes and alpha-1 antitrypsin deficiency.

(6) Mitochondrial inheritance. Also called maternal inheritance, the DNA genomes in the mitochondria (mtDNA) are always transmitted from the mother to her daughters and sons. The fathers do not pass mitochondria or mtDNA to their children. The short, circular, double-stranded mitochondrial DNA (mtDNA) genomes encode a small number of mitochondrial genes that can sometimes cause maternally inherited diseases in humans. The mother transmits the mutant mitochondrial genes to the developing human embryo at fertilization, causing mitochondrial genetic diseases such as Leber hereditary optic neuropathy (LHON).

Inheritance of Complex Multigene Genetic Diseases

Complex or multifactorial genetic disorders such as heart disease, type II diabetes, and obesity do not follow straightforward patterns of genetic inheritance, making it difficult to predict the risk of genetic transmission of these disorders. It is also clear that the environment influences the expression of many genes, potentially altering the normal patterns of inheritance.

Some genetic diseases are caused by chromosome segregation mistakes occurring during the cell division events that created the egg or sperm cells prior to the fertilization event, which leads to a person with the disease who inherits the wrong number of chromosomes in each cell. During normal cell division (mitosis), equal numbers of duplicated chromosomes must be segregated into each of two daughter cells. It is so important to segregate equal numbers of chromosomes to the new daughter cells that the cell has specific proteins and feedback systems designed to inform the cell if the chromosomes accidentally make a mistake and a chromosome miss-segregates (undergoes nondisjunction) during cell division. Mistakes in chromosome segregation can also occur during meiotic cell division, which produces new egg and sperm cells containing half the number of chromosomes (haploid).

Despite the safeguards used by the cell to protect against mistakes in cell division, occasionally a protein or DNA component in the cell functions incorrectly and the new cells inherit an unequal number of chromosomes. The resulting egg or sperm cell will contain an extra chromosome at fertilization, and as a result the new individual will inherit an extra chromosome in all cells. Most mutations that cause the gain or loss of a human chromosome (aneuploidy) are fatal in early embryo development. The few exceptions include Down syndrome and Turner syndrome, which both involve the inheritance of extra chromosomes.

Genetic Testing to Identify Mutant Genes and Genetic Diseases

Different types of DNA testing are used to identify mutations in human genes and to diagnose human genetic diseases. Genetic DNA testing involves directly examining the genes in a patient's chromosome DNA to obtain specific genetic information unique to that individual. In some types of gene testing, scientists actually determine the DNA sequence of a region of the patient's chromosome DNA to find out if the gene in question contains a mutation. The sequence of the patient's gene is compared to the normal (wild-type) allele of the same gene in the human genome sequence database (Unit 2). Recent approaches use the human SNP database to screen for the specific SNP DNA differences between human genomes that are linked to certain genetic alleles and genetic disease mutations. Similar DNA screens are used to identify the 13 different DNA human genetic markers (loci) used by the FBI when identifying humans by DNA fingerprinting (Unit 6).

Other types of genetic tests use indirect methods that rely on analyzing the products of the gene in question. For example, the test might measure the activity of an enzyme expressed in the patient's cells that is compared with the normal activity of the wild-type enzyme, to assess cardiac function. A different test monitors the metabolic by-products of the biochemical reactions altered by the mutant genes under study, or from an expected step in cell and tissue development. A common prenatal test is used to detect the level of the alpha-fetoprotein (AFP) in the mother's blood. AFP is made by the fetus, but unusually high levels of AFP in the mother indicate a possible problem with fetal spine development that warrant further prenatal testing.

Human genetic diseases that are caused by chromosome rearrangements are best identified by studying a karyotype, a display of whole mitotic (metaphase) condensed chromosomes, either stained for cytological analysis or labeled with fluorescent tags for microscopic analysis.

Genetic Testing Applications include the Following:

- screening individuals to identify genetic carriers
- analyzing embryos for mutant genes by preimplantation genetic diagnosis (PGD)
- testing newborn infants for genetic diseases
- screening adults for risk of developing late-onset genetic disorders
- identifying disease genes from patients' symptoms
- determining DNA fingerprinting for forensic identification

UNIT 4 Genetic Testing to Diagnose Genetic Diseases

> **KEY CONCEPTS**
> Gene testing is most often used to diagnosis and to predict genetic diseases, to avoid transmitting mutant genes to biological children and to identify individuals who are at high risk for preventable diseases.

DNA Testing with Molecular DNA Probes

Genetic testing methods directly examine the DNA taken from the cells of a specific individual. Genetic tests conducted at the molecular level can include sequencing at least some of the DNA in an individual's genome to determine which genes have mutations. Testing can also analyze the number of DNA repeats at specific locations in the human genome and also identify specific SNP sites for genotype analysis. Molecular DNA tests typically require the use of specific DNA probes that can base-pair only with complementary DNA sequences in the target chromosome DNA molecule (Figure 4-8). Just like base pairing in the DNA helix, the base pairing between the DNA probes and the target DNA is the result of the formation of multiple hydrogen bonds and is highly specific.

The hydrogen bonds required for the base-pairing of DNA probes are so specific that DNA probes can detect a single base-pair difference between the wild-type and mutant alleles in two human genomes, by screening 6.4 billion DNA base pairs for complementary sequences (Figure 4-8). During a genetic test, the DNA probe will base-pair only to the completely complementary target DNA sequences in the test. A single base-pair

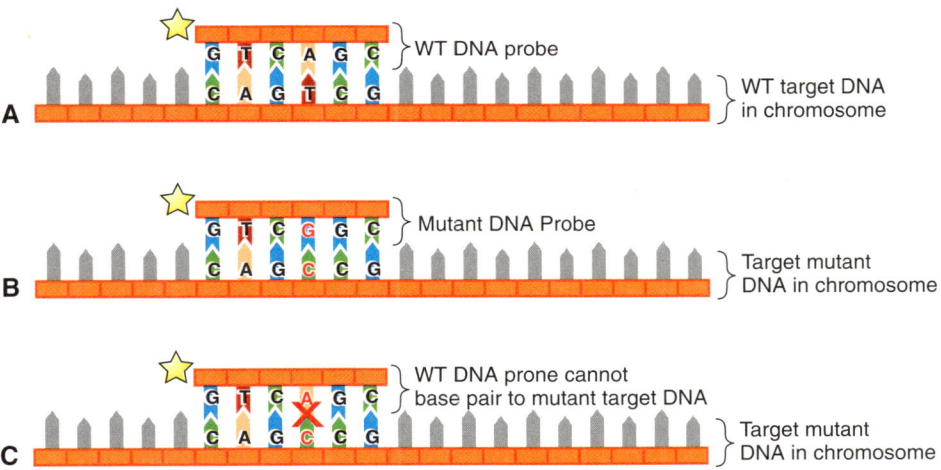

FIGURE 4-8 DNA probes base-pair to specific targets in the chromosome DNA (A) Each DNA probe is a single stranded DNA containing a base sequence that is specifically designed to base pair with a target DNA sequence in the chromosome. Single-stranded DNA probes are often labeled with a fluorescent tag for detection. (B) This mutant DNA probe was designed to base-pair specifically with a mutated DNA sequence in the target chromosome DNA. (C) The wild type DNA probe cannot base pair to the target mutant DNA sequence in the genome because there is a miss-match in the complementary base pairs.

mutation in the chromosome DNA can be detected by using a DNA probe that contains a DNA sequence that is complementary to the DNA mutation. Under appropriate base-pairing (hybridization or annealing) conditions, the mutant DNA probe will base-pair to the mutant gene in the chromosome and will not base-pair to wild-type gene (Figure 4-8).

Genetic testing can reveal inherited mutations that cause various genetic diseases. For example, a mutation in the familial adenomatous polyposis gene can help screen people to find individuals who are at risk for colon cancer. These people should be screened periodically for precancerous colon polyps to avoid potentially lethal colon cancer. Another genetic test is under development to screen for a common genetic iron-storage disease that is difficult to diagnose and that can be fatal without treatment. Scientists are also developing genetic tests to diagnose complex disorders like Alzheimer's disease and some types of cancers. People with a strong family history of a disease should be tested to identify those who have a high genetic risk for developing the disease or disorder. In the post human genome era it will become more common for people to consult human genetic counselors before becoming pregnant to start a family.

> **KEY CONCEPT**
> The power of DNA testing comes from the accuracy of DNA base pairing and the ability of the DNA probes to engage in only highly specific base-pairing interactions. Only a DNA probe with a base sequence that is complementary to the target DNA sequence can successfully base-pair to the target DNA.

Tests for Genetic Diseases Involving whole Chromosomes

The karyotype of a typical human mitotic cell contains 46 condensed chromosomes, which represents the chromosomes in one diploid human cell nucleus before DNA replication (Figure 4-9). Human gamete cells (egg and sperm) are haploid and contain half the number of chromosomes (23). A human karyotype display is an effective way to visualize inherited chromosomes to look for changes in chromosome morphology and chromosome number. This karyotype display clearly shows the extra copy of chromosome 21 that is inherited in Down syndrome, which is one of the few disorders involving the gain (or loss) of entire human chromosomes that is not fatal (Figure 4-9).

The condensed chromosomes in karyotype displays are usually stained to reveal cytological banding patterns or are labeled by means of DNA probes with fluorescent tags used for fluorescence microscopy. Traditional cytogenetic analyses involve black and white karyotype displays of banded metaphase chromosomes that are commonly used to analyze the numbers, shapes, and structures of human chromosomes. The bands on stained chromosomes appear darker or lighter than adjacent bands on the same chromosome, giving each chromosome a unique pattern of light and dark bands that is used to identify the chromosome. Among other physical characteristics, the chromosome banding patterns in a karyotype display can be used to reproducibly identify the chromosome rearrangements indicative of specific human diseases.

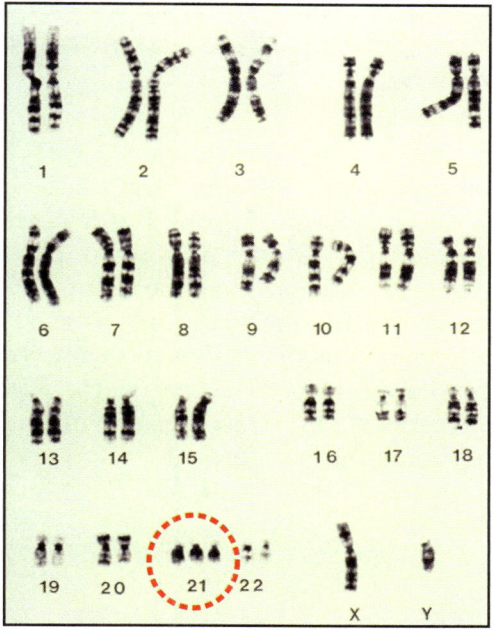

FIGURE 4-9 Human chromosome karyotypes can reveal genetic disease This karyotype is from a male with Down syndrome (trisomy-21) who inherited an extra copy of chromosome 21 (red dotted circle). The normal human chromosome number is 46, which includes 23 from maternal side and 23 from paternal side.

Chromosome banding patterns belong to two groups:

(1) the bands that are distributed along the length of the entire chromosome, which include G-bands, Q-bands, and R-bands, and
(2) the specific bands that appear at a number of restricted sites on the chromosomes, including the bands at the centromeres (C-bands).

The G-bands were originally made by staining chromosomes with Giemsa dye, but they can also be created using other dyes. The dark regions in G-bands contain A+T–rich DNA sequences that are packaged in heterochromatin, specific segments of the chromosome that remain highly condensed throughout the cell cycle and do not become thread-like. The bright regions of the G-bands represent euchromatin, which contains primarily G+C–rich DNA sequences packaged in chromatin that condenses and decondenses with the cell cycle. The R-bands (reverse bands) are the opposite of the G-bands in that the dark regions of R-bands are euchromatic, and the bright regions of the R-bands are heterochromatic.

> **KEY CONCEPT**
> When condensed human chromosomes are stained they reveal different banding patterns that reflect different distributions of DNA sequences rich in either A-T or G-C base pairs and regions of chromatin that exhibit different condensation functions. The stained banding patterns do not necessarily tell us anything about the location or expression of the underlying genes encoded by the chromosome DNA.

One way to Light up Chromosomes is Fluorescence *in Situ* Hybridization

The fluorescence *in situ* hybridization (FISH) method is a very sensitive and accurate way to identify specific target sites of any size in human chromosomes, even a single gene (Figure 4-10). Years before the DNA sequence of the entire human genome became available, FISH was commonly used to confirm the specific chromosome locations of various genes that were previously identified by other methods such as genetic mapping and gene function studies. FISH technology is still used to analyze the fine details of individual chromosome rearrangements because the DNA probes can be designed to span the junctions at the DNA breakpoint and fusion junctions appropriate to the chromosome rearrangement.

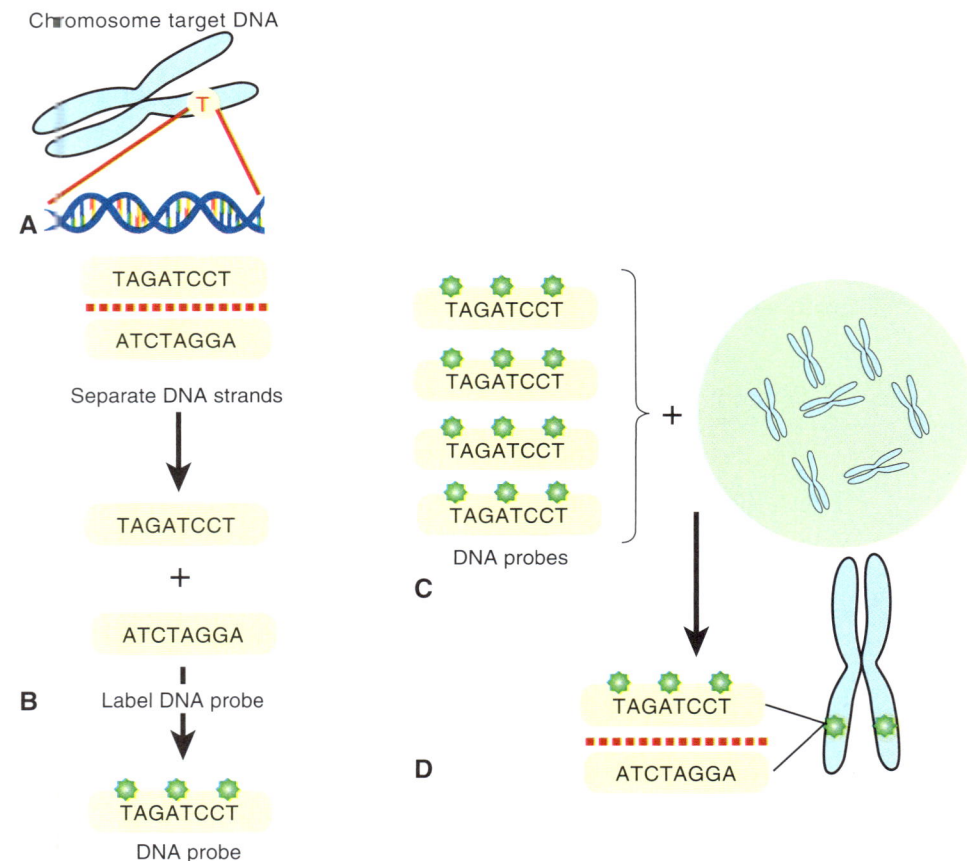

FIGURE 4-10 FISH technology paints chromosome colors. (A) Each DNA probe used in FISH is complementary to a specific target sequence in the chromosome DNA. (B) The FISH DNA probes are labeled with fluorescent tags (green). (C) The condensed metaphase chromosomes are spread on a glass microscope slide and treated to separate the chromosome DNA strands in situ (in place). The single-stranded DNA probe is added to the hybridization (annealing) buffer on the slide and the probe base pairs to the complementary target DNA sequences in the chromosome. (D) The DNA probes that base pair to the target DNA show up as distinct fluorescent signals on the chromosomes visualized in the microscope.

FISH DNA probes are made with sequences complementary to a known chromosome target DNA sequence, and the probes are also labeled with a fluorescent tag. The chromosomes to be analyzed are spread on a glass microscope slide and denatured to separate the chromosome DNA strands in place (*in situ*) (Figure 4-10). When the single-stranded DNA probes are added to the hybridization (annealing) buffer on the microscope slide, the DNA probes hybridize (base pair) to the complementary target DNA sequences in the chromosome DNA. The hybridized DNA probes show up as distinct fluorescent signals on the chromosomes when visualized with a fluorescence microscope (Figure 4-10).

Human chromosomes can be 'painted' by the computer as were the unpaired male chromosomes shown in Figure 4-11. This particular colored pattern results from treating the chromosomes with an enzyme followed by digital imaging and is not a result of FISH. Scientists have also used computer technology and different DNA probes to "paint" all 23 human chromosomes with a range of different colors that provide distinctive spectral karyotypes where each chromosome has a unique color. The spectral images of painted chromosomes are especially useful for characterizing the chromosome mutations and rearrangements that are associated with diagnosing human diseases.

FISH technology can also detect relatively small changes in human chromosomes For example, (Figure 4-12). DNA from one end of chromosome 5 was exchanged with DNA from the opposite end of chromosome 14. A deletion that removes the genes from human chromosome 17 is associated with Smith-Magenis syndrome. This mutation causing mental disability and compulsive behaviors. A small deletion in chromosome 5 causes *cri du chat* syndrome. This mutation causes patients to sound like a crying cat, giving the condition its name, and also causes mental disabilities.

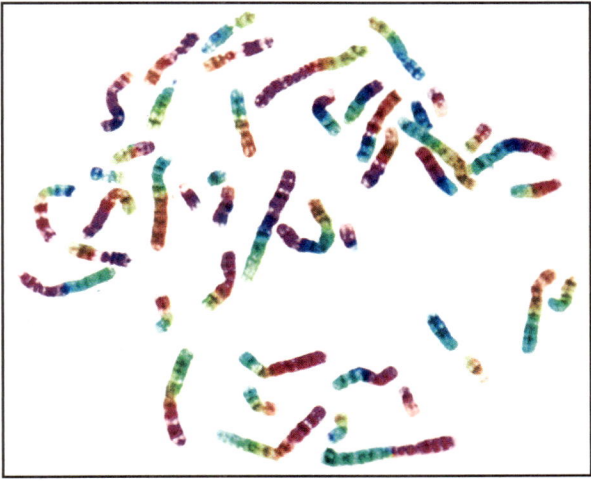

FIGURE 4-11 Painted human chromosomes This normal set of unpaired painted chromosomes was obtained from a normal human male cell in metaphase of mitosis. The colored banding pattern results from treating the chromosomes with an enzyme followed by digital imaging, and is not a result of FISH. Magnification: x2400 at 6x7cm size.

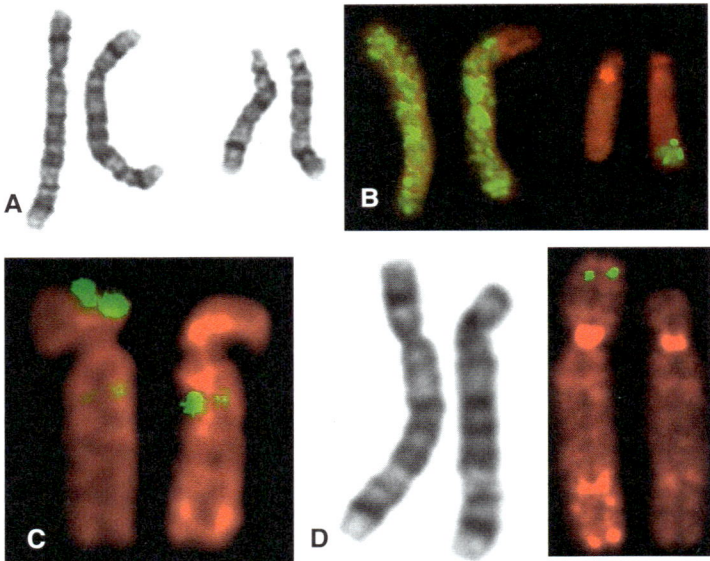

FIGURE 4-12 Chromosome rearrangements painted by FISH (A) A DNA translocation took place between chromosome 5 (pair on left) and chromosome 14 (pair on right). (B) DNA from one end of chromosome 5 (yellow) was exchanged with DNA from the opposite end of chromosome 14 (red). (C) A deletion in chromosome 17 (right) removes the wild type gene (two green dots) on chromosome 17 (left) and causes Smith-Magenis syndrome. (D) A deletion in chromosome 5 (far right) causes cri du chat syndrome. The gene (green/yellow) on the normal chromosome (center right) was deleted from chromosome 5 (right).

Chromosome Rearrangements and Translocations

Human chromosomes undergo chromosome DNA rearrangements and translocations that affect gene expression. An example is a chromosome inversion, where a region of chromosome DNA is duplicated and inserted into the same chromosome DNA in an inverted orientation relative to the arms of the chromosome (Figure 4-13). A chromosome translocation usually involves two different chromosomes, a region of one chromosome is moved into a different chromosome. In some cases the chromosomes swap DNA sequences from different chromosomes (Figure 4-14). The genes encoded within the boundaries of the inverted (or translocated) DNA region are expressed normally, but any gene that spans the junctions between the two different chromosome DNAs will probably (but not always) be disrupted and rendered nonfunctional by the DNA translocation. If the disrupted gene is essential for survival, then cells with the disrupted gene will not grow beyond the point where the essential protein is required for further cell development, and the essential protein is missing in the mutant cells.

A gene that spans the DNA breakpoint of a chromosome translocation might not be disrupted if the coding region remains intact as it crosses the junction of the DNA sequences brought together from the two different chromosomes. The fate of the cells that carry this chromosome translocation depends on the function of the gene interrupted by the translocation. A specific chromosome translocation event creates the Philadelphia (Ph1)

UNIT 4 Genetic Testing to Diagnose Genetic Diseases   149

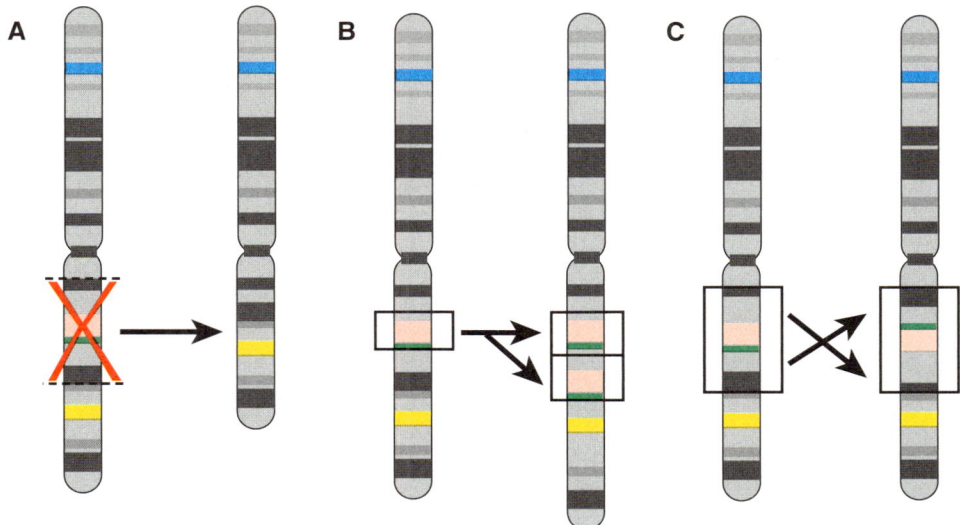

FIGURE 4-13 Chromosome mutations alter entire chromosomes Human chromosomes can be altered by different types of mutations that involve rearranging large regions of DNA. (A) A chromosome deletion completely removes the DNA sequences. (B) Tandem duplication can alter a large region of the chromosome DNA by repetition. (C) An inversion occurs when a region of the chromosome is inverted and reinserted relative to the chromosome arms.

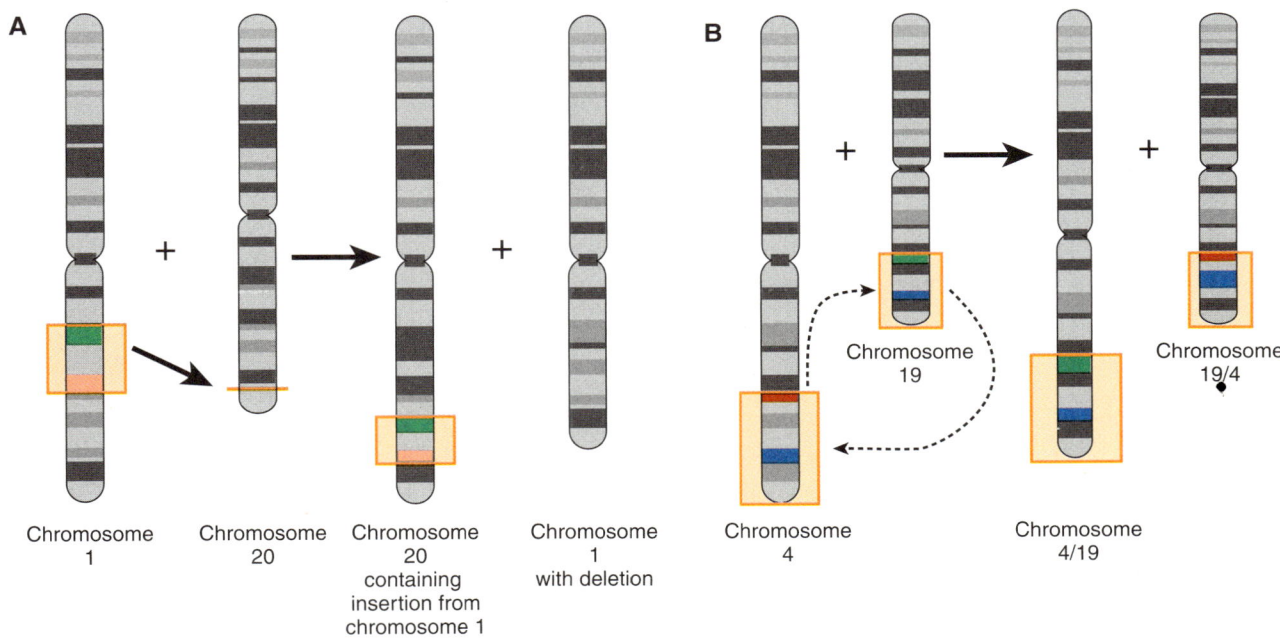

FIGURE 4-14 Chromosome rearrangements can translocate DNA to a different chromosome Chromosomes can acquire many different types of mutations that frequently involve rearranging large regions of DNA. (A) In this translocation mutation, a region of chromosome 1 DNA is inserted into the DNA in chromosome 20. Genes spanning the DNA junctions of this translocation risk disruption. (B) This chromosome translocation involves a swap of DNA sequences between chromosomes 19 and 4.

chromosome, which involves an equal exchange of DNA sequences between human chromosome 22 and chromosome 9. The translocation that creates Philadelphia chromosome 9,22 (Ph1) involves connecting (ligating) the DNA sequences from chromosome 9 to the DNA sequences from chromosome 22 (Figure 4-15). The Ph1 translocation creates a new oncogene by fusing the first half of the *Abl* gene DNA to the second half of the *Bcr* gene DNA. The new *Abl-Bcr* fusion gene codes for a new fusion protein containing the amino-terminus of the *Abl* protein fused to the carboxy-terminus of the *Bcr* protein (Figure 4-15). This chromosome translocation causes the devastating blood cancer, chronic myelogenous leukemia (CML).

Although neither of the original *Abl* or *Bcr* genes was dangerous before the Ph1 chromosome translocation, the new *Abl-Bcr* gene and fusion protein function to promote the development of cancer. The Ph1 chromosome translocation initially occurs in only one blood stem cell in an individual, but the expression of the *Abl-bcr* oncogene fusion protein causes the single mutant cell to grow and divide quickly. In just a short time the mutant cells carrying the Ph1 translocation have outgrown the healthy blood cells, bringing about the disease symptoms. The mutant cells carrying the Ph1 chromosome translocation continue to grow and divide out of control and cause chronic myelogenous leukemia.

Newborn and Prenatal Genetic Screening

Newborn genetic screening is an important way for doctors to identify treatable genetic diseases and disorders, usually by testing a small amount of blood collected from a heel prick immediately after birth. The purpose of newborn testing is to identify certain genetic diseases that can be successfully treated very early, often avoiding long-term consequences and a lifetime of problems. In this way the newborn tests are a preventative health measure because these tests will identify infants that require life-saving intervention and treatment that is known to be effective. In the United States, the genetic testing of newborn infants is administered by individual state governments that decide which tests to perform and how to pay for the screening program. This introduces a level of uncertainty into the effort to include newborn testing as a nationwide preventative health measure. In most cases, the fee charged for screening ranges from about $15 to $60 per newborn, which is often covered by the parent's health insurance and can be supplemented by the health care system available in that state.

Prenatal (pre = before, natal = birth) testing is used to detect genetic diseases before birth and involves the genetic analysis of cells from a growing fetus, including prenatal genetic diagnosis (PGD) (see below). Amniocentesis and chorionic villus sampling (CVS) are the two most common methods of obtaining fetal cells for genetic testing in the lab. The two methods provide similar information, but chorionic villus testing can occur safely at an earlier stage of pregnancy than for amniocentesis. This type of prenatal genetic testing is used routinely to screen for diseases and disorders such as Down syndrome, sickle cell anemia, and CF.

In amniocentesis the doctor uses a large syringe to recover fetal cells and maternal cells from the amniotic fluid surrounding the growing fetus (Figure 4-16). Ultrasound imaging with sound waves allows the doctor to

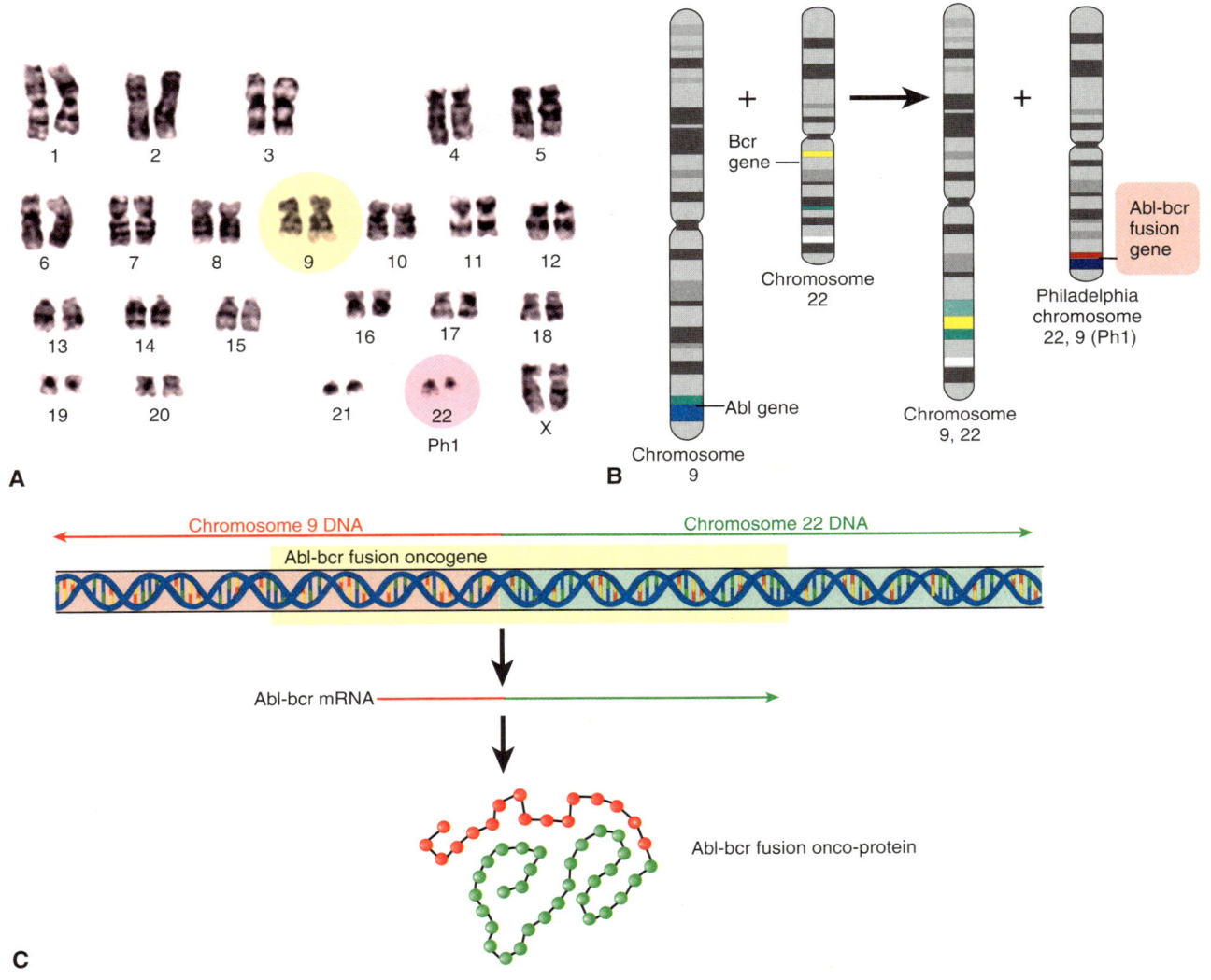

FIGURE 4-15 Philadelphia chromosome translocation creates an onco (cancer) gene
(A) Chromosome karyotype of the Philadelphia chromosome, a translocation between human chromosome 9 and chromosome 22. (B) Diagram showing the region of chr 22 DNA that was translocated to chr 9, and the part of chr 9 DNA that is moved into chr 22. (C) This chromosome DNA translocation creates a new fusion gene (abl-bcr) that is expressed as a hybrid mRNA and produces a fusion protein that causes a blood cancer called chronic myelogenous leukemia (CML).

visualize the positions of the large needle and the developing fetus during amniocentesis to avoid injuring the fetus. The ultrasound images also help to reveal potential developmental problems with the fetus. The chromosomes obtained from the cells in the amniotic fluid are stained to visualize chromosome banding patterns by karyotype analysis and to detect rearranged, missing, and extra chromosomes. DNA testing to detect genetic disease genes can also performed on the DNA obtained from the fetal cells obtained by amniocentesis.

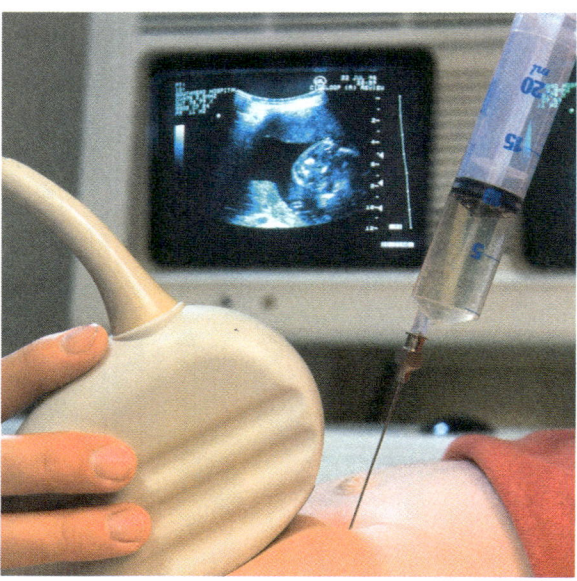

FIGURE 4-16 Amniocentesis is a common pre-natal test Amniocentesis is a medical test used to detect genetic diseases in fetuses before birth. Cells from the fetus and the mother are present in the amniotic fluid surrounding the fetus. A long needle introduced through the pregnant woman's abdomen is used to collect amniotic fluid. The maternal and fetal cells in the fluid sample are tested for genetic diseases by karyotype display and DNA testing.

Preimplantation Genetic Diagnosis (PGD)

Tay-Sachs disease (also known as GM2 gangliosidosis, hexosaminidase A deficiency, or sphingolipidosis) is a devastating, incurable disease caused by a mutation in the human hexA gene. The mutant hexA gene produces defective hexosaminidase enzymes that cannot properly degrade the special type of "fat" made in brain cells called gangliosides. The hexA mutation causes the brain cells to accumulate excess gangliosides, which eventually prevents nerve cells in the brain from transmitting nerve signals.

Tay-Sachs is a particularly insidious disease because infants born with it appear normal at first, but after the first 6 months of age they rapidly lose mental and physical abilities and progressively deteriorate, rarely living past the age of 5 years. Without genetic testing there is no way for parents to know whether they have passed this particular fatal genetic mutation to their child. In some cases the doctor can watch for the appearance of a "cherry red" spot in the retina of the eye, which is an early sign of Tay-Sachs disease (Figure 4-17), but there is no cure.

Genetic testing can be used to screen various adult populations for mutant genes known to convey genetic diseases. This is particularly important because genetic carriers usually have no symptoms of any genetic disease. Population and epidemiology studies show that some genetic diseases are more frequent in certain ethnic populations than in others. The Ashkenazi Jews are a good example since about one in five are genetic

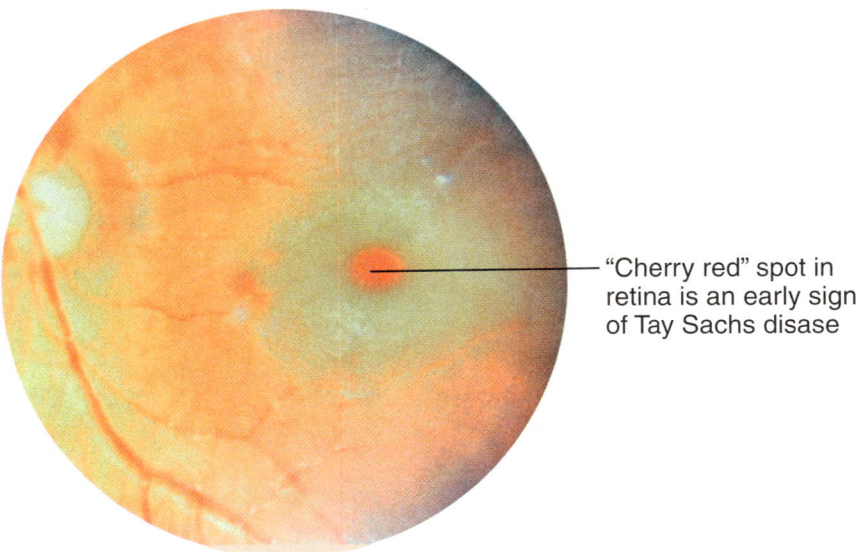

FIGURE 4-17 Tay-Sachs disease causes a red spot in the eye Infants born with Tay Sachs disease appear normal after birth. Without genetic testing the parents do not know whether their child has inherited this fatal gene mutation. The doctor can use an ophthalmoscope to examine the infant's eyes to watch for the appearance of a "cherry red" spot in the retina, an early sign of Tay Sachs disease.

carriers of the hexA gene mutation that causes Tay-Sachs disease. It is essential to screen people of eastern European descent for potential genetic carriers of Tay-Sachs as well as Canavan disease, Niemann-Pick disease type A, and Bloom's syndrome, diseases that are more prevalent in people of eastern European decent.

The devastating real-life consequences of transmitting Tay-Sachs or another fatal genetic disease to a biological child causes more and more people to seek genetic counseling when they are planning a pregnancy. Couples who are at risk of transmitting a fatal mutant gene to a biological child sometimes decide to conceive embryos by *in vitro* fertilization for the purpose of having the embryos individually analyzed by PGD. Although all reproductive technologies are dangerous to both the mother and fetus, the goal of PGD testing is to identify healthy embryos that can be implanted in the mother to achieve a full term pregnancy and give birth to a healthy baby free of a specific genetic disease.

PGD is a sophisticated technique used to test very early embryos for specific genetic mutations. The PGD testing process begins with *in vitro* fertilization with the egg and sperm cells of the prospective biological parents to create human zygotes in the lab. The zygote cells divide in the lab in a Petri dish to the 8-cell embryo stage, at which time a single cell is removed from the embryo and subjected to genetic testing to determine if the embryo carries a particular mutant gene. For example, genetic testing of the mutant hexA allele in the single cell would indicate whether or not the embryo carries Tay-Sachs disease (Figure 4-18).

After a single embryo cell is removed from the 8-cell embryo for testing, the intact 7-cell embryo is returned to the Petri dish in the lab, where it

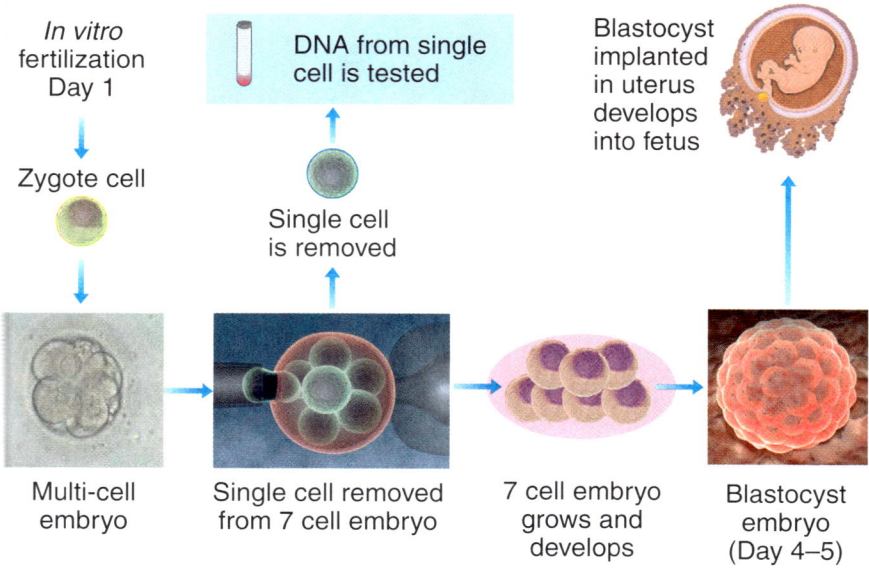

FIGURE 4-18 Single cell removed from embryo used for Preimplantation Genetic Diagnosis
In vitro fertilization is used to create human embryos in the lab for Preimplantation Genetic Diagnosis (PGD). The zygote cell divides to form an 8 cell embryo. One cell is removed from the 8 cell embryo for DNA testing and the 7 cell embryo is returned to the plastic dish. The 7 cell embryo continues to grow and forms a normal, healthy blastocyst that can be implanted in a uterus and carried to term (or stored frozen).

continues to divide and develops into a normal hollow blastocyst embryo (Unit 5) (Figure 4-18). If the DNA testing reveals that the single cell carries a lethal mutant gene, then the corresponding blastocyst in the lab will be discarded. However, if the single cell is genetically healthy, then the corresponding blastocyst embryo will be implanted into the uterus and the pregnancy carried to term. The PGD method offers couples with high genetic risk the option to conceive a child who will be free of inheriting a known lethal genetic mutation.

Genome Testing on a DNA Chip

Clearly DNA testing has become well known to the public as an extremely powerful tool used on TV and the internet to identify criminals, diagnose diseases and predict human genetic traits. Whether or not the public understands the details of DNA testing, in 2010 over 200 human genetic tests were commercially available from genetic online testing services. The largest online genetic testing service at that time was offered by the 23andMe.com company, which offers consumers personal genome testing as a service. Consumers collect saliva samples that are sent to the 23andMe company for DNA analysis.

About 6–8 weeks after submitting a saliva sample, the client can log on to the 23andMe.com web site and access the results of the personal genetic testing. The genotype analysis performed by 23andMe uses customized DNA chip technology that examines an individual's genome DNA sequence at specific SNP sites in the genome, but does not determine the entire DNA sequence of an individual's chromosomes (Figure 4-19). Different human chromosomes have almost identical DNA sequences, except at millions of SNP locations in the human chromosomes where the DNA sequences vary.

A human genotype refers to the wild-type or mutant allele status of selected genes carried in an individual's genome DNA. Human chromosomes carry different combinations of wild type and mutant genes, which taken together determine the genotype of a specific individual. Genotype analysis at 23andMe begins with the client's chromosome DNA, which is extracted from the cheek cells in the saliva sample and amplified by PCR to obtain sufficient copies of the target sequences in the chromosome DNA to be screened for genotype (Unit 2). The PCR-amplified DNA is cut into small DNA fragments that represent selected parts of the human chromosome, which are physically attached to millions of microscopic beads on the surface of the DNA chip (Figure 4-20). Several short, single-stranded DNA probes that are complementary to different SNP DNA sequences are made with each DNA probe containing a different sequence and a different fluorescent tag. When the DNA probes are added to the DNA chip, the different probes base pair to the complementary chromosome DNA attached to the beads. The dots on the DNA chip fluoresce in different colors that indicate the SNP DNA sequence present at each SNP site tested in the genome.

A DNA chip is also an efficient, rapid way to analyze DNA for many different applications. For example, a DNA chip is an effective way to detect the presence of infectious organisms in blood and food using small DNA samples dotted into an array on a glass slide. Each individual dot contains a different DNA sequence that is designed to base pair with one specific gene in the DNA of the pathogenic organism being tested. When the organism's test DNA is added to the chip, it base pairs only with the target dots containing complementary DNA strands. When the chip is scanned with a laser to detect the positive samples containing the base paired DNA, only those dots

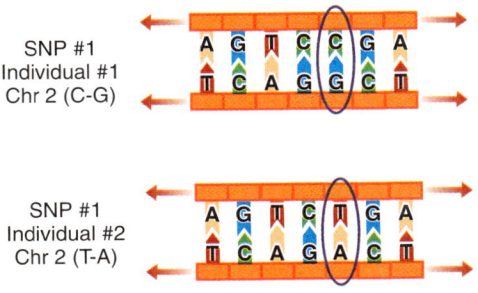

FIGURE 4-19 Single Nucleotide Polymorphisms change one base A Single Nucleotide Polymporphism (SNP) is a single base pair change in the DNA sequence of the same chromosome from two different individuals. The SNP#1 in chromosome 2 from individual #1 is CG and the SNP#1 in chromosome 2 from individual #2 is TA.

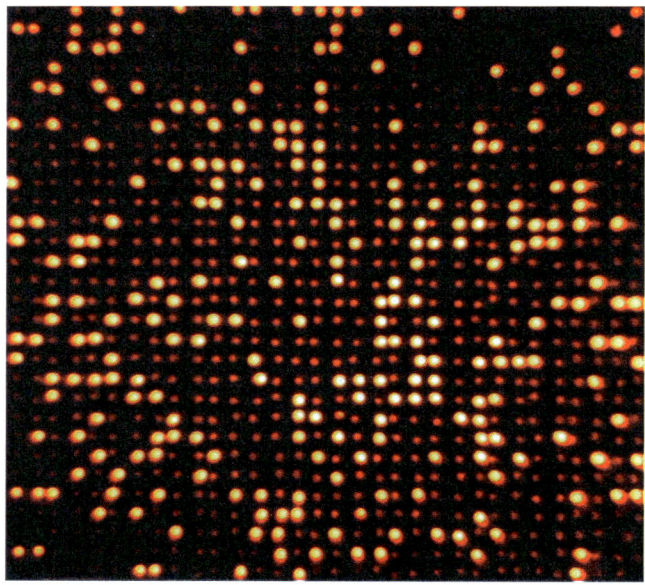

FIGURE 4-20 DNA chip array can determine human genotypes and identify pathogens
This DNA test shows positive test results (yellow dots) that indicate infectious organisms detected in blood and food. Each individual dot on the DNA chip contains a different DNA sequence that is designed to bind to a specific target in the DNA of the organism to be detected. When the test DNA is added to the chip, the complementary DNA strands base pair to the target DNA in the dots and the chip is scanned by a laser to detect the positive signal from the base paired DNA samples (yellow dots). (DNA BioChip shown here is made by BioChip Technologies, Freiburg, Germany.)

with base paired DNA are illuminated as yellow dots. This assay is a very sensitive way to detect very small amounts of contaminating organisms in the food or the blood supply.

Genetic Testing Cautions

Beyond the genetic tests ordered by a physician in a medical setting, consumers can now choose to purchase medical DNA tests from online services that also trace genetic ancestry and build family trees. Experts raised many concerns when commercial genetic testing services became available to the public through online sites. The lack of government regulation of this new industry is a good reason for caution on the part of the consumer. Although online genetic testing services can potentially provide accurate genetic results, it is important for consumers to get a second opinion before making critical medical decisions. The public should also be cautious about unsupported claims on the results of online genetic testing and studies on human genes in general. Some testing services advocate using the test results as tools to customize medical treatments, alter diet, or make lifestyle changes, but it is never wise to use specific genetic information to make general conclusions about people. More questionable claims include genetic testing young children to predict future athletic prowess, intellectual skills, or dress size at age 50 years!

Concerns about individual privacy and the security of genetic information, especially online security, led to the Genetic Information Nondiscrimination Act (GINA), which became law in May 2008 and prohibits U.S. insurance companies and employers from discriminating on the basis of information derived from genetic tests. However there are currently no regulations in the United States to oversee the accuracy and reliability of genetic testing, because genetic testing is considered to be a service and is not regulated by the FDA.

The 23andMe company provides a genetic report to its customers containing genotype information on a predetermined set of specific human genetic alleles. The 23andMe report interprets the results of the SNP screen and provides the consumer with some personal genotype information that is limited because the 23andMe genetic analysis does not include sequencing entire genomes. Of course in any case consumers should remember that the results of these tests cannot be used to predict genetic outcomes or to diagnose a genetic disease online or at home. Researchers can determine if a particular genotype is associated with the risk of developing a known genetic disease, but this type of genotype screening cannot diagnose if an individual has a particular disorder or predict if a person will develop a disease sometime in the future.

Genotype analysis alone is not sufficient to diagnose a disease, because a family medical history and a summary of current symptoms are also very important. Of course an online analysis usually does not include an in-person physical exam, one of the most important aspects of a medical diagnosis. Also medical tests are often needed to rule out other possible causes of a condition and to confirm a diagnosis and treatment plan.

For human genotype testing, the 23andMe genetic testing lab uses a commercially available DNA chip (Illumina HumanHap550+ Genotyping BeadChip), which contains DNA probes representing the possible sequences at 550,000 SNP sites distributed in the human genome. The 23andMe scientists customized the DNA chips by adding tens of thousands of additional SNP sites identified recently in the published, peer-reviewed scientific literature. The genotyping services performed by 23andMe and similar companies are not approved by the U.S. Food and Drug Administration (FDA), and the information provided is not intended for medical use or to diagnose human disease.

Genotype testing and DNA sequencing are very different processes that provide scientists and physicians with different types of genetic information. Genotyping evaluates an individual's genome to determine the specific DNA sequences at a relatively small number of SNP sites in the human genome. Although the DNA sequencing method was used to determine the exact DNA sequence of over 3.2 billion base pairs in the human genome (Unit 2), it is not yet routine to determine the DNA sequence of an entire individual human genome for individual diagnostic purposes. DNA sequencing technologies are rapidly advancing and scientists predict that soon it will be possible to sequence an entire human genome rapidly and accurately for under $1,000 per genome. Current genotyping technologies offer an alternative way to provide efficient and cost-effective genetic information without sequencing entire human genomes.

DNA Tests Reveal Genetic Carriers, Drug Reactions, and Genetic Risk

About one in 29 Caucasians in the United States carries the common genetic mutation that causes CF disease. Cystic fibrosis carriers do not have symptoms of the disease and usually do not know that they are at risk of transmitting a serious genetic disease to their children. A CF carrier does not experience symptoms because the wild-type CF transmembrane conductance regulator (CFTR) gene produces enough wild-type CFTR protein to provide the essential CFTR protein function in cells. Many different cftr mutant alleles can cause CF disease, but most genetic tests screen for the most common cftr allele, F508.

Human genes and the proteins they encode can influence how an individual's body reacts to different drugs, including side effects and sensitivity to certain medications. The effectiveness of different medications varies in different people and can be affected by gene mutation and modified by changes in body metabolism and diet. A good example of the influence of genetic alleles on individual sensitivity to medications is warfarin (Coumadin). This drug is a very effective blood thinner that prevents blood clots in most patients but people who are genetically predisposed to warfarin sensitivity can experience very severe bleeding. An individual's sensitivity to warfarin can be determined before administering the drug, by performing a genotype analysis to screen selected SNP sites in the human genome. Another example is the drug clopidogrel (Plavix®), which helps prevent heart attacks by stopping the blood cells from clumping together in arteries and veins and forming blood clots. However, certain people have a genetic allele that inhibits the metabolism of clopidogrel and significantly decreases the effectiveness of the drug. The cholesterol-lowering statin medications have been shown to be very effective, but therapeutic doses of statins can cause severe muscle pain and weakness in some genetically predisposed people. The potential risk of serious side effects from statin medications can be avoided by SNP genotyping before administering the medication. For genetically susceptible people, the side effects of many otherwise very effective medications can be much worse than the symptoms of the disease.

Why Genetic Risk is Important Information

About one in five Americans will develop type II diabetes by 79 years of age, and the diabetes epidemic in this country and around the world continues to get worse. Research shows that an individual's risk of getting type II diabetes can be reduced by lifestyle changes such as diet, exercise, weight control, and preventive medical care. Many in the medical community have shifted from prevention to the diagnosis and treatment of several age-related diseases that are becoming more prevalent as the overall U.S. population continues to get older.

Genetic Testing Traces Ancient DNA by Molecular Genealogy

Molecular genealogy is a combination of traditional genealogy (tracing one's ancestors by records) and modern DNA analysis (tracing one's genetic ancestry by DNA). Most people recall from grade school having the assignment of building a family tree, which reflects the biological relationships between parent and child extending back over the years. Modern family trees are much more complicated than the simple family trees of the past. Modern blended families have step parents, adopted children and step children.

The written records that are typically used by genealogists to trace an individual's ancestors are often incomplete, but in many cases molecular genealogy can be used to expand the search using DNA testing. The chromosome DNA contains all the genetic information necessary to reveal the location in the world an individual's biological ancestors originated and can identify relatives with a common genetic ancestor.

The online company Ancestry.com provides each client with results from personal genetic testing that reveals a collection of genetic markers with the genetic connections between related people presented as genetic family trees. The Ancestry.com company actually offers the consumer two different types of genetic tests to choose from, which depends on the needs of the client:

(1) Paternity testing analyzes the genetic inheritance of the male Y chromosome, and has several applications including the identification of the biological father of a child, and
(2) Maternal inheritance testing follows the transmission of mitochondrial DNA to sons and daughters through several generations.

When considering the genetic inheritance patterns of the male Y chromosome, it is important to recall that a man (X, Y) inherits his Y chromosome only from his father and donates his Y chromosomes only to his sons. In this way the Y chromosome represents an unbroken genetic link between generations of fathers and sons. The Y chromosome DNA contains rare mutations, which explains why a son's Y chromosome DNA usually has a slightly different DNA sequence than his father's Y chromosome DNA. Females inherit two X chromosomes, one from Mom and one from Dad, and no Y chromosomes. However, a woman can still trace paternal lineage by analyzing DNA from a biological male relative such as a brother, father, or male cousin.

Ancestry.com offers a paternal Y-DNA test that analyzes specific regions of the Y chromosome that contain short, repeated DNA sequences called short tandem repeats (STRs) (Unit 6). All human males have STRs in the Y-chromosome, but some men have a large number of Y chromosome STRs and other men have a small number of Y chromosome STRs (Figure 4-21). In this analysis the number of STRs at specific locations on the Y chromosome is counted and represents the STR profiles of the father and biological son. In genealogy studies these STR profiles can be used to differentiate one paternal lineage from a different paternal lineage. The Ancestry.com Y-DNA test measures the number of STRs at two different locations on the Y chromosome and based on these results Ancestry.com compiles a map showing the migration of the client's ancestors leaving Africa over 100,000 years

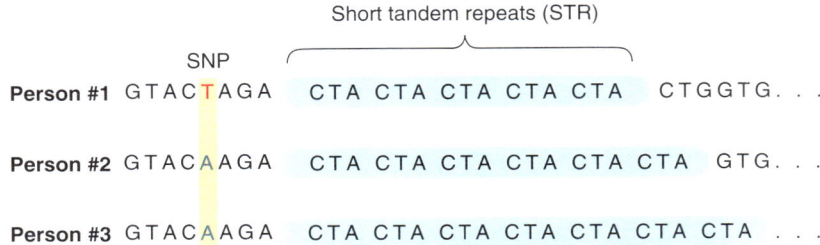

FIGURE 4-21 Short tandem repeats can identify an individual human genome A region of short tandem repeats (STR) from the human genome is shown for three unrelated people. These individual genomes contain different numbers of repeats of the DNA sequence: CTA. This STR is linked to an SNP located in the adjacent DNA. STRs are used for human identification by DNA fingerprinting.

ago, and includes the people who branched out and populated the different regions of the world indicate on the map.

Maternal Lineage Test Results

As humans migrated out of Africa and populated the rest of the world, they adapted to new climates, diets, and living conditions and humans changed and diversified. As tens of thousands of years passed, the ancient human populations became isolated from each other and their genome DNA sequence changed until the members of the different populations became genetically distinct. These ancestral groups of humans can be distinguished by the distribution of SNP changes in chromosome DNAs that are inherited together in DNA "neighborhoods" called haplogroups.

Mitochondrial DNA (mtDNA) is passed from mother to sons and daughters, and is another powerful investigative tool used to trace inheritance and identify ancestors (Figure 4-22).

Drugstore DNA Testing and Access to Genetic Counselors

In May 2010, Pathway Genomics made headlines when the company became the first to offer a DNA test kit for sale in Walgreen's drugstores in the United States. However, days later the FDA opened an investigation and the DNA test kits were removed from store shelves. The Pathway Genomics DNA home testing kits that were removed from drugstore shelves were very similar to several other DNA test kits sold online. Both testing methods included a tube to collect saliva and a prepaid envelope to send the saliva sample to the lab for DNA analysis.

Many doctors doubt the wisdom of selling DNA test kits directly to the public, concerned that consumers might misunderstand or misinterpret the results of the gene tests without access to effective genetic counseling. There is also a concern that many online genetic testing services do not offer genetic counseling when clients receive online test results. Some companies provide access to online genetic counselors to help clients to better understand the

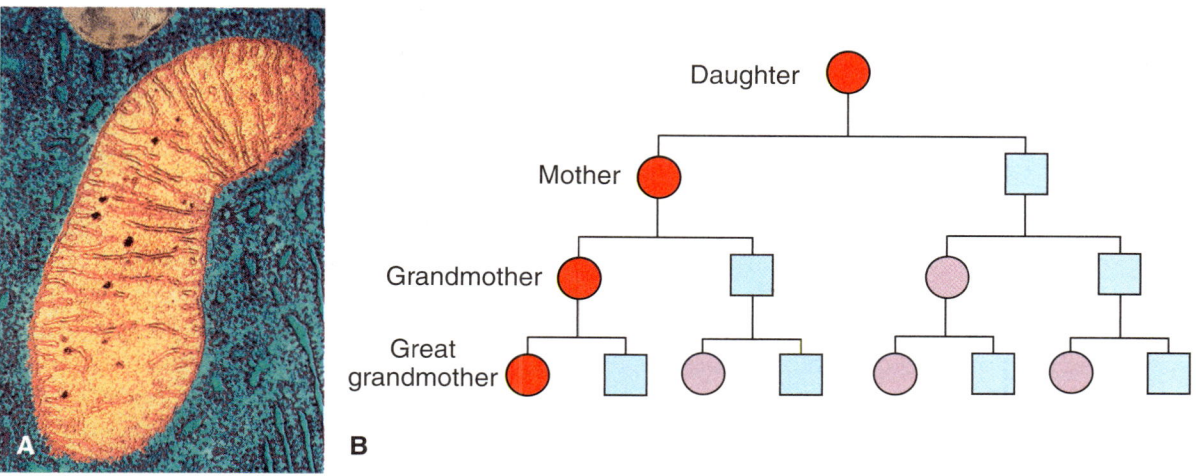

FIGURE 4-22 Mitochondrial DNA is inherited from Mom (A) This transmission electron micrograph (TEM) shows a mitochondrion surrounded in the cell by the rough endoplasmic reticulum (RER) (blue). Proteins translated on the RER ribosomes (black dots) in the mitochondrion are transported and modified for secretion. The folded membranes (cristae) in the mitochondrion contain enzymes involved in respiration, which is the metabolism of sugars and fats to produce energy for the cell in the form of adenosine triphosphate (ATP). (B) Mitochondria and mtDNA are transmitted only from the mother to sons and daughters by maternal inheritance.

results of the genetic tests. The counselors can explain the implications for family members, and provide accurate information needed to make the best decisions. Clients are always advised to have additional genetic and medical testing to confirm any test results.

> **KEY CONCEPT**
> It is important to note that collecting a saliva sample at home using a test kit, which is then sent to a lab for DNA testing is quite different than using a test kit to perform actual enzymatic or chemical DNA test reactions in your bathroom.

Genetic counseling helps people to understand the consequences of the alleles they inherited and to help them make informed decisions about the genetic issues that affect their lives. Genetic counselors are usually medically trained members of healthcare teams that work with individual patients and their families to help decide who should be tested and what tests should be administered. Counselors explain the advantages, limitations and any possible dangers associated with each test. Pre-natal genetic counselors gather a wide range of personal and family information to help doctors assess the risk level of a particular genetic disease that might affect an expected child. Risk factors for genetic disease can include maternal age, family history, ethnicity, distribution of specific genetic markers and alleles, and the results of prenatal tests and fetal ultrasound exams. Genetic counselors also provide information about diseases that "run in the family" and answer questions about the genetic consequences of exposure to chemicals or other agents that act as mutagens and can often cause cancer.

Sickle Cell Anemia Disease

Human RBCs contain hemoglobin protein complexes that carry heme rings and iron (Fe^{2+}), and are responsible for the familiar red color of blood. A normal hemoglobin complex (HbA) contains two alpha-globin proteins and two beta-globin proteins. Each globin protein subunit contains one heme group with an iron atom bound reversibly to an oxygen molecule. Each hemoglobin complex carries and releases four molecules of oxygen as the RBCs travel around the body and release oxygen to cells, tissues, and organs (Figure 4-23).

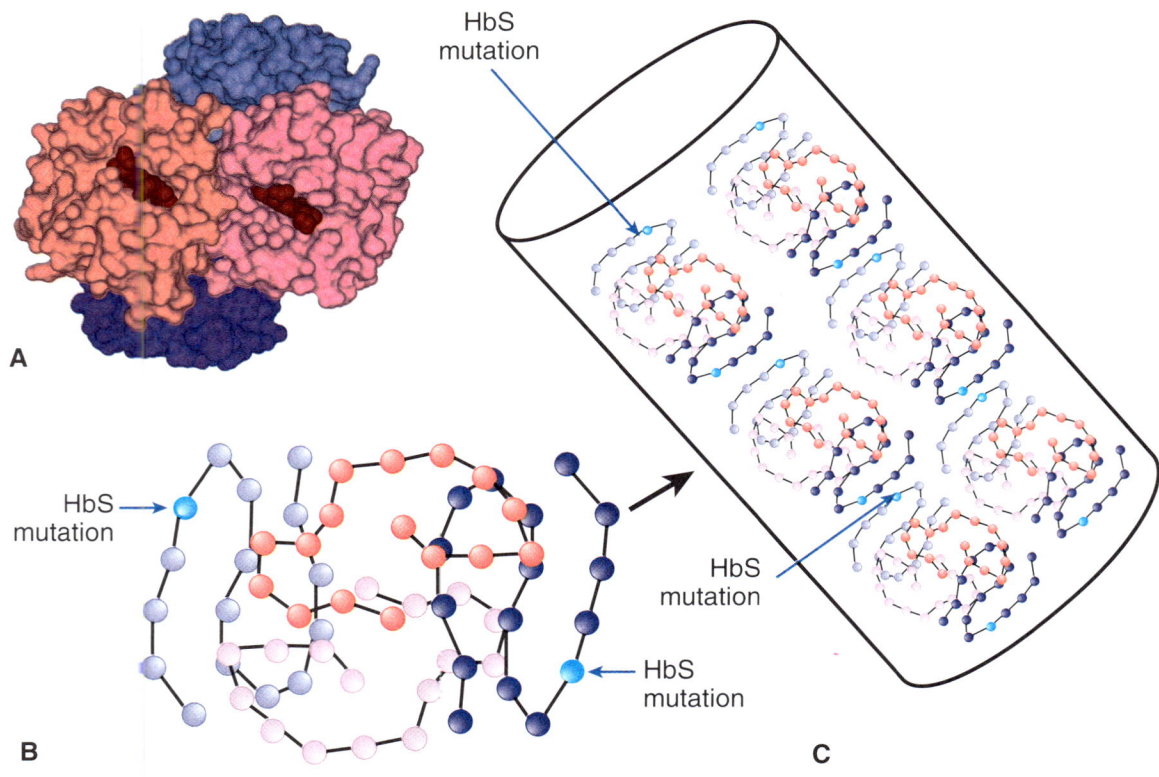

FIGURE 4-23 Sickle cell mutation makes hemoglobin rods that distort the shape of RBCs (A) Normal hemoglobin (HbA) is a complex of 4 proteins (2HbA,2HbB) that carry oxygen in RBCs. The alpha globin and beta globin subunits each carry a heme group with an iron atom bound to oxygen. (B) A single base pair mutation in the beta globin gene causes a single amino acid change in the beta globin protein, which results in the mutant hemoglobin complex, HbS. (C) The HbS mutation replaces glutamic acid with valine in the beta globin protein, causing a hydrophobic patch to form on the surface of the HbS complex. As a result, the mutant complexes stick to each other forming stiff rods that distort the RBCs into sickle shapes, causing sickle cell disease.

Sickle cell anemia disease is caused by an autosomal recessive mutation in the beta-globin gene located near one end (telomere) of human chromosome 11. Two mutant beta-globin proteins and two wild-type alpha-globin proteins assemble to form a mutant hemoglobin complex (HbS). A person who inherits two copies of the mutant beta-globin gene allele will become ill with sickle cell disease, but someone who inherits one mutant beta-globin gene and one normal beta-globin gene is a carrier of sickle cell disease but does not get sick. Although genetic carriers of sickle cell do not have disease symptoms, they can potentially transmit the mutant beta-globin gene to their biological children. The prevalence of people in the population who are genetic carriers of diseases such as sickle cell is a good reason for the public to consider more widespread genetic testing.

Mutant Hemoglobin Complexes Stick together and Distort Cell Shape

The mutation that causes sickle cell disease is an excellent example of a very small change in DNA that has a surprisingly large impact on the structure and function of the cell and on the overall health of the organism. Sickle cell disease is caused by a point mutation in the beta-globin gene that changes a GAG codon, which specifies glutamic acid, to a GTG (GUG) codon, which specifies valine. As a result of this single base-pair mutation, the mutant beta-globin protein contains a hydrophobic (water-hating) valine amino acid in place of a negatively charged (acidic) amino acid, glutamate (also called glutamic acid) (Figure 4-24). This seemingly insignificant change in only one amino acid causes two small hydrophobic "patches" to form on the surface of each mutant HbS hemoglobin complex. The hydrophobic patches cause the HbS hemoglobin complexes in the red blood cells to stick together and form stiff HbS fibers inside the mutant RBCs. The stiff HbS fibers physically distort the shapes of the disk-like RBCs and from abnormal, sickle-shaped RBCs (Figure 4-23, Figure 4-24).

Normal RBCs are flexible disks that can move freely through the small blood vessels and capillaries to deliver oxygen to the tissues and organs in the body. The sickle-shaped mutant RBCs are not flexible and can cause severe pain when the mutant sickle cells get stuck in the smallest blood vessels, the capillaries. In addition the sickle cells cannot efficiently deliver oxygen to organs such as the spleen, liver, kidneys, lungs, and heart causing the organs to starve for oxygen. Normal RBCs circulate in the bloodstream for up to 120 days, but sickle-shaped cells can survive for only 10 to 20 days, causing a shortage of hemoglobin that leads to severe anemia and significantly increases the risk of blood clots and strokes.

About 1 in 500 African-Americans in the United States are born with two copies of the mutant beta-globin gene and suffer from active sickle cell disease symptoms. The mutant beta-globin gene that causes sickle cell occurs most frequently in the genomes of people who are descendents of populations in Africa, India, the Caribbean, the Middle East, and the Mediterranean.

People who are genetic carriers of sickle cell disease have inherited one wild-type and one mutant copy of the beta-globin gene and are also resistant to becoming infected with malaria. This means that people who are

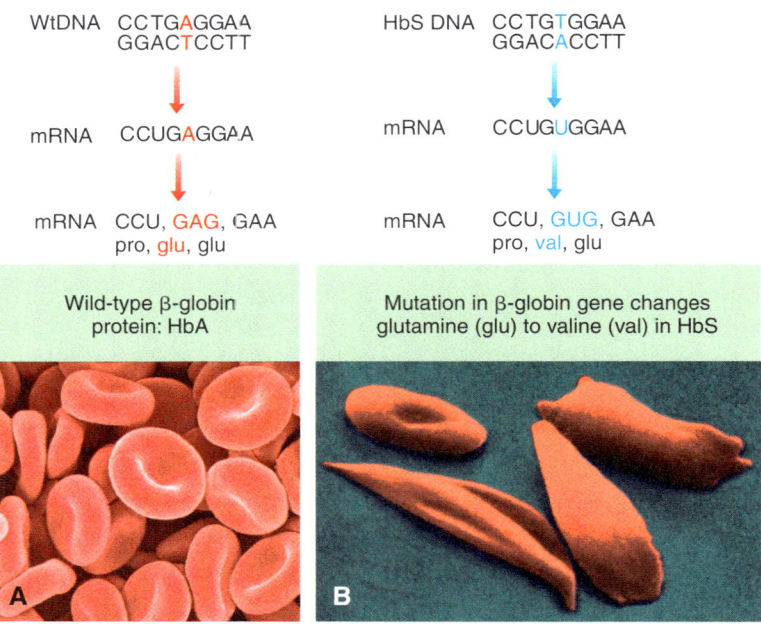

FIGURE 4-24 Tiny point mutation causes big changes (A) The wild-type sequence of part of the ß-globin gene DNA, the corresponding mRNA and amino acid sequences are shown. Normal RBCs have typical disk-like shape. (B) The point mutation in the ß-globin gene alters the mRNA and causes valine (val) to be inserted in place of glutamine (glu) in the mutant hemoglobin S protein (HbS). This causes the formation of sickle-shaped red blood cells in people with sickle cell anemia.

carriers of sickle cell disease have a distinct advantage in many countries where malaria is an epidemic disease and often fatal. The parasite that causes malaria is introduced into the bloodstream when a person is bitten by an infected mosquito. The parasites actually live inside the normal RBCs during part of the parasitic life cycle, but the distorted sickle RBCs do not support the growth of the malaria parasite. This is one reason why sickle cell disease predominantly affects African-Americans who came to the United States as part of the slave trade that flourished in countries with rampant malaria.

> **KEY CONCEPT**
> The point mutation causing sickle cell disease changes only a single amino acid in the mutant beta-globin protein, but the consequence is a large change in the structures and functions of the RBCs.

Cystic Fibrosis is a Common Genetic Disease

CF is a common genetic disease, diagnosed in over 30,000 children in the United States each year. People with CF disease symptoms inherit two mutant cftr genetic alleles and produce only mutant cftr proteins. The normal CFTR gene encodes a large protein of 1,480 amino acids that forms a channel to transport chloride ions (Cl-) across the plasma membrane of the cell (Figure 4-25). Unlike sickle cell disease, there is no single DNA mutation that is responsible for all cases of CF disease. More than 1,000 different cftr mutations have been identified that cause CF disease, but about 70% of the cases of CF disease are caused by one mutation called the F508 cftr allele. The F508 cftr mutation deletes 3 base pairs (CTT) from the gene sequence that removes a single phenylalanine amino acid from the mutant cftr protein (Figure 4-26). A different cftr allele that also causes CF disease contains a mutation that converts a CAG (glutamine) codon to a translational Stop codon (TAG, which is UAG in the mRNA). The result is a prematurely terminated mutant cftr protein that contains only 493 amino acids of the 1,480 amino acids in the normal cftr protein and is functionally defective.

> **KEY CONCEPT**
> The hydrophobic regions of the CFTR protein span the cell membrane and make a channel that controls the flow of salt ions into and out of the cell. This function of the CFTR protein is essential in all human cells and is defective in CF disease.

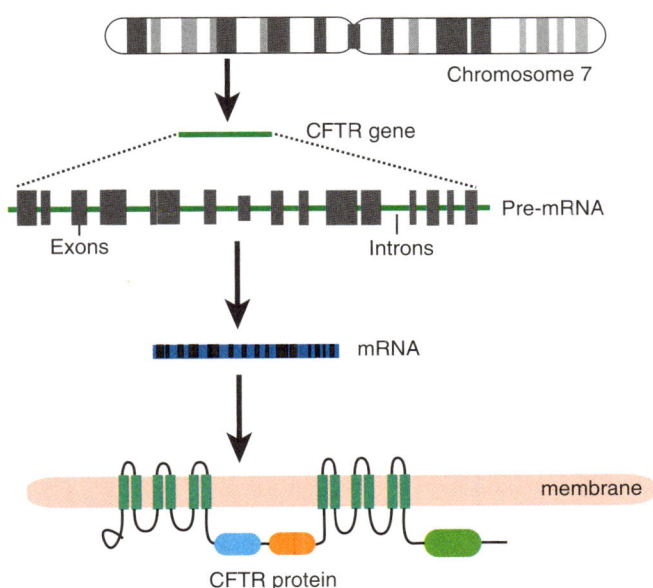

FIGURE 4-25 Cystic fibrosis disease is caused by a mutant membrane protein The DNA mutation that causes cystic fibrosis disease is located on human chromosome 7 and alters the cystic fibrosis transmembrane receptor (CFTR) gene. The pre-mRNA containing introns and exons is processed to generate the final mRNA. The CFTR protein spans the cell membrane and contains several transmembrane domains that are essential for the channel function of the CFTR protein.

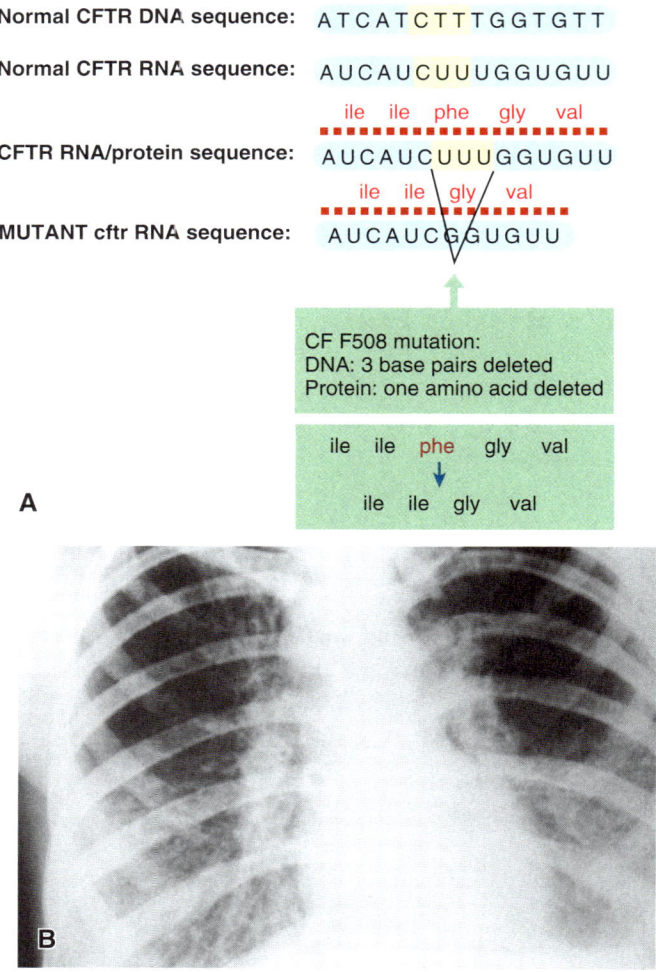

FIGURE 4-26 The CF F508 mutation deletes three base pairs and one amino acid (A) The F508 mutation that causes most cases of cystic fibrosis disease is located in the cystic fibrosis transmembrane receptor (CFTR) gene and protein. The F508 mutation removes 3 base pairs from the gene and deletes one amino acid (phenylalanine) from the protein. The wild type CFTR amino acid sequence of isoleucine, isoleucine, phenylalanine, glycine, valine (ile ile phe gly val) is changed by the cftr CF508 mutation to isoleucine, isoleucine, glycine, valine (ile ile gly val). (B) CF disease seriously damages the lungs, which accumulate thick mucus that prevents the efficient exchange of oxygen and promotes bacterial infections.

The human body perspires as a part of an important mechanism that maintains the body's homeostasis. Sweat is normally made in glands beneath the skin and travels through the ducts to the outside of the body. Normally the CFTR channels transport salt ions across the cell membranes as needed to adjust ion levels, but in the absence of normal CFTR proteins the mutant cftr proteins fail to maintain the proper balance of ions inside and outside the cells. Because the cftr mutant cells cannot absorb salt properly in the absence of CFTR function, people with CF disease produce perspiration containing five times more salt than normal. CF disease causes a continual loss of salt from the body, which potentially alters the balance of ions in the blood and raises the risk of serious cardiac problems, including an abnormal heart rate.

The CFTR protein is essential for the function of most types of human cells and mutations in the cftr gene has widespread consequences that affect many organs and tissues in the body. Unfortunately CF disease causes the most serious damage to the lungs, where the airways and air sacs accumulate thick mucus that prevents the efficient exchange of oxygen (Figure 4-26). These diseased tissues provide a perfect environment for the growth of pathogenic bacteria, which cause repeated respiratory infections in people with CF disease.

More than 12 million people in the United States are genetic carriers of CF disease. Testing to identify genetic carriers is strongly recommended for people who are potentially at risk of conceiving a child with CF. A genetic carrier of CF inherits one normal CFTR gene and one mutant cftr gene (CFTR, cftr) and will not have CF symptoms, The cftr mutation is genetically recessive. When the mutant cftr gene is inherited along with one normal copy of the CFTR gene (cftr, CFTR), the cells can survive on the expression of a single normal copy of the CFTR gene and are not affected by the mutant cftr gene and protein. Genetic carriers of CF disease (cftr, CFTR) can transmit either genetic allele (mutant or wild type) to biological children (Figure 4-27).

Punnett squares are a simple method to predict the probability that biological offspring will inherit certain combinations of mutant and wild-type parental genetic alleles. In Figure 4-27 a Punnett square is used to assess the risk of a child inheriting CF disease, depending on the CF alleles donated by the biological parents. If both parents are CF carriers (CFTR, cftr; CFTR, cftr), then there is a 25% chance (1 in 4) that the parents will have a baby who inherits CF disease with symptoms (cftr, cftr) (Figure 4-27). If only one parent carries a mutant cftr gene (CFTR, cftr), and the other parent is normal (CFTR, CFTR), then there is a 50% chance that their biological child will be a carrier (CFTR, cftr). However, there is no chance that the baby will have CF, since two mutant cftr genes (cftr, cftr) are needed to inherit CF disease. These risks apply to every pregnancy for this couple.

Alzheimer's Disease Involves many Genes

Alzheimer's disease is caused by the destruction of nerves in the parts of the brain that are essential for memory, thought, and language (Figure 4-28). Alzheimer's disease is characterized by progressive dementia and currently affects about 4 million Americans, mostly people over 65 years of age. Microscopic analysis of the brain cells from people with Alzheimer's disease shows the widespread accumulation of abnormal plaques, tangles, and clumps of cellular debris (Figure 4-29). The microtubules in the cytoplasm of Alzheimer's brain cells are destroyed by defective tau proteins, which cause the formation of clusters and tangles (Figure 4-29).

The damaged areas of the brain produce smaller amounts of the neurotransmitters needed to send nerve impulses and transmit signals between nerve endings, decreasing the efficiency of nerve function. Loss of memory is the primary symptom indicating the onset of Alzheimer's disease, which gets worse with time and interferes with routine tasks. As the disease progresses,

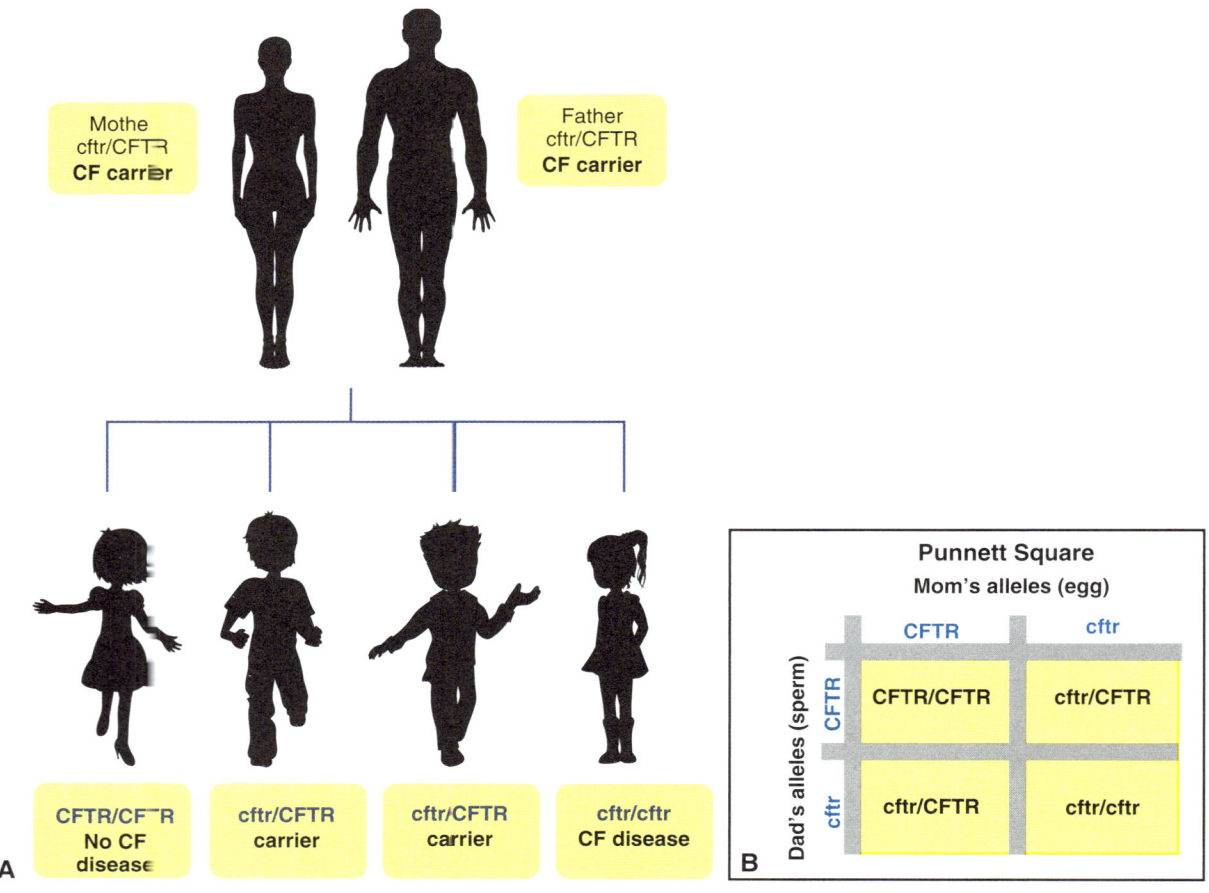

FIGURE 4-27 Genetic carriers can pass mutant CF genes to their children (A) People who are genetic carriers of cystic fibrosis risk passing the disease to their children. There is a 25% chance that the child of genetic carriers will not be affected by CF, a 25% chance that the child will have symptomatic cystic fibrosis disease and a 50% chance that the child will be a genetic carrier of cystic fibrosis disease. (B) Punnett squares predict the probability that biological offspring will inherit certain combinations of parental genetic alleles. The parental alleles are listed across the top and down the left side. This Punnett square assesses the risk of inheriting CF disease. In this case both parents are CF carriers (CFTR, cftr; CFTR, cftr). There is a 25% chance (1 in 4) that the parents will have a child who inherits CF disease with symptoms (cftr/cftr).

Alzheimer's disease causes personality and behavioral changes, and eventually the patient loses verbal skills.

Currently scientists do not understand the molecular mechanism(s) that cause Alzheimer's disease, but research indicates that a combination of multiple genes and environmenal factors is involved in the development of the disease. For example, people who inherit the e4 allele of the APOE gene have an increased risk of developing Alzheimer's disease after age 65.

There is no cure for Alzheimer's disease, but some drugs have been developed that slow the progress of the disease symptoms. Aricept© is a medication that increases the production of neurotransmitters in the brain, and another medicine, Namenda©, protects nerve cells from damage by glutamate.

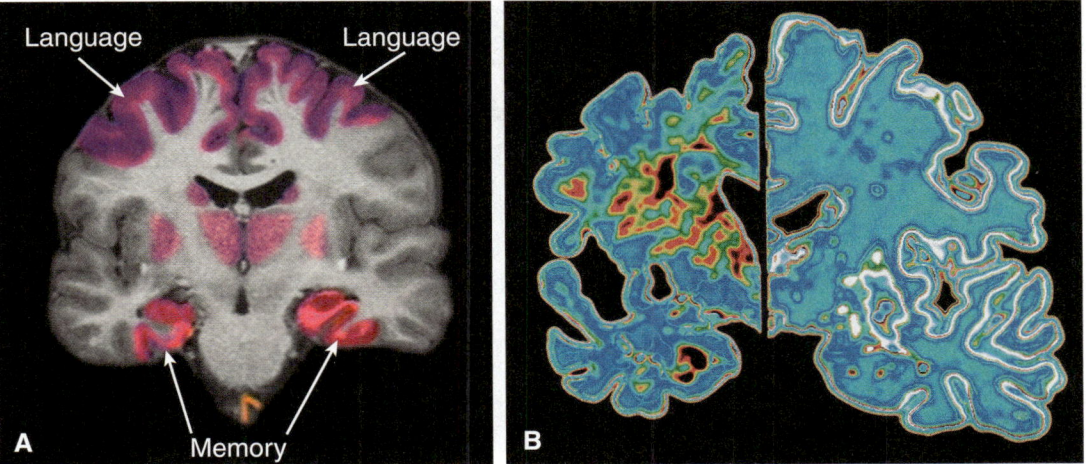

FIGURE 4-28 Alzheimers disease destroys nerve cells in the brain (A) Frontal MRI (magnetic resonance image) of the brain indicates the regions affected by Alzheimers disease including the temporal lobes (dark reddish color), which are involved with memory, and the parietal regions (purple regions on top and on both sides) are also severely affected. These regions are damaged in Alzheimers disease. (B) Computer image of a vertical (coronal) slice through the brain of an Alzheimer patient (left) compared with a normal brain (right). The Alzheimer's disease brain is considerably smaller due to the degeneration of nerve cells. The nerve cells accumulate tangled protein filaments (neurofibrillary tangles) and brain lesions containing beta-amyloid proteins.

Inherited Colon Cancer

Cells in the human body usually reproduce only when the body needs additional cells for a particular purpose. However, in some cases the proteins that regulate cell division are defective, allowing the cells to divide out of control and potentially form tumors. Most types of cells in the human body can develop into cancer cells, including the cells lining the colon, which is the first part of the large intestine. People with a family history of colon cancer have an increased risk of inheriting FAP (familial adenomatous polyposis) or HNPCC (hereditary nonpolyposis colon cancer, also called Lynch syndrome), which are cancers caused by mutations in one of several genes coding for the DNA enzymes that repair damaged chromosome DNA.

FAP is caused by mutations in the APC (adenomatous polyposis coli) gene on human chromosome 5, which predispose people to developing colon cancer. The APC gene codes for a tumor suppressor protein that regulates cell division and prevents the uncontrolled cell growth that leads to cancer development. People who inherit one mutant copy of the APC gene have a high probability of developing colon cancer by age 40. Similarly, people who inherit one HNPCC mutation have an 80% chance of getting colon cancer. HNPCC mutations also increase the risk of developing other types of cancer, including ovarian, stomach, brain, and liver.

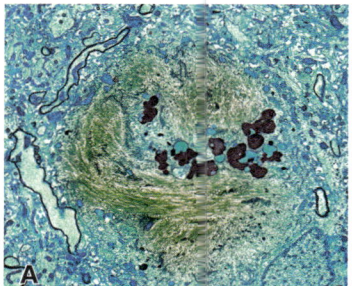 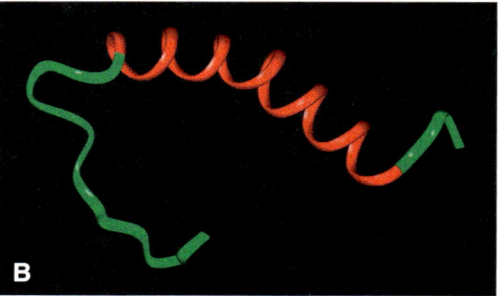 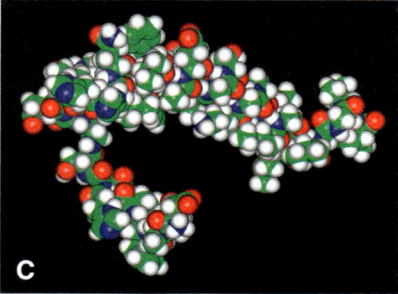

FIGURE 4-29 Alzheimer's disease makes protein aggregates (A) Colored transmission electron micrograph (TEM) of an Alzheimer's brain cell shows neurofibrillary tangles (green) in the cytoplasm (blue), which are aggregates of the tau protein that normally stabilizes microtubule cytoskeletal structures in cells. Similar abnormal tangles are also found in Creutzfeldt-Jakob disease (CJD), and Parkinson's disease. (B) Ribbon model of the ameloid beta peptide found in the plaques in the brains of Alzheimers patients. (C) Molecular space-filled model of the ameloid beta peptide.

Huntington's Disease

Huntington's disease (HD) is a progressive neurodegenerative disorder that destroys cells in the basal ganglia, which is the region of the brain that controls movement, emotion, and cognitive ability. HD is caused by a mutation in the huntingtin protein gene that produces a defective huntingtin protein. Normally the huntingtin protein controls the movement of vesicles that carry cargo to the outside of the cell. HD is a familial genetic disease that is passed from parent to child by the genetic transmission of the HD gene mutation. A parent with HD has a 50% chance of donating a mutant HD gene to each biological child. A child who inherits a mutant HD gene will eventually develop the degenerative symptoms of the disease, although the age of onset and the rate of disease progression are variable. Currently there is no way to stop or even slow the progression of HD.

Fortunately progress in DNA testing makes it possible to accurately diagnose Huntington's disease, and determine the number of triple repeats in an individual mutant HD gene. The National Institutes of Health compared the advances in diagnosis of Huntington's disease to the situation only 50 years ago when Woody Guthrie died.

(http://www.ninds.nih.gov/disorders/huntington/detail_huntington.htm#160623137)

"The great American folk singer and composer Woody Guthrie died on October 3, 1967, after suffering from HD for 13 years. He had been misdiagnosed, considered an alcoholic, and shuttled in and out of mental institutions and hospitals for years before being properly diagnosed. His case, sadly, is not extraordinary, although the diagnosis can be made easily by experienced neurologists."

The wild-type HD gene was identified in 1993 and found to contain many in-frame tandem repeats of the CAG codon (5'-[CAG]$_n$-3'), which encodes a stretch of repeated glutamine amino acids in the protein. Mutant HD genes contain a larger number of CAG repeats than the normal wild type HD allele, and a DNA test is now available that determines the number of CAG repeats in the HD gene in the genome DNA. The coding region of the wild-type huntingtin gene normally contains 10 to 28 CAG repeats that specify a corresponding stretch of glutamine (Gln) residues in the wild-type huntingtin protein. The HD symptoms are caused by a mutant huntingtin gene that encodes more than 40 CAG codons translated into a stretch of more than 40 glutamine amino acids in each mutant huntingtin protein. This stretch of uncharged hydrophilic glutamine amino acids disrupts the normal function of the huntingtin protein.

HD is caused by an autosomal dominant mutation. A person who inherits one copy of a mutant huntingtin gene will develop Huntington's disease, even though this person also inherited a wild-type huntingtin gene with fewer than 28 CAG repeats. In this case the function of the wild-type protein is not sufficient to overcome the detrimental effects of the mutant huntingtin proteins expressed in the same cell.

> **KEY CONCEPT**
> In autosomal dominant inheritance, a parent with one normal and one mutant HD gene has a 50% chance of passing the fatal disease to each biological child.

Breast Cancer Gene Mutations

Cancer develops when normal cells lose the ability to control cell division and continue to divide out of control, forming tumors. The molecular mechanisms responsible for the development of different kinds of cancer cells include mutations in genes that normally control the cell cycle or protein products that function in the maintenance and repair of chromosome DNA. The development of most types of cancer cells requires a series of different DNA mutations that cause the cells to exhibit uncontrolled cell growth and eventually metastasize to other locations in the body. The two genes known to be involved in a genetically inherited form of breast cancer are called the Breast Cancer #1 (BRCA1) and breast cancer #2 (BRCA2) genes, and both encode tumor suppressor proteins. In this case breast cancer develops because the mutations in BRCA1 or BRCA2 cause defects in the process of repairing damaged chromosome DNA. Without the ability to efficiently repair DNA damage, additional mutations accumulate rapidly in the chromosome DNA (Figure 4-30).

The mutations that cause most cancers are not genetically inherited from biological parents but instead are changes in the somatic cell chromosome DNA that accumulate over a person's lifetime. Some cancers tend to "run in the family," but breast and ovarian cancers occur frequently in people who have no family history of cancer. Only 5–10% of all breast and ovarian cancers are caused by inheriting a mutation in either the BRCA1 or BRCA2 gene. Normally the wild type BRCA1 and BRCA2 proteins suppress cancer cell development by preventing uncontrolled cell growth. The BRCA1 and BRCA2 enzymes repair any damaged chromosome DNA before the cell duplicates the chromosomes in preparation for the chromosomes to segregate into daughter cells at cell division. These processes are essential to the cell,

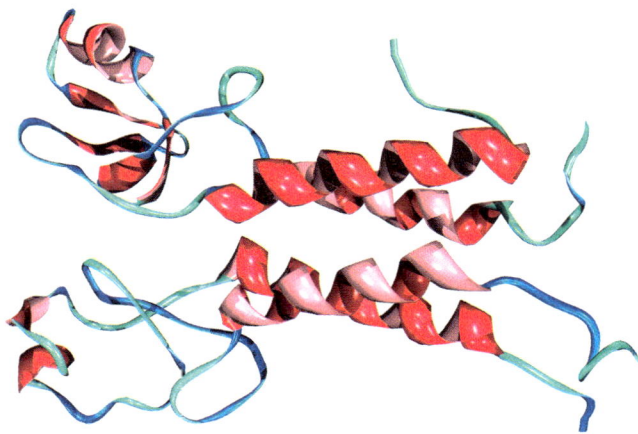

FIGURE 4-30 The human breast cancer gene BRCA1 codes for a tumor suppressor protein
The breast cancer type 1 susceptibility gene (BRCA1) codes for a DNA repair protein that fixes the double-stand breaks in the chromosome DNA. If the DNA damage cannot be repaired, the cell will die. Mutations in the BRCA1 protein prevent the protein from repairing the damaged DNA, causing the cells to divide uncontrollably and form a tumor. BRCA1 mutations also increase the risk of breast, ovarian, and prostate cancers.

which has specific checkpoint mechanisms that force a delay in cell division in cases where the chromosome DNA is not completely replicated or contains unrepaired DNA damage. However, when the mutant BRCA1 or BRCA2 proteins are defective in repairing damaged DNA, the cells attempt to finish cell division with abnormal, damaged chromosome DNA. As a result the affected genome accumulates additional genetic mutations and possibly chromosome rearrangements that lead to the development of cancer cells.

Women who inherit either a mutant BRCA1 or a mutant BRCA2 gene have a greatly increased risk of developing breast cancer at an early age. About 60% of these women will develop breast cancer by age 50, and about 80% will develop breast cancer by age 70. Women who inherit either a mutant BRCA1 or BRCA2 gene also have an increased risk of developing breast or ovarian cancer. The BRCA1 and BRCA2 gene mutations are autosomal dominant, which means that the biological children of a parent (mother or father) who carries either a BRCA1 or BRCA2 mutation will each have a 50% chance of inheriting a mutant BRCA1 or mutant BRCA2 gene. People with strong family histories of breast or ovarian cancer should consider having their genome DNA screened for the BRCA1 or BRCA2 gene mutations. Some women who have inherited these mutations choose to undergo preventative treatments such as breast and ovary removal, while others take tamoxifen and other drugs shown to protect against breast cancer.

Heart Disease is a Major Killer of People in the United States

Multiple genes play a role in many human diseases, including heart disease. The involvement of several genes makes it more difficult to determine the primary factors that potentially put an individual at "high risk" for heart disease.

Studies have shown that a family history of chronic heart disease and relatives with heart attacks at an early age is usually an excellent indicator of elevated risk of heart disease. However, research has also shown that additional risk factors for heart disease include being overweight, smoking cigarettes, and having a sedentary lifestyle (little exercise). People often misunderstand the roles that genes, lifestyle, and environment play in determining future health risks. The term *genetic determinism* refers to the idea that genes alone determine fate without influence from any other factors. But research clearly shows that genes alone do not determine someone's fate, with the exception of autosomal single-gene diseases like Huntington's. Despite research to the

> **KEY CONCEPT**
> Many genes are involved in heart disease, but as yet there is no medical test that can predict a future heart attack from the genetic alleles in an individual's genome. Some behaviors can contribute to heart disease, including smoking, high blood pressure, a high-fat diet, and lack of exercise.

contrary, people who believe in genetic determinism can make harmful decisions and unfair judgments about themselves and others.

We know that heredity and environment influence gene expression, but scientists are just beginning to understand how the environment affects the future genetic risk for generations to come (Unit 6). Individual humans all have the same genes, of course, but everyone carries different forms of these genes, different alleles. Some people carry genetic alleles that put them at increased risk for getting cancer. However, even for these people the risk of getting cancer is much lower if they do not smoke. Several gene alleles make people genetically predisposed to developing type II diabetes. Even for those people, the risk that they will actually develop type II diabetes is much lower if they maintain a normal body weight, eat a healthy diet, exercise, control stress levels, and have access to preventive health care. Environmental factors and mental stress also negatively affect the immune system, weakening the body's immune response and increasing the production of stress hormones such as cortisone.

Recently researchers have identified several human genetic diseases that are caused by incorrectly folded mutant proteins. These genetic diseases are not new, but recent studies have revealed the connections between several genetic diseases and mutant proteins that fold incorrectly. These genetic diseases affect well known people such as Lou Gehrig, who had amyotrophic lateral sclerosis (ALS); President Ronald Reagan, probably the best known person who died from Alzheimer's disease; and actor Michael J. Fox, who has been speaking out to support embryonic stem cell research as an effective treatment for his condition, Parkinson's disease.

Prions are a Bad Influence on 'Good' Proteins

The importance of protein folding to the biological function of the protein is emphasized by the deadly activities of prion proteins, small proteins that cause disease by changing their physical shape. Inside normal cells the normal prion proteins (PrPC and PrP) are benign and do no harm. The problems begin on the very rare occasion that a person ingests meat contaminated with PrPSc prion proteins, and the pathogenic prions enter the body cells.

Prion proteins act alone and they do not contain any type of DNA or RNA genome (Figure 4-31). Prions are not viruses. The PrPSc prion proteins cause mammalian diseases such as bovine spongiform encephalopathy (BSE, "mad cow disease") and Creutzfeldt-Jakob disease (CJD) in humans. Exposure to pathogenic prions is very rare but several countries banned the import of beef from Britain after an outbreak of Mad Cow disease in the U.K. Currently there are no effective treatments for prion diseases, which are always fatal.

Inside normal cells the benign prion proteins (PrPC, PrP) do not cause harm until they come in contact with the pathogenic prion proteins (PrPSc). At this time the normal PrPC proteins are induced to change shape and are converted into the pathogenic shape adopted by the PrPSc disease prion (Figure 4-32). The miss-folded pathogenic PrPSc proteins accumulate and are toxic to the cells, causing various disease symptoms. In the brain cells the miss-folded PrPSc proteins cause the formation of amyloid plaques similar to the plaques and tangles observed in the brains of Alzheimer's patients (Figure 4-29), which is inevitably linked to nerve cell degeneration. Microscopic analysis of the brain tissue from cows with bovine spongiform encephalopathy (BSE) also shows holes that give the brain tissue the characteristic sponge-like appearance common to prion diseases.

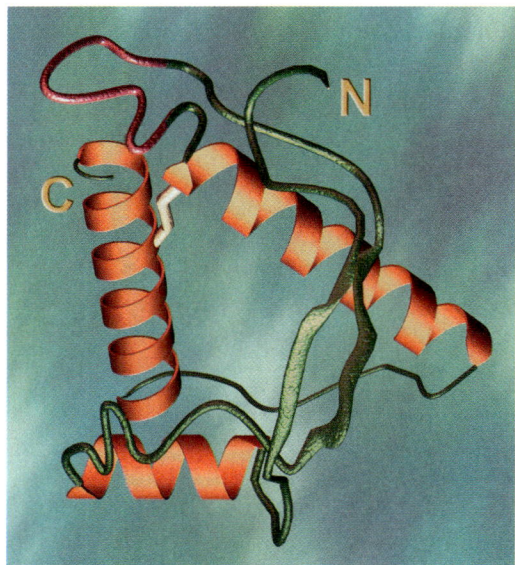

FIGURE 4-31 Prion proteins change shape and cause disease The disease form of the prion protein causes BSE (bovine spongiform encephalopathy) in cows and CJD (Creutzfeldt-Jakob disease) in humans. The disease form of the prion has alpha helices (orange) connected with straight beta strands (green). The carboxyl- (C) and amino- (N) termini of the amino acid strand are indicated.

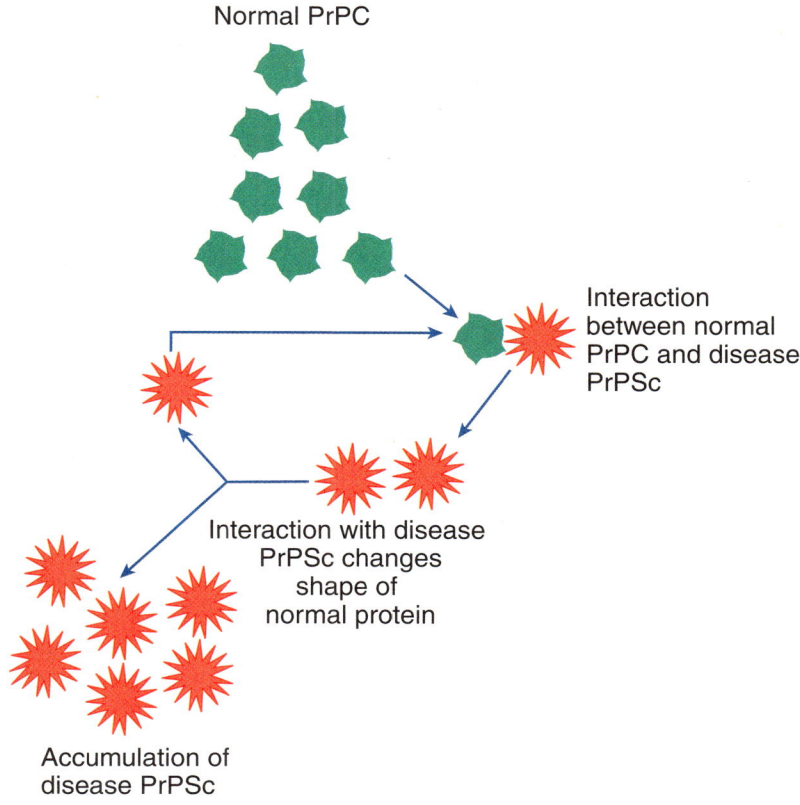

FIGURE 4-32 Prions are a bad influence on other proteins The Prion cycle starts with a normal PrPC protein (also called PrP) that interacts with the disease PrPSc protein. This converts the normal PrPC protein into the shape adopted by the disease prion, PrPSc. The disease PrPSc proteins accumulate and are toxic, causing the disease symptoms.

Unit 4 Questions

1. A gene mutation can potentially change:
   a. The DNA sequence of the gene
   b. The mRNA sequence copied from the gene
   c. The protein sequence coded by the gene
   d. All of the above
   e. None of the above

2. A genetic disease is usually transmitted from person to person by:
   a. Coughing or sneezing
   b. Using a public rest room
   c. Having unprotected sex
   d. All of the above
   e. None of the above

3. The untranslated DNA regions of a eukaryotic gene are:
   a. Translated into RNA but not transcribed into protein
   b. Transcribed into RNA but not translated into protein
   c. Transcribed into protein but not translated into RNA
   d. Translated into protein but not transcribed into RNA

4. Autosomal eukaryotic chromosomes are:
   a. The 22 pairs of chromosomes that are not the sex chromosomes
   b. The chromosomes that specify gender (sex)
   c. The chromosomes carrying all of the dominant genes
   d. The chromosomes carrying all of the recessive mutations

5. A molecular DNA probe has the ability to specifically base pair to:
   a. the other DNA probes in the reaction that have the same DNA sequence
   b. the other DNA probes in the reaction that have completely different DNA sequences
   c. the protein molecules in the reaction that look like DNA on the surface
   a. the other DNA probes in the reaction that have complementary DNA sequences

6. A recessive gene mutation (allele) causes a trait (phenotype or disease) that is exhibited when:
   a. The recessive gene is inherited with a dominant allele of the gene
   b. The recessive gene is inherited with another copy of the same recessive allele
   c. The recessive allele of the gene is inherited by a male
   d. The recessive allele is inherited is inherited by a female

7. The purpose of amniocentesis is to:
   a. Check the health of a baby immediately after it is born
   b. Collect fetal cells for genetic testing of the fetus
   c. Perform *in vitro* fertilization in the lab
   d. Implant blastocyst embryos into the uterus

8. A chromosome rearrangement can involve which of the following changes:
   a. A region of DNA is moved from one chromosome to another chromosome.
   b. A region of DNA is removed (deleted) from a chromosome and discarded.
   c. A region of DNA is copied and re-inserted into the same chromosome, duplicating the DNA.
   d. all of the above
   e. none of the above

9. In preimplantation genetic diagnosis, which cell(s) undergo DNA testing?
   a. Cell(s) from an 8-cell *in vitro* embryo are collected before the blastocyst stage
   b. Cell(s) from the pregnant mother's blood are collected
   c. Cells from the outside surface of the blastocyst embryo formed by *in vitro* fertilization
   d. Sperm cell(s) from the biological father

10. A person who is the carrier of a genetic disease:
    a. Always experiences the symptoms of the disease
    b. Has inherited two mutant genetic alleles that both cause the same disease
    c. Has inherited one dominant mutant allele and one normal allele that cause the disease
    d. Has inherited one recessive mutant allele and one normal allele that cause the disease

UNIT 5

Embryonic Stem Cells and Animal Nuclear Cloning

Human Development and Embryology

As a scientist (and a person) I find the entire process of human embryogenesis to be an awesome and amazing molecular miracle! And embryonic stem cells (ESCs) are the most miraculous type of cell because ESCs have the power to change into any type of specialized cell needed to make a human body! (Figure 5-1). As a result of access to powerful DNA and molecular genetic technologies it is now possible for scientists to better understand the complex molecular mechanisms at work that allow the ESCs to establish and maintain a pluripotent state. As we shall soon discover, scientists have recently learned to convert adult skin cells into un-differentiated cells (iPS cells) that mimic ESCs. The iPS cells are pluripotent and can be induced to generate highly specialized nerve, muscle and cardiac cells.

The cells that make up a human embryo start with a single fertilized cell and a new genome. The two parental genomes, one from Mom and one from Dad, are donated to the embryo by the 'adult' egg and 'adult' sperm cells and the genomes come together at fertilization to create the new zygote genome, called the epigenome. The epigenome is pre-programmed to execute the genetic instructions needed to direct the growth and differentiation of the different embryo cells throughout embryo development. Each embryo cell contains 23 pairs of chromosomes as expected but as we will see, the epigenome is different from the typical adult cell genome (Unit 6).

At about 4-5 days after conception in humans, a small number of very powerful ESCs are growing deep inside a hollow blastocyst embryo. At the appropriate time in development the embryo cells began change and migrate to new locations, altering the shape of the embryo. The embryonic stem cells in the hollow blastocyst embryo have the amazing ability to develop into any type of specialized cell in the human body. Human ESCs could potentially be used to generate specialized tissues and organs for transplant and other regenerative therapies used to repair diseased or injured tissues.

It is important to explore the basic biology behind human embryonic stem cell development to understand how the ESCs generate different types of specialized human cells and tissues in the embryo.

Embryonic Stem Cells are at the Top of the Human Stem Cell Hierarchy

In the developing human embryo, the embryonic stem cells have unique characteristics that distinguish the ESCs from the other embryo cells (Figure 5-1).
Embryonic stem cells share the following characteristics:

- Are undifferentiated (unspecialized) cells
- Can undergo continual cell division (self-renewal)
- Can generate differentiated cells with specialized functions

Embryonic stem cells differ significantly from other cells in the embryo including other stem cells because ESCs can generate all of the different cells and tissues required in the human body (except the placenta). The small number of embryonic stem cells inside the blastocyst can generate over 200 different types of specialized human cells that function in the human body (Unit 1).

During embryo development, the stem cells must respond to internal signals from proteins expressed by the cells and to external signals received from the environment (nutrients, hormones), including changes in environmental conditions (temperature) that can trigger some cells to differentiate into specialized cell types. Most of the cells in the adult human body are fully differentiated and perform specialized functions that are specific to a particular type of cell. However, even after birth and

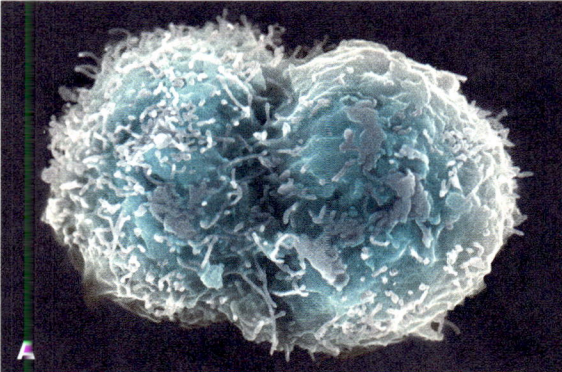 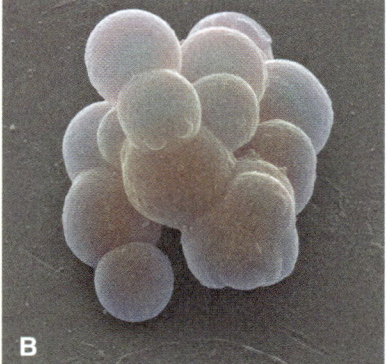

FIGURE 5-1 Embryonic stem cells have not differentiated but can develop into any type of body cell (A) This stem cell is dividing to form two stem cells. Embryonic stem cells can make more stem cells or can differentiate into any other type of cell in the body. (B) These multipotent stem cells obtained from umbilical cord blood can differentiate into precursor cells that generate all of the specialized types of human blood cells.

into adulthood, the human body maintains a small number of progenitor stem cells that are multipotent and can regenerate specialized cells and tissues in the body.

Human stem cell development usually involves cell division, which is the key process used to generate additional numbers of identical or different cells. Like all eukaryotic cells, human cells divide using two different universal mechanisms, mitosis and meiosis:

(1) In mitosis, a single parent cell (haploid or diploid) undergoes cell division and produces two identical daughter cells. This cell division does *not* change the total number of chromosomes in each cell (ploidy).

(2) In meiosis, a single parent cell undergoes two successive cell divisions and produces four daughter cells. In this case the number of chromosomes inherited by the daughter cells is half the number of chromosomes in the parent cell. The parent cell is diploid (46 chromosomes) but each of the two daughter cells is haploid (23 chromosomes).

When a human cell divides by mitosis it makes two identical cells that are identical to the parent cell (Figure 5-2). However, when an embryonic stem cell undergoes mitotic cell division, it can potentially differentiate into a completely different type of stem cell with a specialized function. The ability to differentiate into a specialized cell type is central to the potential of stem cell developmental.

The Origin of Human Embryonic Stem Cells

Human embryo development at the molecular level begins when a human egg is fertilized by a sperm (Figure 5-3). Under natural circumstances fertilization usually takes place in the female reproductive tract, but the process can also occur by *in vitro* fertilization when the sperm and eggs are mixed together in

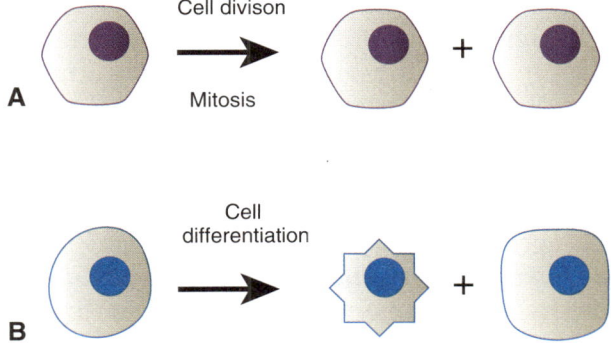

FIGURE 5-2 Stem cells can divide by mitosis or differentiate and specialize (A) Cells typically divide by mitosis to make additional cells that are identical to the parent cell. (B) Cell division and differentiation is one way that stem cells can respond to signals (gene expression, environment, nutrients, hormones) and produce specialized cells.

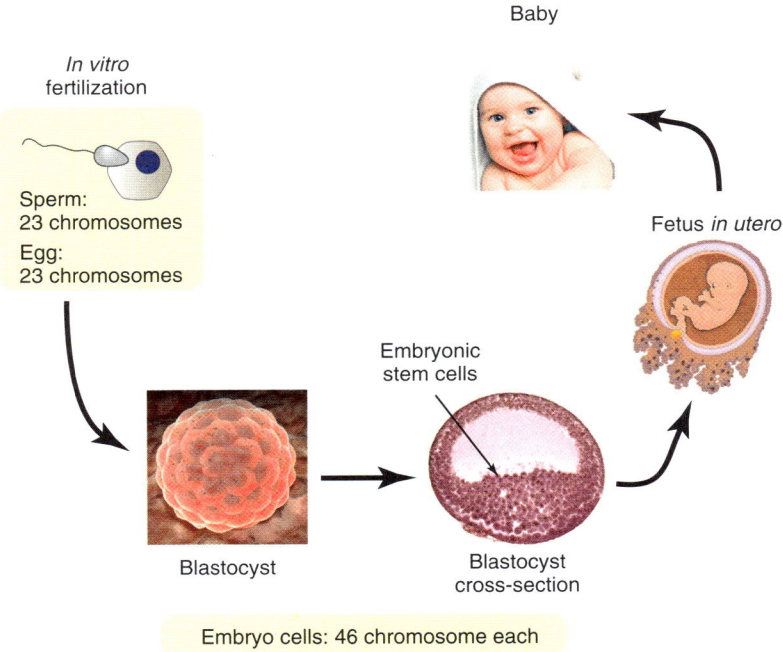

FIGURE 5-3 Fertilization generates the blastocyst and embryonic stem cells Fertilization of the egg cell by sperm is the initial step in human embryo development (embryogenesis). The fertilized zygote develops into the blastocyst embryo, which is a key stage in mammalian embryo development. The inner cell mass (ICM) in the blastocyst generates a small number of embryonic stem cells (ESCs). These ESCs are the pluripotent stem cells that give rise to all of the different types of specialized cells in the human body.

a plastic laboratory dish. The egg cell nucleus containing 23 maternal chromosomes (from the mother) and the sperm cell nucleus containing 23 paternal chromosomes (from the father) (Unit 4) come together during fertilization and create the single nucleus in the diploid zygote cell that contains all 46 human chromosomes. At fertilization the mitochondria and the other components located in the cytoplasm of the egg cell become part of the cytoplasm in the zygote by maternal inheritance, but the sperm contributes only the sperm nucleus containing the sperm DNA genome (Unit 6).

In early embryogenesis the zygote cell divides by mitosis to produce two identical embryo cells, which divide into four embryo cells, which divide into an 8-cell embryo (Figure 5-3). The cells in the embryo continue to divide and by the 4th or 5th day after fertilization about 120 cells have formed a hollow ball called a blastocyst embryo (Figure 5-3). In natural conception, the blastocyst forms while the embryo is traveling through the fallopian tube, long before it reaches the uterus. However, whether fertilized *in vivo* or *in vitro*, for pregnancy to occur, the blastocyst embryo *must* successfully implant into the wall of the uterus of a surrogate female for further embryo development. If the blastocyst fails to implant in the uterine wall, the body will naturally discard the embryo by spontaneous abortion.

The human blastocyst embryo hosts the small numbers of amazing embryonic stem cells (ESCs) that grow in the inner cell mass (ICM) inside hollow blastocyst (Figure 5-3). Embryonic stem cells also rank at the top of the stem cell hierarchy because ESCs are pluripotent and can produce any type of specialized cell in the human body (except the placenta) (Figure 5-4). ESCs can make more ESCs, and can also make multipotent progenitor stem cells that can generate only a limited number of differentiated types of cells. In comparison, ESCs are pluripotent, and gametes (eggs and sperm) are developmentally totipotent because the union of sperm and egg cells gives rise to all the cells and tissues in the human body, including the placenta.

Stem Cell Renewal

When ESCs undergo mitotic cell division to make more identical ESCs, the process is called stem cell renewal (Figure 5-4). However, the pluripotent ESC can also divide asymmetrically and produce two different cells, an identical pluripotent ESC and a multipotent progenitor stem cell (Figure 5-4). The multipotent progenitor stem cells have more limited developmental potential and can generate a smaller number of specialized cell types compared to pluripotent ESCs. The ESCs can generate each of the specialized cell types in the human body (except placenta) (Figures 5-4 and Figure 5-5).

When a multipotent progenitor stem cell divides it can produce another progenitor stem cell or it can generate a differentiated cell that performs a specialized function in the body (Figure 5-5). Specialized cells exhibit the structural and biochemical properties that are characteristic of fully

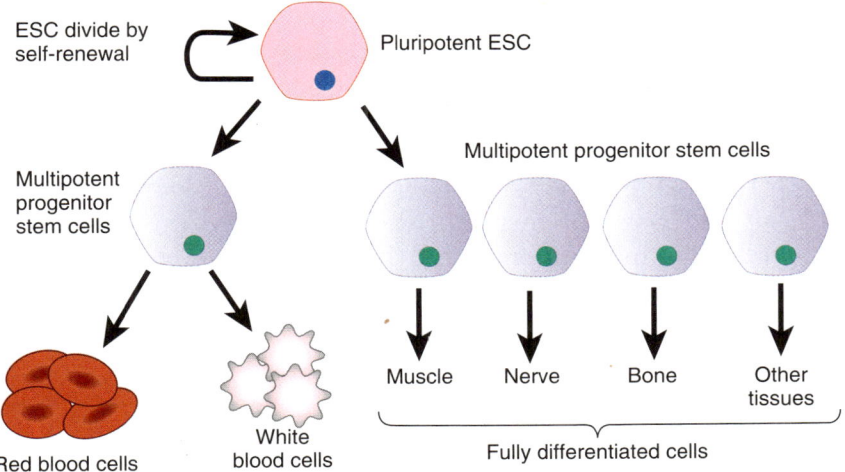

FIGURE 5-4 ESCs have the most potential and are at the top of the stem cell hierarchy
The hollow blastocyst embryo contains the inner cell mass (ICM), which generates a small number of embryonic stem cells (ESCs). The ESCs are pluripotent and can develop into any cell type in the human body. The ESCs produce multi-potent progenitor stem cells that can generate all of the different types of specialized cells and tissues needed in the body, including blood cells, muscle, nerve, etc.

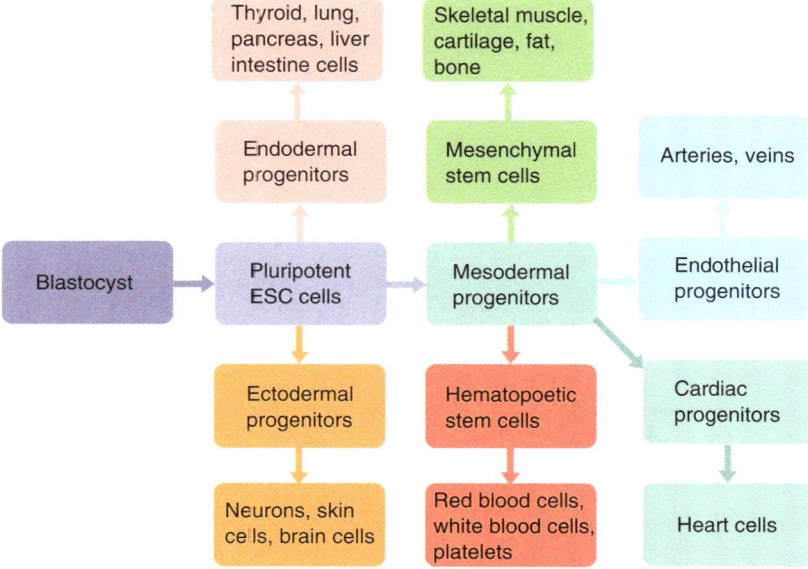

FIGURE 5-5 Flow chart of specialized human cells generated by ESCs The ICM cells generate a small number of embryonic stem cells (ESCs) in the blastocyst. The ESCs produce multipotent progenitor stem cells, which generate the three 'germ' layers in the human embryo and give rise to all of the different types of specialized cells and tissues needed in the human body.

differentiated cells. In the past the development of specialized cells was called "terminal differentiation" because scientists thought that after a cell had become specialized with a specific function, the cell could not de-differentiate, or develop in reverse, and revert to an unspecialized state. However, very exciting recent research shows that specialized cells are not terminally differentiated because under certain conditions these highly developed cells can be induced to de-differentiate and regain their pluripotent developmental potential (see iPS cells below).

To be able to better understand the functions of specialized human cells and the issue of terminal differentiation, it is important to look at the molecular mechanisms and the cellular components involved in the transition from ESC to highly differentiated nerve, kidney, or skin cells. The genetic pathways leading to cell development in the human embryo involve

> **KEY CONCEPT**
> ESCs are pluripotent and can give rise to all the specialized cell types in the human body (except the placenta). To accomplish this feat, the ESCs undergo asymmetric cell division and produces a multipotent progenitor stem cell that in turn gives rise to a highly differentiated type of cell in the body.

dramatic changes in the overall shape and size of the growing embryo. During gastrulation the ESCs inside the blastocyst embryo divide and migrate to new positions to determine the shape of the developing gastrula embryo. The ESCs generate the progenitor stem cells, which in turn give rise to the three different germ cell layers in the developing embryo, the endoderm, ectoderm, and mesoderm (Figure 5-6). The endoderm layer produces specialized cells in the thyroid, lung, pancreas, liver, and intestine; the ectoderm layer generates the adult skin cells, neurons, and brain cells; and the mesoderm layer develops into mesenchymal stem cells that generate skeletal muscle, cartilage, fat cells, and bone (Figure 5-6).

As the human embryo develops, the multipotent progenitor stem cells give rise to all the necessary types of specialized cells that are destined to generate the future organ and tissue systems in the fetus, baby, and adult (Figures 5-6). The stem cells give rise to increasingly specialized cells that are predetermined to perform specific jobs in the developing embryo, in the fetus, or after birth. For example, the endoderm layer of the human embryo that made the pancreatic bud structure in the embryo can generate only specialized cells in the pancreas and at this developmental stage cannot spontaneously change programs and develop into another cell type.

> **KEY CONCEPT**
> Inside the hollow blastocyst embryo, the inner cell mass generates a small number of embryonic stem cells (ESCs) that have the developmental potential to generate all of the different types of specialized human cells needed in the adult human body.

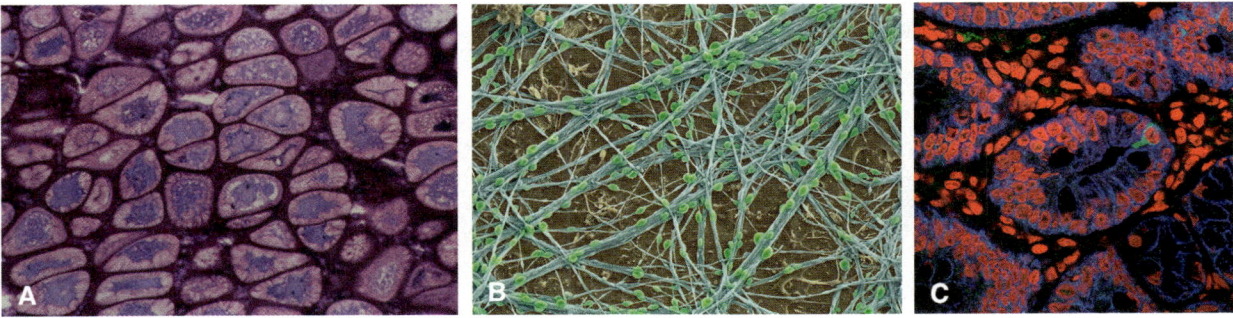

FIGURE 5-6 ESCs generate stem cells in the three embryonic germ layers ESCs generate progenitor stem cells that give rise to the three different germ cell layers in the developing embryo, the endoderm, ectoderm, and mesoderm (A) The mesoderm stem cells develop into cartilage as well as fat cells and bone. (B) The ectoderm germ cells give rise to nerve and brain cells. (C) The endoderm layer develops into the cells in the adult small intestine.

Adult Progenitor Stem Cells (ASCs) are Multipotent

After birth the human body maintains a small number of multipotent adult progenitor stem cells, which each have the ability to regenerate and replace some types of specialized cells and tissues in the adult body. For example, multipotent ASCs produce specialized skin cells to replace damaged cells at the site of a cut or other injury to the skin. Many types of progenitor ASCs have been identified in the adult body, and adult progenitor stem cells have been isolated and characterized from human blood, cornea, retina, heart, fat, skin, dental pulp, bone marrow, blood vessels, skeletal muscle, and intestines (Figure 5-7).

The potential for successful cell replacement therapies in regenerative medicine depends on the use of adult stem cell lines that have been fully characterized, and includes cells, tissues, and organs derived from adult

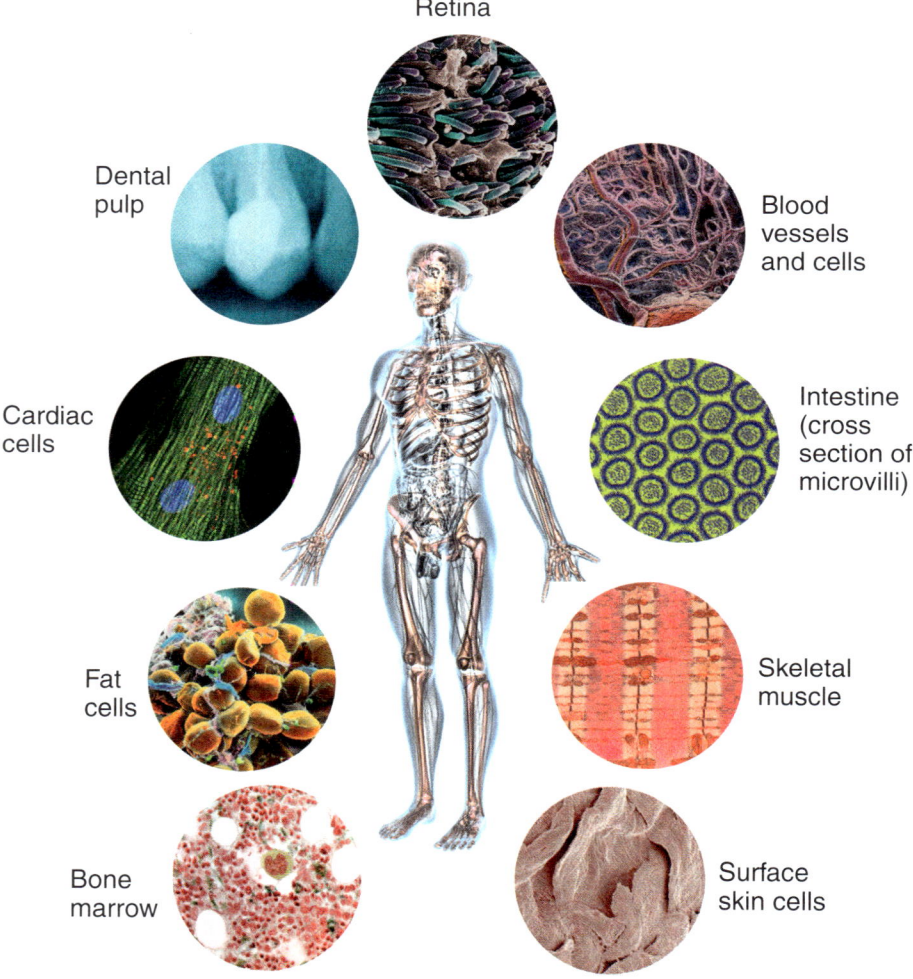

FIGURE 5-7 The human body contains several types of adult stem cells Adult stem cells (ASCs) have been identified in tissues from several locations in the human body, including blood, cornea, retina, heart, fat, skin, dental pulp, bone marrow, blood vessels, skeletal muscle, and intestines.

stem cells. Research using ASCs avoids the ethical complications surrounding human embryonic stem cell research because studies on human ASCs do not involve human ESC or embryonic stem cell derivatives in cell replacement therapies. Unlike some ESCs growing in culture in lab, adult stem cells do not spontaneously differentiate into random cell types or grow out of control in the lab.

In the human body, some types of cells have short life spans and must be periodically replaced with newly differentiated cells of the same cell type in the body. For example, RBCs circulate for a short time (120 days) in the adult bloodstream and must be replaced with new RBCs. The multipotent hematopoietic stem cells (HSCs) live permanently in the marrow of the long bones in adults and produce two lineages of blood stem cells: the lymphoid progenitor stem cells, which generate cells in the immune system (T- and B-lymphocytes), and the myeloid progenitor stem cells, which develop into blood cells such as macrophages, granulocytes, platelets, and erythrocytes (RBCs) (Figure 5-8).

Newly made skin cells continue to develop all over the human body, starting in the deeper layers of the skin and the skin cells are more specialized in the

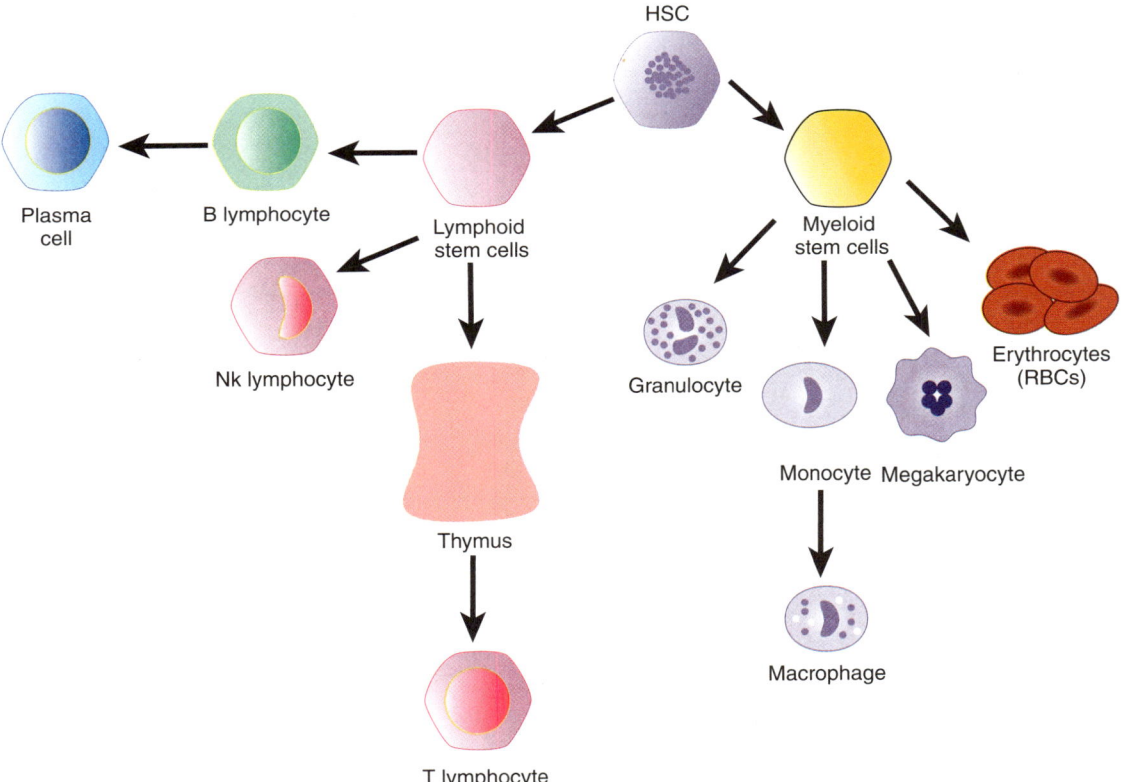

FIGURE 5-8 Multipotent hematopoietic stem cells produce all types of blood cells
Hematopoietic stem cells (HSCs) grow in the bone marrow of the long bones of adults where they generate two lines of blood stem cells, lymphoid and myeloid. The lymphoid progenitor stem cells produce the cells in the immune system (T- and B-lymphocytes), and the myeloid progenitor stem cells develop into blood cells such as macrophages, granulocytes, platelets, and erythrocytes (RBCs).

upper layers. The keratinized skin cells on the surface are discarded every day because the new specialized skin cells are waiting to take over.

Specialized Cells in the Human Body

The human body contains hundreds of different types of specialized cells with distinct shapes and morphologies that are indicative of the essential biochemical functions performed by the tissues containing the specialized cells. The cardiac muscle cells in the heart maintain a consistent beat in the human body, and the individual cardiac cells growing in plastic dishes in the lab also maintain a rhythmic heart beat. Other muscle cells capable of contraction are similar to cardiac muscle cells but are required for the voluntary movement of the striated muscle cells.

The human body contains many types of nerve cells with different highly developed and specialized structures (Figure 5-9). Many nerve cells contain long projections (axons) that are necessary for the nerve cells to transmit impulses over long distances between adjacent nerve cells. Purkinje nerve cells have bulb-like cell bodies and branched processes called dendrites, which conduct impulses to the cell body. These specialized nerve cells are located in the cerebellum, the part of the human brain that controls muscle coordination and balance.

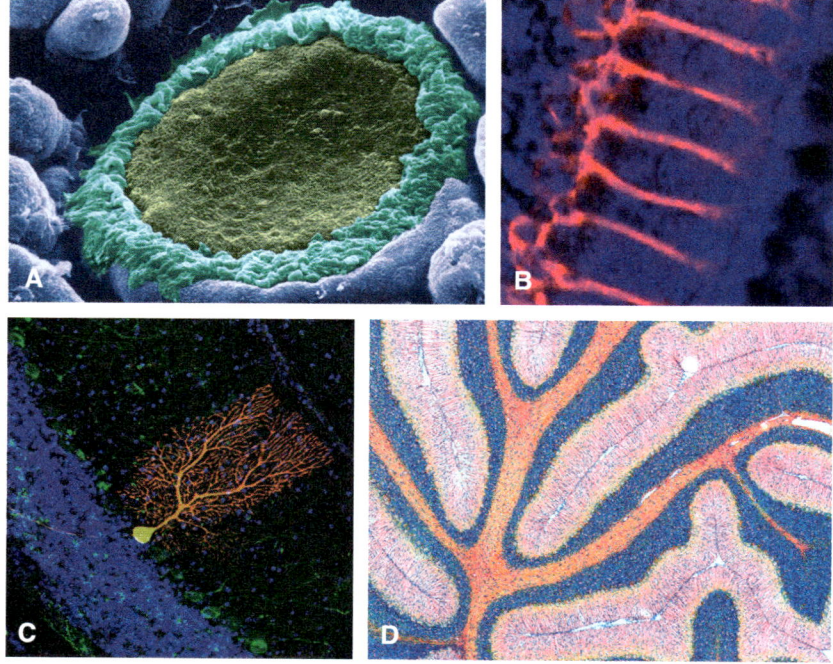

FIGURE 5-9 Differentiated human cells perform very specialized functions (A) Taste buds (blue, green) on the tongue contain cells that detect taste molecules and send nerve signals to the brain. (B) Inside the human inner ear are tiny sensory hair cells (blue) that contain glutamate-aspartate transporter proteins (pink). (C) The Purkinje nerve cells in the brain have a bulb-like cell body (yellow) and numerous branched processes called dendrites (orange). (D) This highly-folded tissue in the cerebellum of the human brain contains neurofilaments (blue), which are most numerous in the inner white matter of the cerebellum and cell nuclei (red), which are most abundant in the grey matter. The glial cells produce the glial fibrillary acidic protein (GFAP) (green).

Although they are too small to see with the naked eye, the surface of the human tongue contains taste bud cells that detect sensory molecules in the mouth (Figure 5-9). Taste buds contain highly developed sensory nerve cells that respond to chemicals in the mouth by releasing the appropriate chemicals and transmitting sensory nerve signals. Also too small to see with the naked eyes are the tiny hair cells in the cochlea of the inner ear that are embedded in endolymph, a fluid that conducts waves in response to sound waves (Figure 5-9). The physical movement of the ear hairs is converted into electrical nerve signals that are transmitted to the brain by nerves that use the neurotransmitter, glutamate. Excess glutamate levels are toxic to cells, so the hair cells in the inner ear produce glutamate-aspartate transporter (GLAST) enzymes that metabolize the excess glutamate and protect the hair cells from injury caused by loud auditory impulses.

> **KEY CONCEPT**
> The stem cells in adult tissues and organs in the body routinely replace some types of old or damaged cells with new cells that can perform the specialized functions.

The Beginning of Human Embryonic Stem Cell (hESC) Research

In their natural setting in an embryo, embryonic stem cells divide to make more cells by self-renewal. However, when scientists first tried to grow ESCs in the lab they found that human embryonic stem cells were very difficult to grow. It was not until 1998 that Dr. James Thomson and his research team (University of Wisconsin) reported for the first time that they had successfully grown human embryonic stem cells (hESCs) in the lab. Thomson's research team accomplished this feat by distributing the hESCs recovered from the inner cell mass of human blastocyst embryos (donated for research) in culture dishes without feeder cells (Figure 5-10). Earlier attempts at culturing hESCs required the use of a feeder layer of normal mouse cells growing on the bottom of the plate, which supported the growth of a layer of hESCs on top. The ability to grow ESCs in culture dishes without feeder cell layers was a major advance in the field that eliminated many problems.

The supply of hESCs grown in the lab can be expanded by collecting the hESCs growing in a single plastic dish and gently diluting the cells into fresh liquid medium in several plates (Figure 5-10). The hESCs on the fresh plates divide and grow and at the appropriate time the cells are collected, diluted and plated into fresh medium. It is essential to grow hESCs in the lab in order to study ESCs and better understand how ESCs become pluripotent and generate specialized cells. Research on hESCs revealed that these unique cells express transcription factors and reprogramming proteins that function to maintain a pluripotent developmental state.

Scientists developed ESCs that could be cultured for long periods in the lab without the cells differentiating spontaneously or accumulating chromosome abnormalities. These ESCs represent genetically stable embryonic stem cell lines that can be propagated, stored frozen and shared with various research groups. Access to well-characterized ES cell lines is an extremely

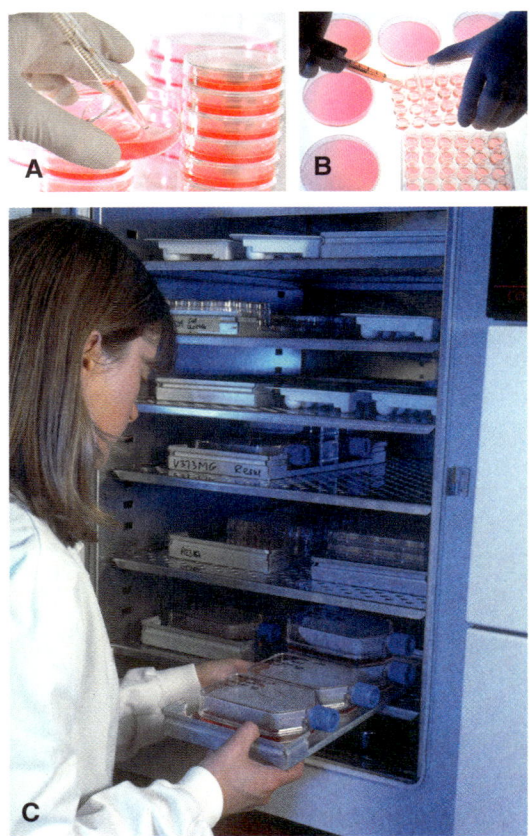

FIGURE 5-10 Human embryonic stem cells are grown in the lab (A) Stem cells growing in the lab must be periodically divided, diluted and distributed into additional plates in fresh medium. (B) For some types of experiments the stem cells are plated into multi-well sample trays. Gloves are worn and sterile conditions are observed to avoid contaminating the stem cells. Bacteria can overgrow a plate of stem cells in just a few hours. (C) Cultures of human stem cells are grown in a temperature-controlled incubator in the lab.

important step toward the development of stem cell based treatments for Alzheimer's disease, Parkinson's disease, diabetes, arthritis, and trauma such as spinal cord injury, stroke, and burns. Standardized stem cell lines are also important for applications in biotechnology such as increasing the accuracy of the testing performed for drug design and development, allowing researchers to avoid having to test drugs and chemicals on live animals.

The Promise of Human Embryonic Stem Cell Research

Embryonic stem cell research has already had major impact on science and the future of hESC research holds much promise for many areas of medicine. Scientists are developing techniques to induce pluripotent embryonic stem cells to differentiate into cells with specialized functions (Figure 5-11). In order to generate specialized tissues for medical transplant therapies, scientists must identify the specific molecular signals that reproducibly induce hESCs to differentiate into predetermined types of specialized cells.

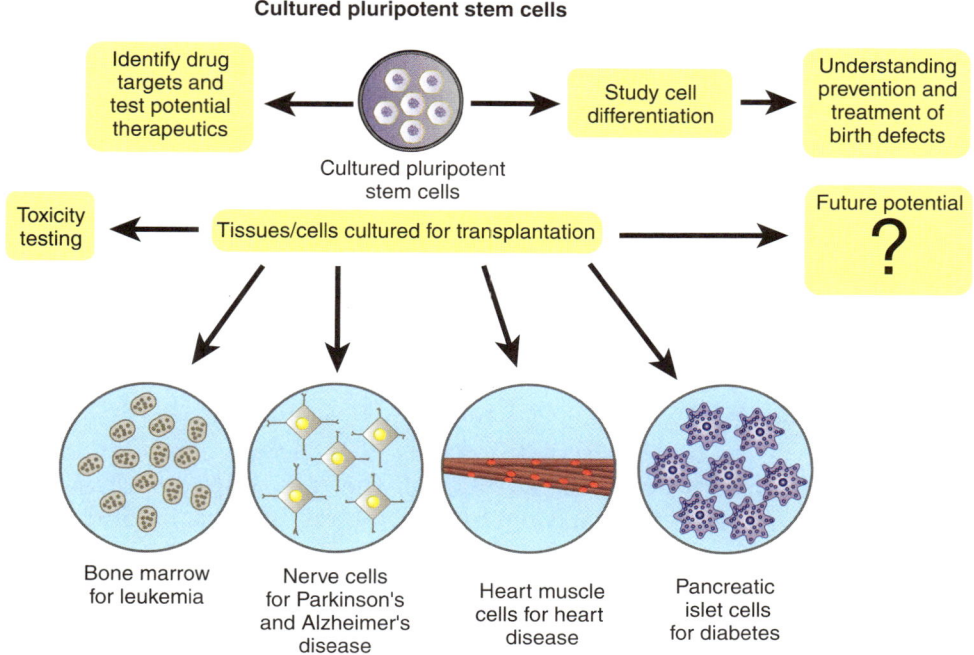

FIGURE 5-11 The potential promise of stem cell research Stem cell research has already had a huge impact on many areas of science, medicine and biotechnology. Stem cells have many applications but the most potential for stem cell research lies in generating specialized cells for tissue regeneration therapies. Stem cells can be used to produce new pancreatic cells and finally cure diabetes, make cardiac cells to restore the function of damaged heart muscle cells and regenerate nerve cells to treat neurological disorders like Parkinson's and Alzheimer's disease.

Researchers are working to provide reliable sources of specialized human cells for various types of medical treatments and other applications (Figure 5-11). Tissue regeneration and transplant therapies could potentially produce pancreatic cells that would finally cure diabetes and provide heart muscle cells to restore the damaged cardiac muscle after a heart attack. Hopefully transplanted spinal nerve cells will soon be used to regenerate damaged spinal cord cells in humans as was demonstrated in mice (see below). Specialized stem cell transplants could also benefit people with neurodegenerative diseases by replacing damaged nerve cells and restoring neurological function. Some degenerative diseases and disorders could potentially be effectively treated with transplants of either undifferentiated hESCs or partially differentiated progenitor stem cells. Eventually scientists plan to generate specialized cells using a patient's own stem cells, which have the advantage that the transplanted tissue will not be rejected by the patient's immune system.

To realize the amazing potential of embryonic stem cells, it is important for scientists to better understand the genes and proteins involved in the process of cell differentiation that generates different types of cells. Studying the molecular mechanisms responsible for the development of stem cells has helped researchers to create novel approaches to cell replacement and regeneration therapies. The ability to culture specialized cell lines allows researchers to develop differentiated cells with the features needed for many applications such as testing toxins and identifying new cellular drug targets. Stem cell research and information from the human genome sequence have also helped to better prevent, detect and treat birth defects.

Scientists continue to identify the signals that can induce ESCs to change developmental programs and generate an array of specialized cells. Internal signals controlled by the cell include the expression of specific genes and proteins such as transcription factors, genome reprogramming proteins, and other regulatory proteins. External signals also influence cell development including molecules secreted by other cells, chemicals in the local environment and physical contact between ESCs. Cell-to-cell communication can involve signaling proteins that are released from one cell and bind to specific protein receptors on the surfaces of the recipient cells (Figure 5-12). Receptor proteins are not secreted by the cell but are usually anchored to the outer surface of the plasma membrane, where they bind directly to the signaling molecules outside the cell.

Scientists are working to develop an effective embryonic stem cell therapy for cardiac disease. If successful the impact would be enormous because heart disease is the number one cause of death in the United States. Recent studies show that hESCs and some types of adult stem cells can be induced

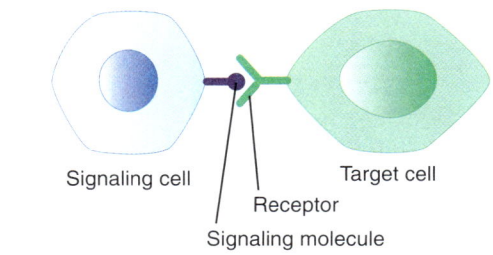

FIGURE 5-12 Cells communicate by cell-cell signaling (A) Cell-to-cell communication can involve signaling proteins that are secreted from one cell and travel to bind to specific protein receptors on the surfaces of the recipient cells, which transmits the signal. (B) Some cell communication involves physical contact between adjacent cells which brings together the adjacent cells.

to differentiate into heart muscle cells called cardiomyocytes. The hESCs are used to replace the weak and diseased heart cells with transplanted functional cardiac cells generated from progenitor stem cells in the lab. Cardiac cells have also been produced from other types of stem cells including myoblasts (muscle stem cells), mesenchymal cells, endothelial progenitor cells, and umbilical cord blood cells (Figure 5-6).

> **KEY CONCEPT**
> Embryonic stem cells give rise to multipotent progenitor stem cells that give rise to subsets of specialized cell types in the human body.

Federal Funding for Human Stem Cell Research Banned in 2001

The potential benefits and medical advances expected from stem cell research did not sway people from raising ethical objections to the use of stem cells obtained from human embryos. In 2001 President George W. Bush prohibited the use of federal funding to support embryonic stem cell research except for work performed on a small number of approved stem cell lines. This ban prevented any stem cell research supported by federal funds from being performed at universities and research institutions that typically sponsor the majority of biomedical research in this country.

Many U.S. scientists worked with the limited number of human embryonic stem cell lines available for research under the federal ban, but over time it became increasingly apparent that there were many problems associated with using the approved stem cell lines, and most could not be used for research or medical purposes. The ban on federal funding support for stem cell research caused many excellent scientists to pursue stem cell research abroad. Other researchers joined privately funded biotechnology companies that do not rely on federal funds to support embryonic stem cell research.

After President Obama took office, he issued an executive order to change the federal ban on stem cell research ban. Opponents of embryonic stem cell research sued the National Institutes of Health (NIH) to maintain the ban, but a federal appeals court ruled in 2011 that the research could continue until the court case is heard. Whatever the outcome of the federal court case, U.S. researchers and taxpayers have already lost over a decade of progress in stem cell research compared to the private companies and foreign competitors working to exploit human stem cells.

Many celebrities have been advocates for embryonic stem cell research, including actor Michael J. Fox who has Parkinson's disease and former First Lady Nancy Reagan. Former President Ronald Reagan died from Alzheimer's disease in 2004 (Unit 4). Stem cell research offers new hope for effective treatments for these and other degenerative neurological diseases.

Alternatives to Human Embryonic Stem Cells

The ban on stem cell research greatly expanded the search for less controversial sources of human ESCs that were appropriate for use in research and medical applications. The lucrative future potential of commercial applications from stem cell research and the prospect of new medical treatments inspired many scientists in both the public and private sector to search for new sources of human embryonic stem cells. Scientists have also identified new sources of adult stem cells that have the potential to develop into specialized cells.

Human ESCs are Found in Amniotic Fluid and in the Umbilical Cord

In 2003, researchers discovered rare embryonic stem cells in the amniotic fluid surrounding the fetus as early as 10 weeks after conception (Unit 4). Further tests confirmed that the hESCs found in human amniotic fluid express the Oct4 transcription/reprogramming protein, one of several transcription factor proteins required for ESCs to maintain an undifferentiated, pluripotent state. Additional studies show that the ESCs collected from amniotic fluid can generate specialized brain and muscle tissues that are not be rejected when used in a transplant treatment. Amniotic stem cell lines can also be frozen for future use and provide a personalized lifetime bank containing potentially any tissue.

The umbilical cord is the physical lifeline between the developing embryo or fetus and the mother and carries the maternal blood supply containing oxygen and nutrients throughout pregnancy (Figure 5-13). Umbilical cord cells were first used in 1988 in transplants to treat leukemia, and now umbilical cord cell transplants are a routine effective treatment for childhood blood cancers. The small numbers of stem cells collected from umbilical cord blood are cultured in the lab to generate large numbers of healthy progenitor blood stem cells. A typical stem cell treatment consists of about one million cultured blood stem cells that are introduced into the patient by intravenous injection. The multipotent progenitor blood stem cells take up residence in the patient's bone marrow, where they give rise to healthy platelets, red and white blood cells, bone cells, and cartilage cells (Figure 5-8).

During pregnancy the placenta attaches the developing embryo to the uterus and to its source of nutrients and oxygen. Fetal blood contains precursor stem cells that can differentiate into different types of blood cells. After birth the blood in the placenta and the umbilical cord can be collected and frozen for long term storage (Figure 5-13). Human progenitor stem cells obtained from umbilical cord blood can give rise to a large number of specialized cells with many applications in medicine.

Umbilical stem cell transfusions can effectively treat leukemia and other immune system diseases in adults and children without a significant risk of rejection by the host. ESCs from several different sources have the developmental potential to differentiate into many types of cells that could potentially be used to replace damaged or diseased tissues. Stem cells obtained from placental tissues have been used to generate pancreatic islet cells, hematopoietic stem cells, cardiomyocytes, neurons, and hepatocytes (liver cells).

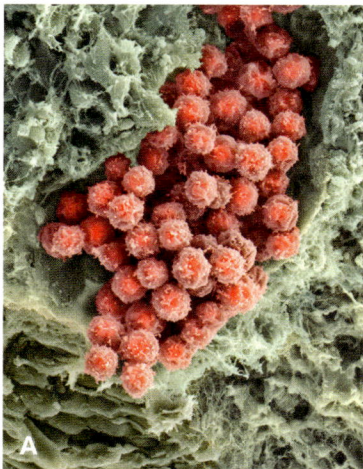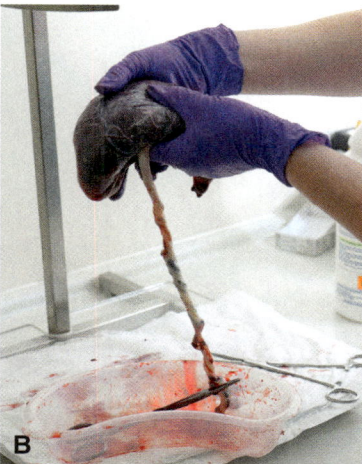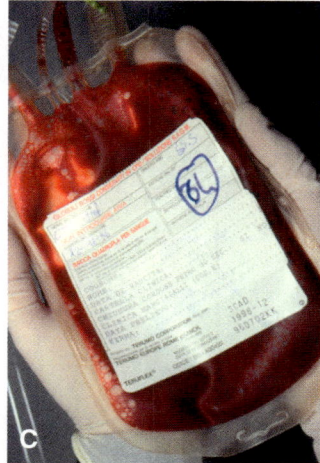

FIGURE 5-13 Umbilical cord blood contains multipotent stem cells (A) Fetal blood contains stem cells, which differentiate into progenitor stem cells that form different types of blood cells. (B) After birth the blood can be drained from the placenta and the umbilical cord, collected and frozen for long term storage. (C) Blood transfusions using umbilical stem cells can effectively treat leukemia and immune system diseases. The cells derived from umbilical cord cells do not usually trigger rejection by the host. (Photographed at the Milan Cord Blood Bank in Italy.)

Bone Marrow Stem Cell Transplant

For over 30 years, human blood cell cancers have been treated with bone marrow (stem cell) transplantation and chemotherapy. The hematopoietic stem cells located in the bone marrow of the long bones in adults generate many different types of specialized blood cells that function in the body. In someone with the blood cancer leukemia, the white blood cells grow out of control and become the most prominent cells in the bloodstream (Figure 5-14). In some cases the patient's own bone marrow stem cells can be harvested, treated to kill the cancer cells and then reintroduced into the patient by intravenous injection (Figure 5-15). The transplanted blood marrow stem cells return to the bone marrow and produce healthy red and white blood cells in the patient. Some patients require a bone marrow transplant from a genetically unrelated donor, but there is always a risk that the patient will reject the transplanted cells even when there is a genetic match between the appropriate transplant alleles from the donor and recipient.

Ethical "Extra Embryos" and Extra ESCs?

In 1978 the first "test tube baby", Louise Joy Brown, was born in Great Britain and by the year 2010 over 4 million infertile couples have conceived children with the help of *in vitro* fertilization (IVF). Recent advances in human reproductive technology now permit couples who are at risk of transmitting a fatal genetic disease to a child can use a combination of IVF and preimplantation genetic diagnosis (PGD) to avoid the risk of passing a lethal mutant disease

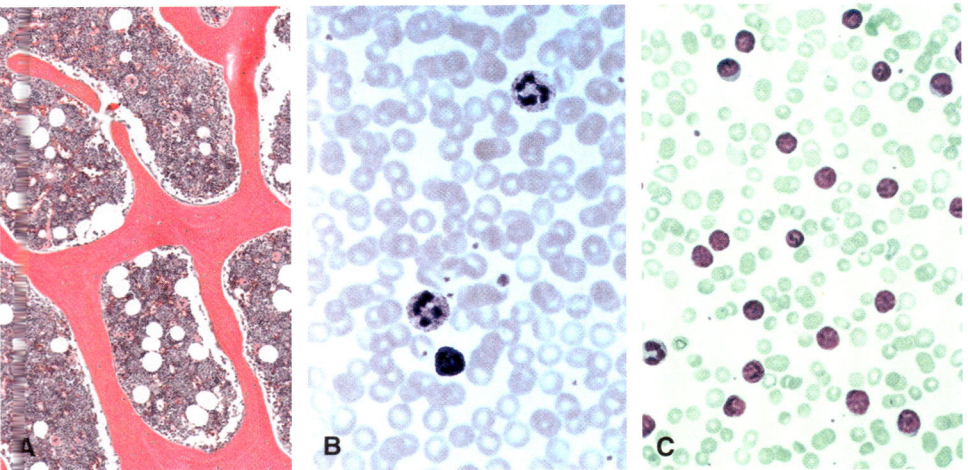

FIGURE 5-14 Bone marrow stem cells generate new blood cells (A) Healthy bone marrow contains the erythroid and myeloid stem cell lines that generate new blood cells. (B) The main cellular components of healthy human blood are erythrocytes (pink) and monocytes (purple). (C) This blood sample came from a patient with leukemia and shows the many white blood cells with large nuclei (dark purple), which are a hallmark of this type of blood cancer.

gene to offspring (Unit 4). The process of preimplantation genetic diagnosis could be used to generate 'extra' embryos and stem cells for future medical purposes.

In PGD a single cell is removed from an 8-cell embryo that is tested for the presence of a specific mutant gene(s) or chromosome abnormality. If the 7-cell embryo does not carry the mutant gene in question, the 7-cell embryo continues to grow and develop into a normal blastocyst embryo that implants into the uterus. The single cell removed from the 8-cell embryo is viable

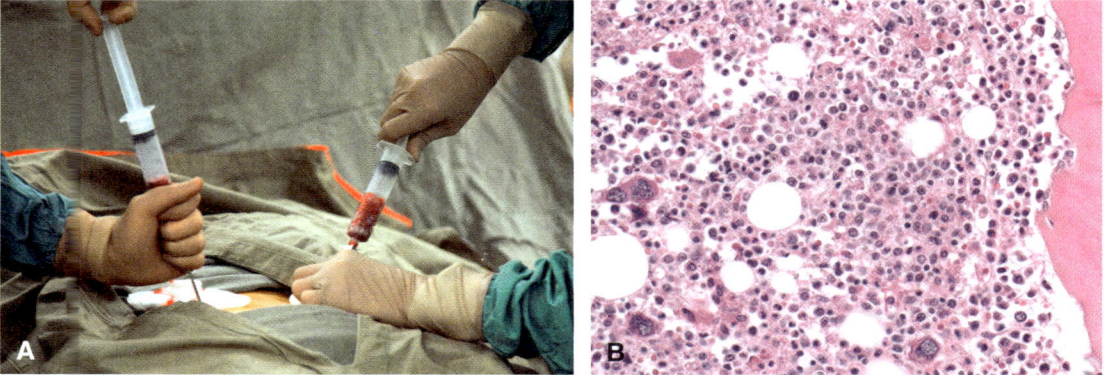

FIGURE 5-15 Bone marrow transplant used to treat blood cancer (A) Bone marrow stem cells are removed from the patient and treated to kill the rapidly dividing cancer cells. The treated bone marrow cells are transplanted back into the patient where they grow and produce healthy red and white blood cells. (B) These are bone marrow cells from a leukemia biopsy (bone is pink).

and can develop into a fully formed normal blastocyst embryo, offering a possible strategy to generate an "extra" blastocyst embryo that could be frozen for future pregnancies or to provide genetically matched stem cells from future treatments. Human PGD testing has been available for many years, but the idea of growing "extra" human embryos as sources of personalized ESCs would be extremely controversial is raised here in the context of ethical questions and discussion.

Induced Pluripotent Stem Cells

In 2007 scientists developed a new type of human stem cells called induced pluripotent stem (iPS) cells. This novel approach generates stem cells that very closely resemble authentic ESCs and have the enormous advantage that iPS cells do not involve human embryos (Figure 5-16). To better understand how stem cells function, the scientists studied the genes expressed by the embryonic stem cells and found that ESCs produce specific transcription factor proteins and reprogramming proteins. The researchers stained the ESCs with different antibodies and detected stem cell–specific proteins, including, Nanog, Sox2, Kif4, and Oct4. These proteins function to control gene

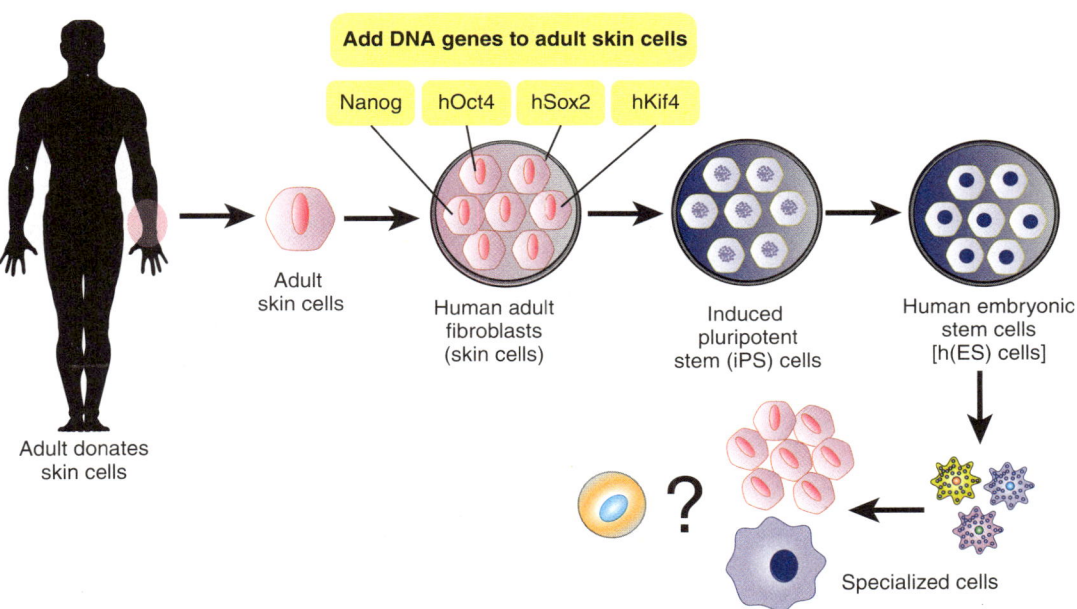

FIGURE 5-16 Adult skin cells induced to become pluripotent stem cells DNA strands coding for the transcription factor (TF) and re-programming (remodeling) proteins Oct4, Sox2, Kif4, and Myc2 were added to fully differentiated adult skin cells growing in plastic dishes. The TF protein genes were expressed from the added DNA, altering the expression of genes in the skin cells and changing the developmental state of the skin cells. The genes needed to create an embryonic stem cell were activated and genes that maintain the functions of an adult skin cell were repressed. Amazingly the induced pluripotent stem (iPS) cells made from adult skin cells adopt the physical and functional characteristics of normal human ESCs. In addition, iPS cells can differentiate into specialized cells such as nerve and muscle, etc.

expression in stem cells and to maintain the pluripotent developmental potential of the ESCs.

The research results have implicated a small number of TF and reprogramming proteins that function to establish and maintain the pluripotent state of the human embryonic stem cells. Based on this evidence, the scientists treated adult skin cells growing in the lab with DNA molecules that code for the four human proteins, Sox2, Kif4, Myc2, and Oct4 (Figure 5-16). The genes were introduced into the cells on a virus DNA vector that carried the genes into the nucleus for expression. These proteins made in the adult skin cells caused the skin cells to express different genes and begin to change shape and de-differentiate. The altered skin cells then started to divide and began to generate the unspecialized induced pluripotent stem (iPS) cells that closely resemble authentic embryonic stem cells.

Pluripotent iPS Cells can Generate Specialized Cells

Most scientists were very surprised that researchers could successfully convert adult skin cells into undifferentiated iPS cells just by adding specific DNA genes. Then the Su-Chun Zhang research group (University of Wisconsin–Madison) went the next amazing step and successfully converted the iPS cells into specialized human nerve cells. Researchers are working to determine the complete range of specialized cell types that can be generated from the iPS cells. Although human iPS cells closely resemble authentic human ESCs, additional research is needed to determine if the induced pluripotent stem cells have the same unlimited developmental potential as authentic embryonic stem cells. Additional research is underway to determine the complete range of specialized cell types that can be generated from different iPS cells.

In the future, researchers plan to use iPS cells made from a patient's adult skin cells to generate personalized, genetically matched specialized cells and tissues that can be transplanted without risk of rejection by the host patient. Researchers need to better understand the molecular mechanisms responsible for the conversion of adult skin cells into undifferentiated iPS cells. It is also important to understand the fundamental changes that take place when a genome is reprogrammed to direct the development of specialized cells or to convert differentiated cells into unspecialized iPS cells (Unit 6).

The future potential of iPS cells will not be fully realized until researchers have characterized iPS cells from different sources of adult cells and generate standardized iPS cell lines that produce the appropriate specialized donor cells. This relatively new technology has not yet been tested sufficiently for routine use of iPS cells in medical therapies, but further studies are underway to develop future applications based on iPS cells.

Further research on the molecular mechanisms that drive cell differentiation and genome reprogramming will continue to reveal the fundamental differences between iPS cells and ESCs. Researchers and clinicians will determine how these differences might impact the use of iPS cells in transplant therapies and treatments. Researchers will continue to explore the mechanisms involved in converting adult somatic cells into iPS cells, but for the foreseeable future, iPS cells will not replace authentic hESCs as the focus of basic ESC research.

Scientists Engineer Safer iPS Cells

The first iPS cells were produced using a DNA vector to carry the appropriate genes into the nuclei of the adult skin cells, and the vector DNA integrated randomly into the skin cell genome DNA. The reprogramming proteins expressed in the skin cells interact with other proteins in the cells and profoundly alter the shapes and functions of the iPS cells.

There were some drawbacks of the iPS cells. The random insertion of the vector DNA into the skin cell genome DNA can potentially disrupt an essential gene with serious consequences. Then scientists discovered that the continued expression of the reprogramming genes in the iPS cells can promote cancer development in the stem cells. This complication threatened to significantly limit the usefulness of the iPS cells for therapeutic applications. However, scientists realized that once the iPS cells were reprogrammed and exhibiting pluripotent characteristics, the iPS cells no longer needed continued expression of the reprogramming genes to generate new specialized cells. The scientists designed a modification of the bacterial Cre enzyme/loxP system that works in eukaryotic cells and permits the reprogramming genes to be efficiently removed from the iPS cell genome (Figure 5-17).

The reprogramming genes inserted into the iPS cell genome are flanked on either side by loxP sites, short DNA sequences that are recognized by the bacterial Cre enzyme. The Cre enzyme that is made in human iPS cells cuts the genome DNA only at the two loxP sites, deleting the chromosome DNA between the two loxP sites and removing the reprogramming genes from the iPS cell genome (Figure 5-17). Researchers now routinely use the Cre enzyme/loxP system to remove reprogramming genes (and other genes) from eukaryotic chromosome DNA.

> **KEY CONCEPT**
> The ability of human embryonic stem cells to make specialized cells and tissues that could cure diabetes, save eyesight and help paralyzed people to walk is very addictive despite the controversy surrounding the use of human embryos. Now innovative and creative molecular researchers have created induced pluripotent stem (iPS) cells that function just like embryonic stem cells, but without the stem cells!

hESCs are a Promising Potential Treatment for Parkinson's Disease

Parkinson's disease destroys nerve cells in the *substantia nigra* located deep within the human brain (Figure 5-18). The nerve cells destroyed by Parkinson's disease normally produce dopamine, a neurotransmitter that carries nerve signals across the gaps (synapses), where the ends of two nerves

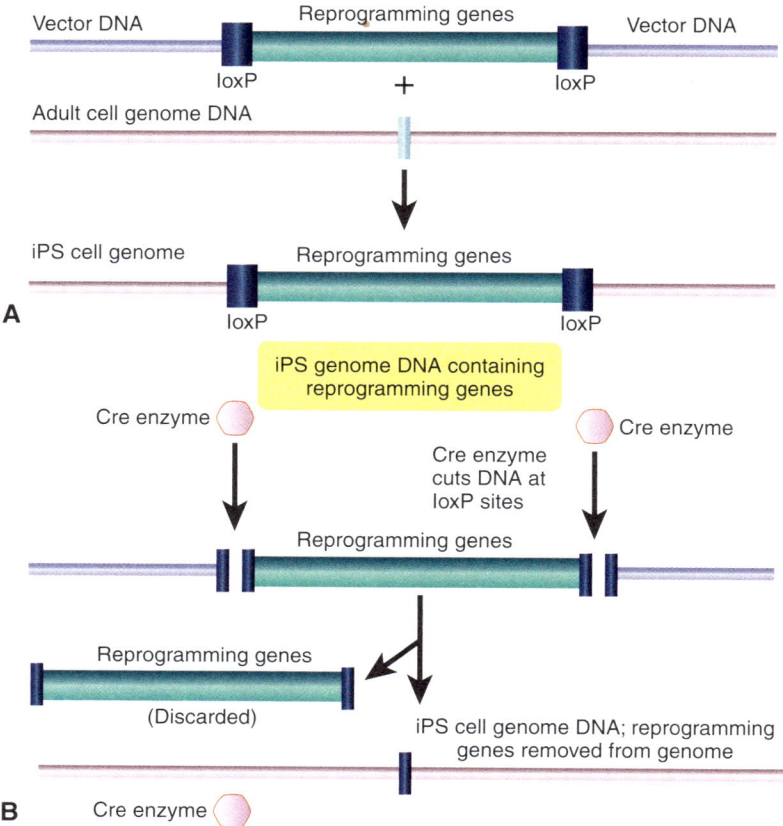

FIGURE 5-17 Reprogramming genes are removed from iPS cells The bacterial Cre enzyme/loxP system was modified to work in eukaryotic cells and can be used to delete any gene from any genome. In this case it was used to delete the reprogramming genes from the iPS cell genome. (A) The reprogramming genes that were inserted into the human genome are flanked on either side by short DNA sequences called loxP sites. (B) The Cre enzymes are expressed in the iPS cells and cut the human chromosome DNA only at the two loxP sites, deleting the DNA between the two loxP sites that contains the reprogramming genes. The chromosome DNA is ligated at the site of the deletion.

come together (Figure 5-19). When the nerve signal reaches a synapse, it triggers the release of neurotransmitters from the end of the presynaptic nerve cell. The neurotransmitters travel across the gap and bind to receptors on the surface of the postsynaptic nerve cell, transmitting the signal from one nerve to the next nerve cell.

The neurological symptoms of Parkinson's disease are caused by insufficient levels of the neurotransmitter dopamine, which inhibits the transmission of nerve impulses across synapses. In addition, aggregates called Lewy bodies form in the nerve cells in the brains of Parkinson's disease patients. Lewy bodies contain tangled filaments of alpha-synuclein proteins that contribute to nerve degeneration. The symptoms of Parkinson's disease begin with a mild mask-like face, general stiffness, and shaking and tremors in the arms and legs; these worsen with time until eventually the patient becomes immobile, often leading to death. Many therapies to treat Parkinson's disease have been developed, including drugs, which

UNIT 5 Embryonic Stem Cells and Animal Nuclear Cloning

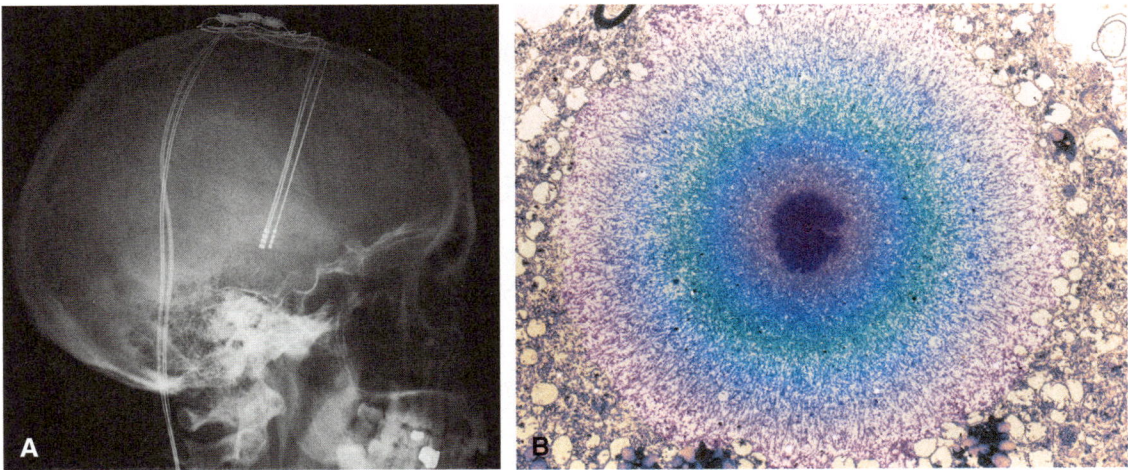

FIGURE 5-18 Parkinson's disease damages one region of the brain Deep brain stimulation (DBS) is used to treat Parkinson's disease because this disease destroys the nerves in a localized region deep inside the brain called the substantia nigra. During DBS the neurosurgeon places electrodes deep inside the brain and adjusts the placement of the electrodes with X-rays as shown in the image. The electrodes in the brain connect to wires on the scalp (top) that lead to a battery. Electrical impulses are sent through the wires to stimulate different parts of the brain. (B) This Lewy body formed in a nerve cell in the brain, and is a diagnostic feature of Parkinson's disease. Lewy bodies contain mostly filaments made of alpha-synuclein protein (blue/pink). The accumulation of Lewy bodies is associated with progressive nerve degeneration.

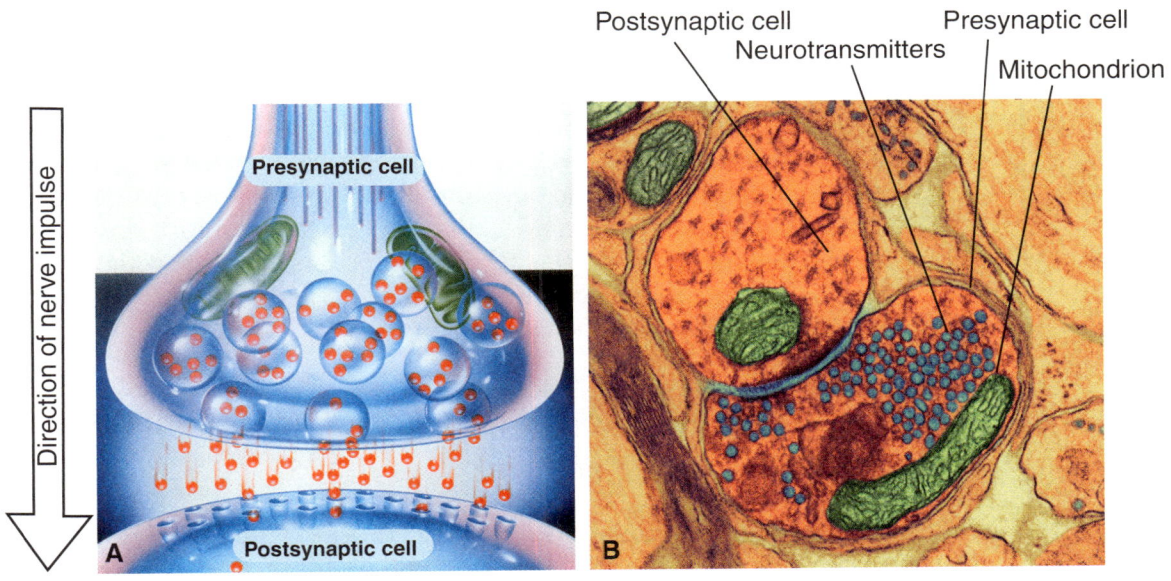

FIGURE 5-19 Neurotransmitter chemicals transmit signals across nerve synapses! (A) The synapse is the gap at the junction between two nerve cells. The nerve impulse is traveling from the nerve at the top (pre-synaptic) to the nerve at the bottom (post-synaptic). When the nerve signal reaches a synapse it triggers the release of neurotransmitters contained in vesicles (blue) at the end of the presynaptic cell. The neurotransmitters are released and travel across the gap, binding to receptors on the surface of the post-synaptic cell. (B) The nerve impulse travels in one direction across the synapse from the pre-synaptic cell (red, lower right) to the post-synaptic cell (red, upper left). The nerve signal triggers the release of neurotransmitters contained in vesicles (blue) from the presynaptic cell. The mitochondria (green) provide energy for the cells.

are important to alleviate symptoms, and deep brain stimulation, which is delivered by electrodes implanted in the brain.

The early experimental treatments for Parkinson's disease involved transplanting human fetal brain cells into the brains of adult Parkinson's patients. In these trials, the fetal cells survived for some time after the transplant and secreted dopamine, which temporarily improved the Parkinson's symptoms. The success of this approach was encouraging, although some patients suffered side effects from the fetal cell transplants, such as excessive uncontrolled motions (dyskinesia). Like embryonic stem cell research, fetal cell research also raises ethical and social issues and is another source of public and scientific controversy.

The prospect of developing effective nerve cell transplant therapies to treat Parkinson's disease is especially promising because the nerve damage in Parkinson's disease is localized to only one region of the brain. The most straight-forward transplant treatment for Parkinson's disease involves replacing the damaged nerve cells in the brain with transplanted healthy nerve cells, followed by tests to confirm that the transplanted nerve cells are making dopamine.

Researchers collected adult skin fibroblasts from Parkinson's disease patients and converted the adult skin cells into iPS cells. The reprogramming genes were removed from the iPS cell genome by means of the Cre/loxP system, and the "safe" iPS cells were induced and generated nerve cells producing dopamine. The next step involved transplanting the specialized nerve cells into the brains of Parkinson's patients to replace the neurons destroyed by the disease. Over time the transplanted cells should grow and eventually re-establish the supply of dopamine needed to relieve the characteristic symptoms of shaking, tremors, and balance problems.

Recently researchers studying the *substantia nigra* regions of the brains of healthy elderly people and Parkinson's patients found that the cells from both sources have mutant mitochondrial genome DNA with large deletions. The frequency of the mtDNA mutations increases with the age of the patient and contributes to the death of nerve cells in certain degenerative neurological diseases. Researchers do not yet understand the mechanisms responsible for creating these mtDNA deletions or why cells containing these mtDNA mutations accumulate in the region of the human brain destroyed by Parkinson's disease.

Stem Cell Transplant can Successfully Treat Spinal Cord Paralysis in Rats

For decades scientific dogma has stated that damaged nerve cells could not regenerate and regain nerve function. The prospect of transplanting viable nerve cells into a human patient to treat paralysis has long been considered impossible because fully differentiated cells cannot develop in reverse and become an embryonic cell. However, recent research on human neurodegenerative diseases shows that transplanted nerve cells can survive and function temporarily in the human brain. Now stem cell technology provides medicine with the ability to generate any type of genetically matched cells and tissues for use in human transplantation therapy.

To better understand how hESC therapy might be used to treat spinal cord injuries in humans, it is important to know how the brain and spinal cord work together in the central nervous system (CNS). In vertebrates, the spinal cord is a long, bundle of nerves and support cells that starts in the medulla region of the brain and is threaded through the openings in the vertebrae of the backbone along the length of the vertebral column. The spinal cord transmits nerve signals between the brain and the rest of the body by managing three major types of nerve signals: the motor information transmitted from the brain and down the spinal cord to the limbs, the sensory information that is sent from the body up the cord to the brain, and the nerve reflexes that are coordinated by the spinal cord. Transmitting nerve impulses along the spinal cord requires that the neurons transmit the signal along the spinal cord and the oligodendrocyte cells (glial cells) protect the nerve cord and ensure that the nerve impulse is properly transmitted.

The oligodendrocyte cells perform this essential support for the neurons by synthesizing the myelin sheath that surrounds the axon projections extending from the neurons (Figure 5-20). Myelin is the layer of special fat deposited by the Schwann cells that wrap around the nerve fiber. The layers of myelin insulate the axon and increase the speed of the nerve impulses as they are transmitted along the long projections (axons) of adjacent nerve cells. The myelin sheath acts like the insulation around an electrical wire and is necessary for the efficient transmission of nerve impulses along the spinal cord. A spinal cord injury not only damages nerve cells but also kills the glial cells, causing the axons to lose the myelin sheath protection. Even nerves that are located long distances from the site of the injury can lose myelin, extending the symptoms of paralysis far from the primary site of the injury.

Dr. Hans Keirstead (University of California, Irvine) is a leading scientist in stem cell research who pioneered the use of embryonic stem cells to treat spinal cord injuries. Keirstead and other researchers used lab rats to explore the possibility that damaged spinal cords can be repaired by injecting new oligodendrocyte cells into the damaged spinal cord tissues. The transplanted

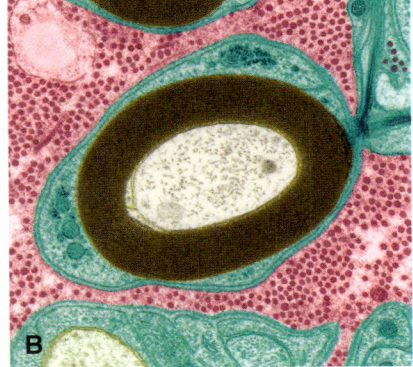

FIGURE 5-20 Axon nerve fibers are covered with myelin insulation (A) Axon nerve fibers are completely covered with myelin (pink) to insulate the fiber and increase the speed of nerve impulse transmission. (B) A cross section of a nerve fiber shows the myelin sheath (brown), a fatty layer surrounding the axon (white). The myelin sheath is deposited on the nerves by the Schwann cells (green). The nerve fibers are surrounded by a delicate connective tissue containing collagen (pink).

oligodendrocyte cells function to remyelinate the nerve cell axons and restore nerve transmission. In 2005, Keirstead used hESCs to generate progenitor human oligodendrocyte cells, which were then used to treat the spinal cord injuries in paralyzed rats.

In 2006 Keirstead used IVF to produce rat blastocyst embryos and to grow the rat ESCs in the lab. The motor neuron cells that were generated by the ESCs in response to certain drugs were transplanted into the injured spinal cords of paralyzed lab rats. Keirstead's research team performed tests to find out whether the transplanted motor nerve cells would actually grow and replace the damaged spinal cord cells. A few months after transplanting the cells into the paralyzed rats, the researchers looked for signs that the donated stem cell–derived neurons had survived transplantation and were functioning. They found clear evidence that the transplanted nerves had made connections to the rat's muscle cells, and had restored partial function to the paralyzed hind legs. Additional analysis on the experimental rats showed that some transplanted nerve cells had differentiated into functioning motor neuron cells. The transplanted nerve axons grew into the lower legs of the host animals and made functional connections with the leg muscle cells, as confirmed by a 50% increase in strength tests. Additional evidence from electrophysiology and behavioral studies showed that the treated rats had partially recovered from hind limb paralysis and confirmed that the transplanted ESC-derived neurons made functional connections with the nervous system. Compared with humans, the rat is a relatively small test animal, so the tests to determine if the transplanted nerve cells can connect over longer distances must await research on large-animal models.

The First Human Stem Cell Clinical Trials Begin in the United States

Stem cell research has the potential to provide very effective cell replacement treatments based on understanding the molecular mechanisms that convert pluripotent embryonic stem cells into a large array of highly specialized human cells. Scientists are learning to control the molecular processes behind stem cell differentiation in order to develop more effective cell transplant therapies for human diseases, injuries, and disorders. The current demand for human organs for transplantation far exceeds the number of donated organs, making it even more important for scientists to learn how to efficiently convert stem cells into the desired type of specialized cells and tissues in the lab. It is hoped that research on pluripotent stem cells will offer future sources of human transplant tissues developed to specifically treat chronic diseases such as diabetes and degenerative neurological diseases such as Alzheimer's disease and Parkinson's disease.

In January 2009 the U.S. FDA approved a 21,000-page application submitted by Keirstead and Geron, a biopharmaceutical company, and launched the first Phase I human clinical trials using hESCs to treat spinal cord injuries. The clinical protocol involves injecting the hESC-derived oligodendrocyte precursor cells into the site of injury in the patient's spinal cord. In animals, this approach resulted in a significant increase in the re-myelination of the nerve axons, and the paralyzed rats exhibited improved movements after only one dose of oligodendrocyte precursor cells. Research shows that both mesenchymal stem cells and neural stem cells can generate the oligodendrocyte precursor cells that hold such promise for successful future spinal cord transplant therapies.

The approved Phase I clinical trials to treat spinal cord damage with stem cell transplants will take place at several U.S. medical centers and involve only 8 patients who must undergo stem cell transplantation within 2 weeks of spinal cord injury. Phase I clinical human trials are designed to evaluate the safety of a particular treatment or drug, so the hESC Phase I trial is not expected to cure spinal cord injuries. However, patients might experience some improvements many months after treatment.

Scientists at Geron Corporation began to study human embryonic stem cells in 1999, just before the ban on federal funding for stem cell research took effect in 2001. The hESC research performed by the Geron scientists was supported by funds that were not restricted by the federal ban. The hESCs used in the clinical stem cell trials in 2011 were obtained from human blastocyst embryos donated by couples pursuing infertility treatments (Unit 4). The hESCs collected from blastocyst embryos were treated in the lab to convert them into the appropriate progenitor neural stem cells needed for transplantation into the injury site. The progenitor neural stem cells populate the injured tissues and generate new nerve cells in the region of the damaged spinal cord. The transplant patients in the clinical trials will receive immune-suppressive drugs to guard against transplant rejection and will be monitored for medical follow-up for 14 years.

Clinical Trials to Treat Blindness in Humans with hESCs

Advanced Cell Technology (ACT) is the only other U.S. biotechnology company given FDA approval to conduct clinical trials involving human embryonic stem cells. ACT scientists used hESCs to generate the specialized retinal pigment epithelium (RPE) cells needed to treat Stargardt's macular dystrophy (SMD). SMD is a genetic disease affecting children who inherit two mutant copies of the ABCA4 gene and become blind as the disease progressively destroys photoreceptor cells in the retina (Figure 5-21). The layers of retinal tissue at the back of the eye contain the rod and cone photoreceptor cells. The light entering the eye, which is focused and inverted by the cornea and the lens is projected onto the retina at the back of the eye (Figure 5-21). When rhodopsin encounters light, the protein changes shape and sends a signal through the layers of retinal cells to the optic nerve and the brain.

The ACT Phase I/II human clinical trial will assess safety, check for possible side effects, and will involve only 12 patients at three U.S. hospitals. The human stem cell lines for this clinical trial were derived from a single cell taken from a three-day-old human embryo. The ACT clinical trial will test the effectiveness of the RPE cell transplants derived from the reprogrammed iPS cells.

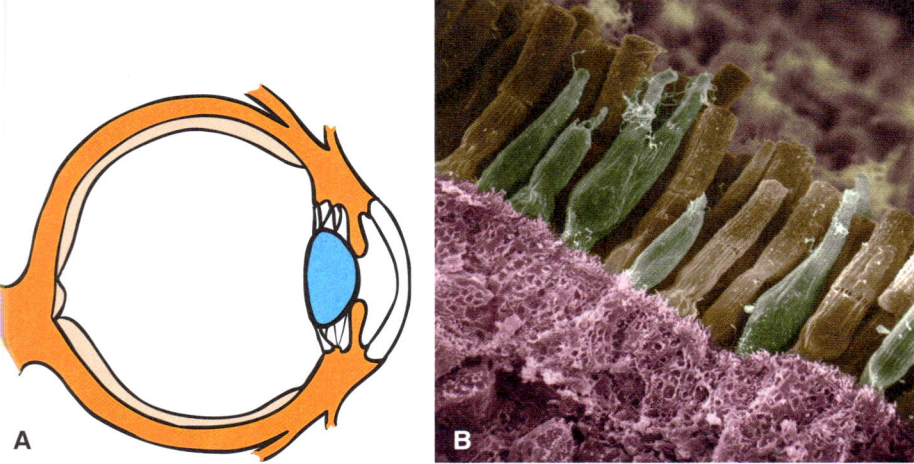

FIGURE 5-21 Retinal stem cells differentiate into photoreceptor cells (A) Rods and cones are part of the retina tissue on the back of the eye. (B) Rods (brown) are long nerve cells that permit vision in dim light, and cones (green) are shorter cells that detect color. These photoreceptor cells send signals to the brain through the optic nerve.

Human Cloning and Somatic Cell Nuclear Transfer (SCNT)

The general term *cloning* refers to the process of making an identical copy of the original. In DNA science there are three important and very different types of cloning:

(1) Recombinant DNA cloning,
(2) Reproductive animal cloning and
(3) Therapeutic animal cloning.

A basic understanding of the differences among these cloning methods is essential to be able to understand the applications of the different cloning methods to solving problems in research and medicine.

Recombinant DNA cloning refers to the process of inserting a DNA fragment from one biological source (for example, a plant gene or a human gene) into a DNA vector to propagate the cloned DNA (see Figure 1-22). Many types of vectors are available for DNA cloning, and many commonly used vectors are based on the circular double-stranded plasmid DNAs that are found naturally in bacterial cells. When a gene is cloned, the DNA fragment containing the gene of interest is released from chromosome DNA by cleavage with a restriction enzyme, and then the DNA fragment is inserted into the vector DNA, cut only once with the same restriction enzyme. The DNA fragment to be inserted into the vector is then joined to the vector DNA with a ligase enzyme. The final product is a recombinant DNA molecule that contains a bacterial DNA plasmid vector and a DNA insert with eukaryotic exons (Unit 1).

Recombinant DNA cloning experiments usually have one of two overall goals. The first goal is to make many identical copies of cloned DNA by growing the recombinant plasmids in bacterial host cells, which was the only method available to produce many copies of cloned DNA until PCR amplification was introduced.

The second goal involves cloning DNA into a type of vector engineered to express the cloned genes and proteins in the host cells. In cases where a eukaryotic gene is to be expressed in prokaryotic host cells, a cDNA copy containing only exons is made from the eukaryotic gene. The cDNA copy is inserted into a vector engineered with a prokaryotic-specific promoter in front of the cDNA to express the cloned coding region in bacterial cells and a transcription stop signal located after the coding region that ensures efficient expression (Unit 3). In most cases a cDNA copy is used to express eukaryotic genes because a cDNA copy of an mRNA contains just exons and lacks the introns in the gene. As a result, the cDNA is commonly used to express a eukaryotic gene and protein in eukaryotic host cells that might not correctly splice the pre-mRNA from the cloned gene. The cDNA copy of the eukaryotic gene is inserted into a vector flanked on one end by an appropriate eukaryotic promoter and on the other end by a poly(A) addition signal (AATAAA) to ensure that the RNA copies expressed from the cloned gene will be processed correctly and efficiently transported to the cytoplasm for protein translation.

Animal Cloning Differs from Recombinant DNA Cloning

Animal cloning refers to the process of making a genetically identical copy of a living creature using somatic cell nuclear (SCNT) (Figure 5-22). The SCNT method involves the physical transfer of a nucleus from a donor cell to a recipient cell that lacks a nucleus. It is very important to understand the different purposes, applications, and ethical concerns associated with the different methods of animal cloning, therapeutic and reproductive cloning.

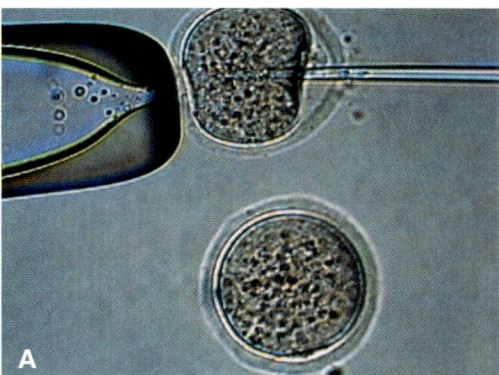 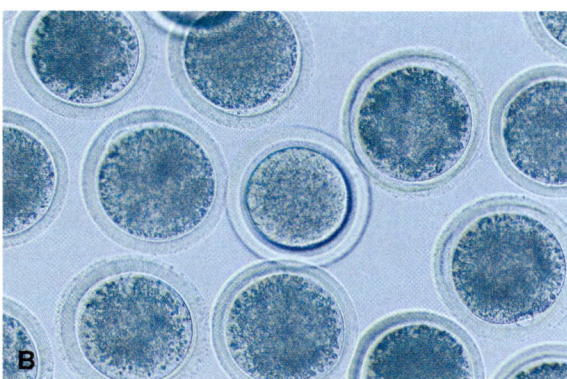

FIGURE 5-22 Nucleus transplanted into "empty" egg cell (A) Light micrograph of an "empty" mouse egg gently held in place by a glass vacuum pipette (left). The nucleus was previously removed from the empty recipient egg cell. The tiny donor nucleus is barely visible in the needle inserted into the empty egg (right). The donor nucleus is injected into the empty egg through the needle. The cloned egg cell will develop in the lab into a blastocyst embryo that contains pluripotent embryonic stem cells. (B) These egg cells were collected from a cow (Bos taurus) used by a company in Argentina to clone cows that produce human growth hormone (hGH) in their milk.

Human cloning is not legal in the United States and earlier claims of success from other countries have not panned out. The inefficiency of the animal cloning process and the lack of scientific understanding about the genetic and cellular mechanisms underlying reproductive cloning have made U.S. scientists take a strong stand against cloning humans. The American Medical Association and the American Association for the Advancement of Science have issued formal public statements against human reproductive cloning, and the U.S. Congress has legislated a ban on human cloning.

SCNT is Central to:

- therapeutic cloning and
- reproductive cloning.

Therapeutic cloning involves using hESCs to generate specialized cells for transplantation that are a genetic match to a specific patient (Figure 5-23). This approach would avoid the risk of tissue rejection. In contrast, as indicated by the name, reproductive cloning has a very different goal, the live birth of a cloned animal that is genetically identical to the animal that donated the nucleus.

The therapeutic and reproductive cloning processes both begin with SCNT when the nucleus of the unfertilized egg cell is removed and discarded to prepare the recipient "empty" egg cell for nuclear transfer (Figure 5-24). The donor nucleus from an adult body cell is then transferred into the empty egg cell,

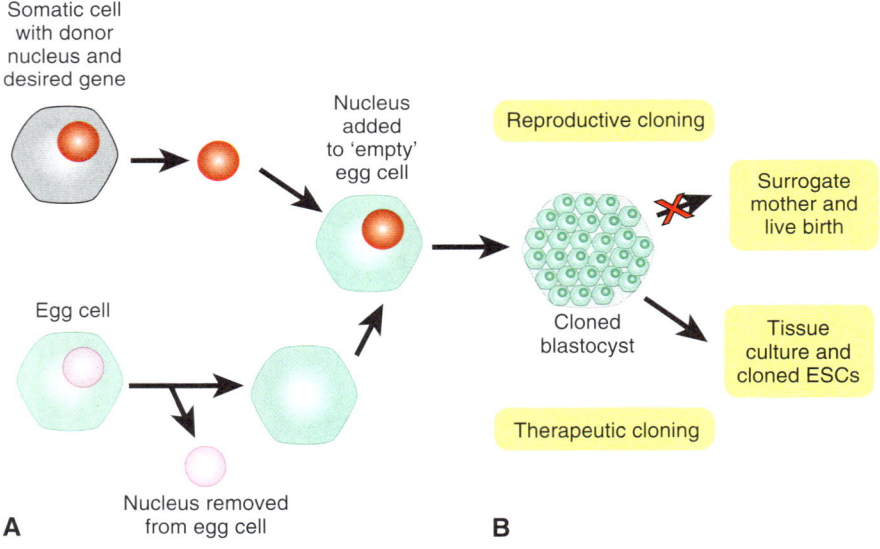

FIGURE 5-23 Reproductive and therapeutic cloning are very different (A) Reproductive cloning and therapeutic cloning have very different outcomes but they begin the same way. The nucleus (dark pink) from a somatic body cell is transferred into an 'empty' egg cell that has already had the nucleus removed. The resulting 'cloned egg cell' grows in the lab into a 'cloned' blastocyst embryo. (B) In therapeutic cloning the blastocyst is used to obtain the hESCs that are a genetic match to the patient and can be induced to develop into specialized cell types for personalized treatments. Alternatively for reproductive cloning the human cloned blastocyst would be implanted into a surrogate female to carry the pregnancy. Of course, human reproductive cloning is not legal in the U.S.

which is now a "cloned" egg cell because it contains a genetically unrelated donor nucleus.

The cloned egg cell containing the transferred adult nucleus begins to divide, making 2 cells, then 4 cells, then 8 cells as the cloned zygote develops into an early embryo and by the 4th or 5th day after fertilization, a cloned blastocyst embryo has formed (Figure 5-24). The inner cell mass in the blastocyst embryo generates a small number of pluripotent embryonic stem cells, which then give rise to the three embryonic germ layers that in turn will generate all of the organ systems and tissues in the human body (Figure 5-24).

Therapeutic and Reproductive Cloning Diverge at the Blastocyst

The fate of a cloned blastocyst embryo depends entirely on the longer-term goal of the therapeutic or reproductive cloning procedure. If the goal is reproductive cloning, the cloned blastocyst embryo will be implanted in a surrogate female and the pregnancy carried to term.

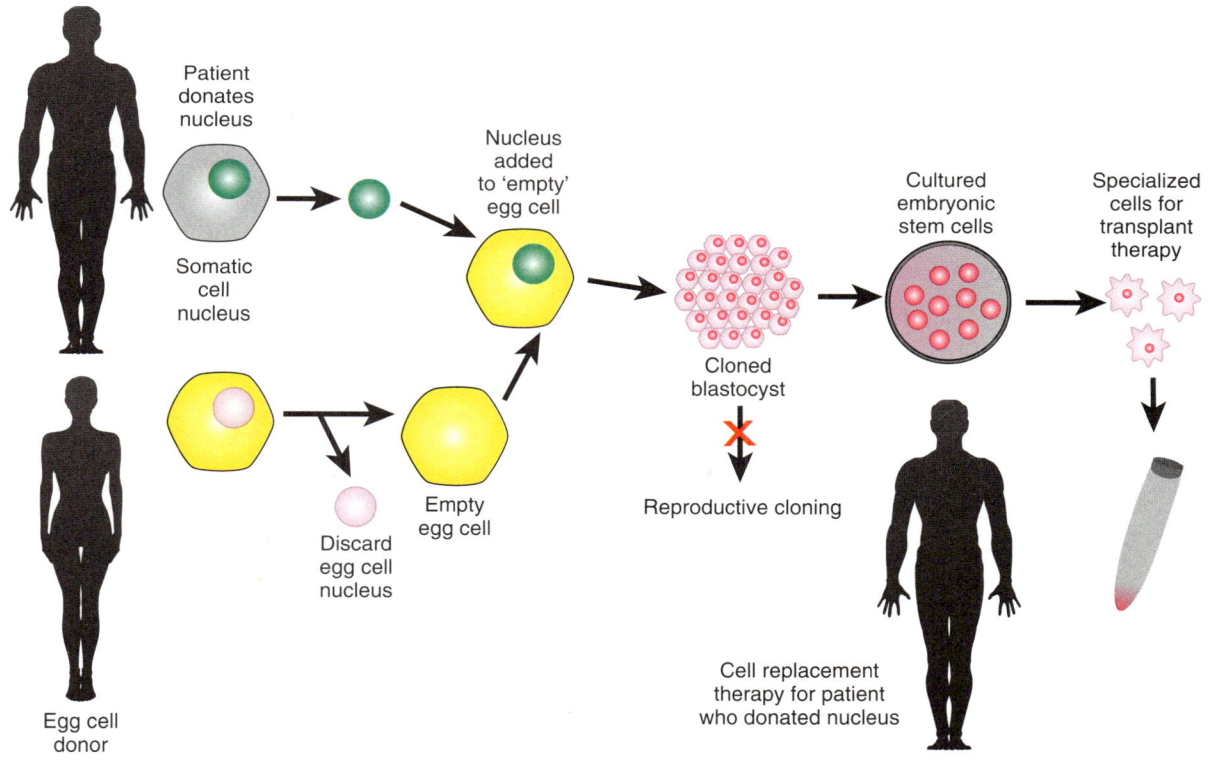

FIGURE 5-24 SCNT, therapeutic cloning and cell replacement therapy In therapeutic cloning the 'cloned' blastocyst is used to obtain cloned hESCs that are a genetic match to the patient. The hESCs can be stored or induced to develop into specialized cell types to treat the genetically matched nuclear donor. Alternatively the human cloned blastocyst could be implanted into a surrogate female to achieve pregnancy. Human reproductive cloning is not legal in the U.S.

> **KEY CONCEPT**
> Reproductive cloning requires that the blastocyst be implanted into the uterus of a surrogate female. Of course, human reproductive cloning is not legal in the United States.

The Goal of Human Therapeutic Cloning

Therapeutic cloning is very different from reproductive cloning, although the blastocyst embryos are made in the same way in the two methods. In therapeutic cloning the embryos are never implanted in surrogate females; instead, they are used to obtain cloned embryonic stem cells (Figure 5-24). These cloned ESCs could potentially be used to generate specialized cell types that are genetically identical to the patient who donated the somatic cell nucleus and could be used in transplant therapies without any risk of tissue rejection by the patient.

The goal of therapeutic cloning is to create ESCs with the characteristics required for specific applications in medical research and clinical therapies. Human embryonic stem cells are of key importance because these pluripotent cells can generate any type of specialized cells in the human body. Eventually therapeutic cloning will be used to routinely generate human ESCs with the potential to develop into any type of specialized tissues for transplantation into the genetically matched recipient patient. Although this approach is not yet available in clinics, in the future this approach could significantly reduce the risk of transplant rejection and decrease the number of people with transplants who would otherwise require a lifetime of taking immunosuppressive medications.

Cloning Live Animals, Hello Dolly, The Cloned Sheep!

Dolly was the first mammal ever cloned from the nucleus of an adult cell (Figure 5-25). Dolly was created by British scientist Ian Wilmut at the Roslin Institute in Edinburgh, Scotland in 1997 using SCNT and reproductive cloning technology. Dolly was cloned from a donor nucleus retrieved from the udder skin cell of a 6-year-old Finn Dorset sheep. The donor nucleus was transferred into an empty egg cell from a Scottish Blackface ewe (a female sheep). The resulting cloned zygote developed into a blastocyst embryo in the lab, and was then implanted into the uterus of an unrelated surrogate female sheep to develop until birth. The process used to clone Dolly and other cloned animals is not very efficient. In fact, Dolly was the only live lamb born from the many donor nuclei that were transferred into 277 empty sheep eggs, allowed to develop into blastocyst embryos, and implanted into surrogate female sheep.

Dolly was the genetic clone of the adult Finn Dorset sheep that donated the skin cell nucleus. However, animals cloned by SCNT, including Dolly, are not completely identical to the animal that donated the nucleus. In addition

 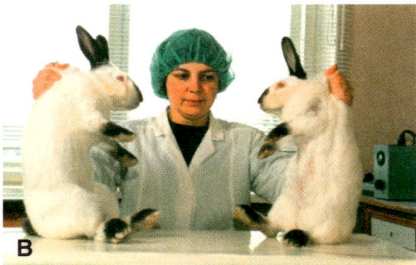 

FIGURE 5-25 Professor Ian Wilmut created Dolly, the first cloned animal (A) British scientist Ian Wilmut is shown with Dolly the first cloned animal. (B) Genetically consistent cloned rabbits are important for use as animal models to study human diseases. (C) This veterinary scientist is shown with two of five genetically-identical cloned sheep.

to the genes that are inherited on the chromosomes in the donor nucleus, a cloned animal also inherits the cytoplasm from the recipient egg, including the mitochondria, and any associated mitochondrial genetic diseases.

Dolly was a cloned animal and she gave birth to six lambs conceived the "old-fashioned way". However, Dolly also developed the types of illnesses usually associated with advancing age, including cancer and arthritis. Unfortunately, Dolly died in 2003 when she was only 6 years old, an early death for Finn Dorset sheep, which usually live to be about 12 years old. Dolly's early death raised additional questions about the long term health of cloned animals. Dolly was cloned using the nucleus of a fully differentiated adult skin cell carrying an "old" genome, suggesting that Dolly was old before her time because she had been cloned with an "old" adult genome. Researchers are working to identify the factors that affect the health of cloned animals, including a better understanding of the fate of the donor nucleus and DNA genome in the recipient cytoplasm. Scientists have begun to unravel the genome reprogramming process that occurs after fertilization or nuclear transfer, and involves epigenetic mechanisms such as DNA and protein methylation (Unit 6).

The SCNT cloning process represents a substantial transition for the adult nucleus, which is removed from the cytoplasm of a fully differentiated cell and transferred into an empty cytoplasm. The nucleus of an adult skin cell (including humans and sheep), typically expresses those genes required for the cell to exhibit the characteristics and functions of an adult fibroblast skin cell. When the adult skin cell nucleus is transferred into the cytoplasm of an empty egg cell, the human nucleus brings with it the genetic potential to direct the development of all types of cells needed in the human body.

The human genome contains all the necessary genes to instruct the development and growth of a human being from fertilization until death. However, despite having access to all of the human genes, at that point the donated adult nucleus must repress expression of the skin-cell-specific genes and activate expression of the genes needed to begin embryo development. These genome-wide changes involve nuclear reprogramming and genome modification. It is amazing to consider that the nucleus survives the physical trauma of extraction and transfer into the empty egg cell and then the genome is successfully reprogrammed to change the genetic instructions to those required to direct embryo development. At the same time the adult skin cell must abandon its previous plan of increasing signs of old age like cell shrinkage and eventual loss of anchoring molecules allowing the dead skin cells to flake off of the surface of the body. Scientists are actively studying the molecular details of the mechanisms involved in genome reprogramming.

> **KEY CONCEPT**
> Dolly was the first animal to be cloned using reproductive cloning, which involved transferring a nucleus from an adult sheep's udder cell to an empty sheep egg cell. As a result Dolly was said to be 'cloned with an adult nucleus'. Some scientists proposed that the age of the adult nucleus has consequences in reproductive cloning and explains why Dolly dies at an early age.

Over 20 different animal species have been cloned using the reproductive cloning method since Dolly was born (Figure 5-25). In 1998 scientists in Hawaii used this approach to produce 50 genetically identical, cloned mice starting with a single mouse nucleus. Access to research organisms like cloned mice that are genetically identical except for a single gene mutation is ideal for research. Although it might seem unnecessary to clone rabbits because they reproduce so well on their own, genetically characterized, cloned rabbits also offer important consistency for research on human diseases (Figure 5-25). In 2006 Iowa researchers successfully cloned ferrets to use as model research animals that are especially important to aid in the study of certain respiratory diseases in children.

The prospect of producing genetically modified animals and plants to help alleviate the worldwide food shortage is a controversial idea. In 1999 a bull steer that is normally bred for meat was cloned for the first time, focusing the international debate on the safety of genetically modified meat from cloned animals. In 2005 Chinese scientists cloned the first water buffalo with the goal of using cloned animals as a future source of meat and milk. The Bio Sidus biotechnology company in Argentina successfully cloned the first domestic cow in 2002 and more recently cloned genetically modified cows that carry the human gene needed to make human growth hormone (hGH) and secrete the hGH hormone in their milk (Figure 5-26).

Normally a mating between a horse and a donkey produces a sterile mule. In 2003 the first cloned mule was born, named Idaho Gem. The first horse was cloned in Italy in 2003 and is shown in Figure 5-27 along with her identical twin and surrogate mother, a mare called Stella Cometa. The same scientists used a nucleus from Fontanella Zapping, a champion Holstein Fresian dairy cow, to clone a calf, Zapping in 2002. The first cloned cat called Copy Cat was born in 2002 and a few years later South Korean scientists cloned the first dog, an Afghan hound named Snuppy. Commercial services

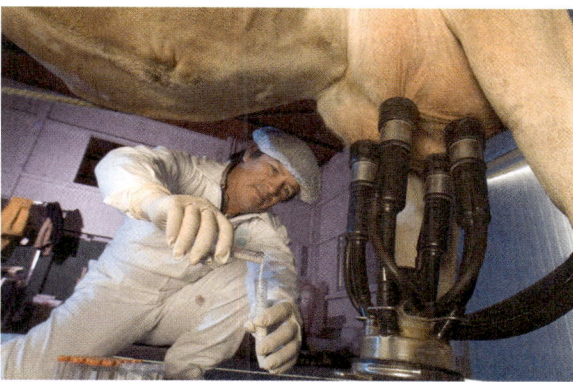

FIGURE 5-26 Grow really strong with genetically modified cow's milk Researchers transfer milk samples into test tubes after the milk is retrieved from the cow's udder. This genetically modified Jersey dairy cow (Bos taurus) is part of a project using cloned cows to produce recombinant products in their milk, in this case, human growth hormone (hGH). This research is underway at the first company to clone a cow, BioSidus, a biotechnology company in Argentina.

FIGURE 5-27 My best friend is a clone! (A) The first cloned horse (Equus caballus) Prometea stands in front of her surrogate mother, a mare called Stella Cometa. (B) Prometea's best friend is a cloned cow named Zapping (left), whose genetic and nuclear mother is Fontanella Zapping, a champion Holstein Fresian dairy cow.

now offer to clone anyone's domestic pets for a price, including dogs, cats, guinea pigs, and hamsters.

Cloned Pigs Make Organs for Use in Human Transplantation

An important future goal of therapeutic cloning is to generate tissues and organs for effective human transplant therapies. In brief, the nucleus from the patient's cell is transferred into an enucleated egg, and the resulting cloned egg divides to make a blastocyst embryo containing ESCs. The cloned

ESCs make specialized cells and tissues that are a genetic match to the patient and as a result could be transplanted into the patient without risking rejection. In the future, scientists hope to produce genetically matched organs from cloned human embryos that would significantly decrease the requirement for organ donations. Of course, there are many obstacles to overcome in order for cloned organs to become reality, including the need for more efficient nuclear transfer techniques and improved methods to grow functional organs in the lab.

The cloned Christmas piglets, born on Christmas Day 2001, represent an important first step toward engineering animals that grow animal organs specifically for transplantation into human patients. Because of the chronic shortage of human organ donations for medical transplantation, scientists have searched for alternative sources of organs. Pigs are a potential source of organs for transplantation, in part because pig organs are similar in size, structure, and development to human organs. The use of pig organs also offers scientists a strategy to avoid tissue rejection, which is a major complication of tissue and organ transplants. When a human organ is donated to a recipient human patient, the transplanted organ is often rejected by the recipient's (host's) immune system because the donated tissue or organ is seen as "foreign" by the recipient's body (graft-versus-host rejection). The Christmas piglets represent a major advance because they were engineered to lack a gene encoding a pig protein that causes rejection when the pig tissues are transplanted into a human recipient.

Scientists created "knock-out" pigs lacking a specific pig gene that causes the pig tissues to be rejected by the human immune system. The scientists first removed this gene in pig cells grown in the lab and then used the engineered pig cells to generate the cloned animals used to grow the pig organs for transplantation. Like humans, pigs inherit two copies of every gene; in 2002, a British biotechnology company produced genetically engineered double-knock-out pigs lacking both copies of the gene involved in transplant rejection.

Animal Cloning to Save Endangered Species

Reproductive cloning and interspecies nuclear transfer methods were first used to clone endangered animals in 2001 by Advanced Cell Technologies (ACT). The process of cloning endangered and extinct animals faces significant challenges, including the fact that the egg cell and the surrogate female carrying the pregnancy are both different species from the animal to be cloned. ACT scientists used interspecies nuclear transfer to clone an endangered gaur ox (*Bos gaurus*). The nucleus from a male gaur ox was transferred into an empty egg cell and the resulting blastocyst embryos were implanted into the uterus of a surrogate mother, a female domestic cow that carried the gaur ox fetus to term. The gaur ox calf called Noah was born in 2001 but unfortunately died within days of birth from an infection that was not related to the cloning. In 2004 a different group of scientists successfully cloned an endangered African wildcat using a similar approach with a domestic cat as the surrogate mother. A rare species of wild sheep, the mouflon (*Ovis orientalis musimon*), which lives only on the islands of Sardinia and Corsica was cloned by replacing the nuclei of egg cells from a domestic sheep with nuclei from mouflon somatic body cells. The cloned blastocyst embryos were implanted into the uterus of surrogate sheep and the first cloned mouflon sheep was born in 2000 at the University of Teramo in Italy. In 2007, South Korean scientists reported the successful cloning of several grey wolves, another endangered species.

Plants are extremely important experimental systems used for agricultural research, and scientists have developed several model plant systems for study. An international standard for plant genetic research is *Arabidopsis thaliana*, which is also called Thale cress and mustard weed (Figure 5-28). *Arabidopsis* plants are appropriate for plant research because they grow rapidly, produce large amounts of seeds, and are easily genetically manipulated. In addition the *Arabidopsis* genome has been sequenced and the genes involved in many plant functions have been identified (Figure 5-28).

Ethical Questions, Concerns, and Risks of Animal Cloning

Reproductive cloning is a very inefficient process, often requiring more than 100 nuclear transfer events to produce a single viable cloned animal. The health of the cloned animals is also a concern, since some have been born with compromised immune systems, infections, and tumors. Research in Japan showed that some cloned mice are not healthy and often die before the normal lifespan, and a healthy appearance in young cloned animals does not necessarily predict long-term survival. In 2002 researchers reported that about 4% of genes in the cloned mice did not function correctly.

There are many ethical, legal, and social issues surrounding the general concept of animal cloning. Many people do not like the idea of using animals

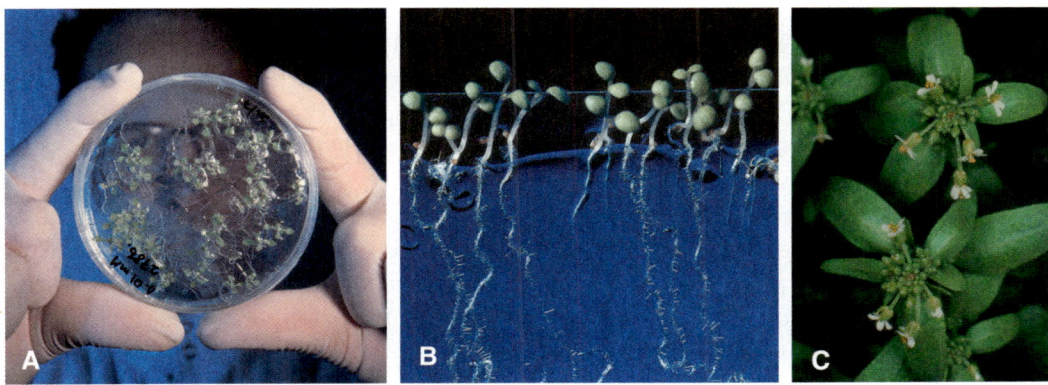

FIGURE 5-28 A tiny weed is an international genetic standard (A) This agar petri dish plate contains tiny Arabidopsis thaliana seedlings carrying random genome mutations. The mutant seedlings will be tested for traits that will help identify the plant genes involved in certain processes. (B) A. thaliana seedlings have very fine root hairs that absorb water and nutrients from the soil into the roots. (C) A. thaliana is a model plant used as an internationally genetic standard for plant genetic research.

engineered to be "incubators" for growing replacement parts for humans. However, others are persuaded that animal cloning is an acceptable solution when they consider that nuclear transfer methods can also be used to try to save endangered animals.

Epigenetic Genome and Nuclear Reprogramming

The nucleus is much more than just a storage compartment meant to protect and replicate the chromosome DNA. During a typical eukaryotic cell cycle, the nucleus undergoes many dramatic changes in shape, structure, and composition that accompany the various stages of the cell cycle. Of course, the dynamic nucleus is involved in many functions, including gene expression and regulation, DNA replication, and accurate chromosome distribution to daughter cells each time the cell divides (segregation). In most eukaryotic cells (except budding yeast cells) the nuclear membrane disassembles during mitosis when the spindle fiber apparatus assembles from tubulin proteins in the cell. The spindle fibers attach to the centromeres on the chromosomes, which help to move the chromosomes to opposite poles so that the two daughter cells will inherit equal numbers of chromosomes and equal numbers of genes. At the end of cell division, the nuclear membrane reassembles around both groups of chromosomes to create a nucleus in each of the two new daughter cells.

Some of the health problems encountered with cloned animals result from changes that occur when the donor genome and nucleus are reprogrammed in the recipient cell cytoplasm. Normally embryo development begins when the sperm and egg genomes come together at fertilization. However, the situation in animal cloning is very different, because the cloned embryo has received an adult nucleus that is expressing genes to maintain a fully differentiated state. Reprogramming the new genome in the recipient cytoplasm is an essential process that is controlled in part by epigenetic mechanisms that do not change the base sequence of the genes.

During the differentiation of unspecialized stem cells into highly differentiated cells, the nuclei must accommodate major changes in gene expression patterns as well as changes in cell structure and function. A similar situation exists during fertilization when the egg and sperm nuclei fuse together into the zygote nucleus with 46 human chromosomes. The normal zygote nucleus is preprogrammed to express the genes needed for embryo development. In comparison, the transplanted adult nucleus is very different from the zygote nucleus, even though the DNA sequences of the genomes are identical. The adult skin cell nucleus is programmed to make skin proteins, but it is surrounded by a cytoplasm on the brink of starting embryogenesis.

The adult nucleus must be reprogrammed to suppress the expression of the adult skin cell genes and turn on expression of the genes needed for embryo development. Reprogramming the adult nucleus and genome during cloning is inefficient, as is the reprogramming process that transforms specialized adult cells into induced pluripotent stem (iPS) cells. The idea of reprogramming specialized adult cells into pluripotent embryonic cells goes against central dogma, which stated for decades that terminally differentiated cells and tissues—such as the pancreas, heart, bone, liver, and nerve—cannot revert into less specialized types of cells. However, now scientists have shown that terminally differentiated cells can be induced to become "young" unspecialized iPS cells, which has inspired a great deal of excitement and interest in the area of regenerative research and transplantation medicine.

Scientists have proposed that defects in the genome reprogramming process probably contribute to the long-term health problems sometimes observed in cloned animals. The apparent early demise of Dolly the cloned sheep has raised questions about the molecular reprogramming mechanisms at work in the transplanted nucleus in the cytoplasm of the empty egg. The adult skin cell nucleus expresses the genes for skin cells, but once the adult nucleus is physically transferred into an empty egg cytoplasm, the adult genome must be reprogrammed to take on the task of directing embryo development.

The molecular components and biochemical mechanisms responsible for reprogramming a transplanted adult nucleus continue to be the subject of intense research. Reprogramming the human genome requires epigenetic changes in the genome that alter gene expression, and the epigenetic changes can potentially persist for many generations (Unit 6).

Unit 5 Questions

1. When embryonic stem cells undergo the process of "self-renewal" they make more:
   a. highly developed cells for specialized tissues
   b. embryonic stem cells with unlimited developmental potential
   c. stem cells that will generate different blood cells
   d. cells by meiosis with half the number of chromosomes in each cell

2. In the case of embryonic stem cells (ESCs), the term pluripotent describes that the ESCs can:
   a. Divide over and over again to make many (pluri-) copies of the same ESCs
   b. Dominate (-potent) the other cells by growing on top of nearby cells in the developing embryo
   c. Generate all the different types of specialized cells, tissues, and organs in the human body
   d. Generate the different types of specialized cells and tissues needed to form the placenta

3. Which of the following are true characteristics of a human blastocyst embryo?
   a. The blastocyst embryo is hollow
   b. The blastocyst embryo is formed on the 4th to 5th day after fertilization
   c. The blastocyst embryo contains the ESCs growing in the ICM
   d. All of the above
   e. None of the above

4. Which of these events must occur before a pregnancy can take place?
   a. The sperm cell must fertilize the egg cell.
   b. The blastocyst embryo must be implanted in the uterus.
   c. The zygote must begin cell division to make an embryo.
   d. a and c only.
   e. a, b, and c.

5. Some people consider embryonic stem cell research to be very controversial because:
   a. This research requires scientists to destroy the unfertilized egg cells donated by infertile couples
   b. This research involves the creation of embryos that are half-human and half-mouse
   c. This research involves ESCs obtained from donated human embryos
   d. This research involves making extra copies of cloned human embryos for unauthorized public use

6. Human stem cells have been identified from sources other than human embryos, including:
   a. Umbilical cord blood
   b. Amniotic fluid
   c. Skin cells that have been reprogrammed into stem cells
   d. All of the above
   e. None of the above

7. Nerve cells in the human body contain an outside coating or sheath called:
   a. Axon
   b. Myosin
   c. Mitochondrion
   d. Myelin

8. Therapeutic cloning and reproductive cloning methods diverge at which stage (step)?
   a. The decision to either implant the blastocyst embryo or harvest ESCs
   b. Fertilization, when the sperm penetrates the egg cell
   c. Gastrulation, when the three germ layers emerge in the embryo
   d. At the point of transition from an embryo to a fetus in the uterus

9. Somatic cell nuclear transfer (SCNT) refers to:
   a. The development of the body cells that do not contain nuclei
   b. The ability of bacterial cells to take up DNA from the surroundings
   c. The process by which viruses introduce DNA into cells in the body
   d. The process normally used by ESCs to develop into more highly specialized cells
   e. The process of transferring the nucleus from a body cell into an egg cell that lacks a nucleus

10. Which type of process involves cloning DNA into vectors that are grown in bacterial cells?
    a. Reproductive cloning
    b. Therapeutic cloning
    c. Recombinant DNA cloning
    d. Stem cell cloning

UNIT 6

DNA Forensics and Epigenetic Reprogramming

Skin Fingerprints and DNA Fingerprints: What is the Difference?

People are usually very familiar with the idea of skin fingerprints, the swirls, ridges, and whorls of skin on the end of our fingers used to make ink prints that are unique to each individual person (Figure 6-1). These skin fingerprints can be used to identify people because each individual skin fingerprint has distinguishing patterns. Skin fingerprints have been used to successfully track criminals and identify murder victims for over 100 years. To identify people, law enforcement agencies maintain large collections of known skin fingerprints in databases. In the past it was necessary to compare the known and unknown skin fingerprints by eye, a tedious and time-consuming task. Now digitized skin fingerprints are stored in computer databases and unknown prints are compared to known fingerprints by computer. Skin fingerprints have some obvious limitations compared to DNA fingerprints. For example,

FIGURE 6-1 Do genetically identical twins have identical fingerprints? (A) For decades people have been identified using fingerprints made with ink. (B) Fingerprints are an example of the biometric data used to identify individuals. (C) These identical twins and have identical DNA genomes. Do they have identical fingerprints?

221

human decomposition will rapidly obliterate the skin on the finger tips, destroying fingerprints. Also skin fingerprints are not visible if the person wears gloves or wipes away the fingerprints, although it is now possible to lift fingerprints from the inner surfaces of latex gloves.

Do Genetically Identical Twins Have Identical Skin Fingerprints?

Skin fingerprints are used to identify people because individual humans have unique fingerprint patterns, but do identical twins have identical skin fingerprints (Figure 6-1)? If genetic mechanisms completely control the development of skin fingerprints, the answer to this question is yes, but if other factors in addition to genetics control the development of skin fingerprints, it is possible that identical twins have different skin fingerprints.

The development of genetically identical twins starts after an egg is fertilized by a sperm, when the fertilized egg is programmed to divide into a two-cell embryo but instead the fertilized egg divides into two physically separate, genetically identical cells. Both cells start to divide and they give rise to two genetically identical embryos that develop into two separate, genetically identical twins (Figure 6-2). These are called monozygotic identical twins because they are the products of one (mono) zygote (fertilized egg) that split into two zygotes with identical human DNA genomes.

The swirls and whorls in human skin fingerprints develop as the embryo grows in the uterus and are influenced not only by genes but also by the local environment of each embryo in the uterus, including nutrition, blood pressure, the position of the embryo, and the rate that each individual embryo's fingers grow in the first three months. The skin fingerprints of identical twins are similar, but not identical, and they have small differences in the whorls and ridges that are sufficient to allow their skin fingerprints to be used to distinguish between genetically identical twins.

> **KEY CONCEPT**
> Human skin fingerprints and some other features of the developing embryo and fetus are not entirely determined by genetics but are also influenced by environmental factors.

Humans usually inherit only two copies of each gene, one from Mom and one from Dad. However, some genes exist in more than one copy per genome because of copy number variation (CNV) (Unit 2). The scientists that sequenced the human genome discovered that the copy number of some genes varies widely among human genomes and that copy number variation was even observed between the genomes of genetically identical twins. Genes on a chromosome exhibit CNV when a limited region of chromosome DNA is copied by mistake, either once or several times during normal DNA replication. The affected genes exist in the genome at more than just two copies per diploid cell, and as a result the genes

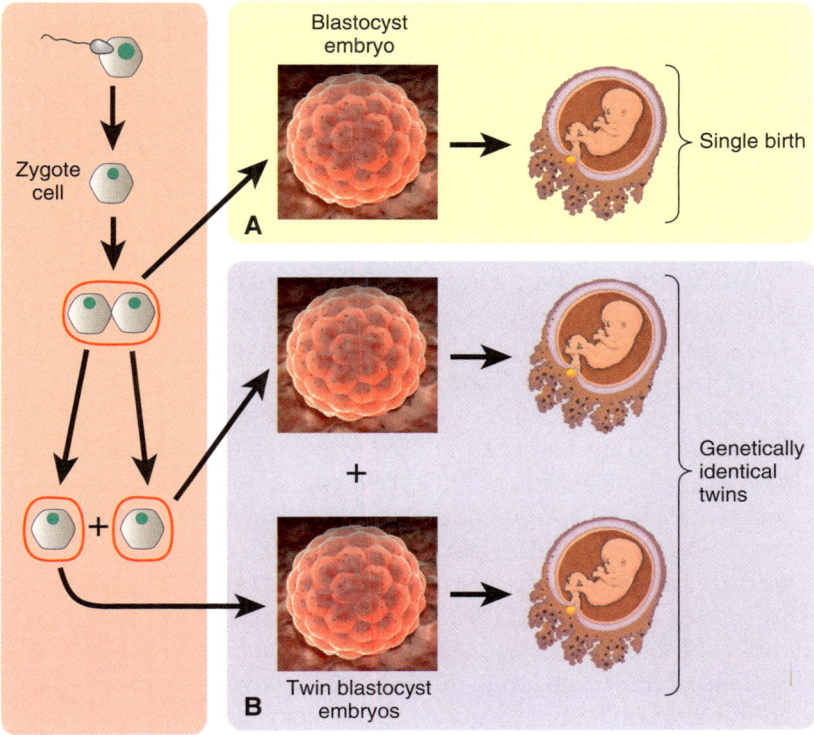

FIGURE 6-2 Identical twins are produced from one fertilized egg (A) The zygote cell formed by fertilization contains the unique genome of a new individual. Normally the zygote develops into a single blastocyst that produces one baby. (B) Genetically identical twins that are generated from one fertilized zygote cell that splits to give rise to two blastocysts and eventually two babies are monozygotic twins.

present at more than two copies per cell are likely to be expressed at higher levels that the other genes. It is possible that the replication error leading to copy number variation might occur in the genome of one of the twin embryos but not in the genome of the other twin. The identical twins have identical DNA genomes except for the regions duplicated by CNV, which would lead to overexpression at the protein level.

DNA polymorphisms and SNPs often occur naturally in regions of the chromosome DNA that contain stretches of repeated DNA sequences. During DNA replication, when the DNA strands containing repeated sequences base-pair, the repeated DNA bases sometimes slip, making short single stranded loops that can potentially add or remove repeated DNA sequences in the final replicated chromosome. These replication mistakes occur naturally in dividing cells but do not usually alter the coding regions of genes or affect the functions of proteins. This is because the huge numbers of repeated sequences in the human genome do not typically contain genes or code for proteins. These types of replication events do help to expand and propagate base pair changes in the polymorphic repeats that are passed to the next generation.

DNA Fingerprinting Technology Relies on DNA Genome Differences

It is now possible to directly compare the DNA sequences of different human genomes to identify specific individuals in the human population. Genome sequencing has shown that human genomes have almost identical and DNA sequences, and even related genomes contain single base pair changes. DNA fingerprinting technology depends on detecting a handful of relatively rare single base pair differences among the 3.2 billion base pairs present in each human DNA genome, including single nucleotide polymorphisms (SNPs) (Unit 2). The DNA map shown in Figure 6-3 is from a person who inherited two copies of chromosome 7 that have different SNP base pairs at some positions in the DNA, including the C to T SNP shown. This particular SNP site can be used to distinguish between these two copies of human chromosome #7. SNPs are located in the protein-coding regions and in the noncoding DNA regions and most SNPs do not alter gene function. However, studies show that specific SNP variations can predispose people to certain diseases, while other SNPs can influence an individual's response to certain drugs and sensitivity to environmental factors such as bacteria, viruses, toxins, and chemicals.

The SNPs in chromosome DNA are evolutionarily stable from generation to generation, which makes SNPs very useful markers for genetic research, including population studies. SNP markers are very powerful genetic tools that have been used to rapidly identify human genes involved in specific genetic diseases, even in cases where the SNP DNA sequence does not alter the function of the disease gene. In cases where a particular SNP marker and a mutant disease gene are always inherited together, there is a physical DNA link between the mutant allele and the SNP marker (they are on the same chromosome DNA molecule) (Figure 6-3). The mutant gene and the SNP marker are located close together on the same chromosome 7 DNA molecule (Figure 6-3). The other copy of chromosome 7 contains the wild-type (normal) SNP sequence located on the same DNA helix as the normal gene allele.

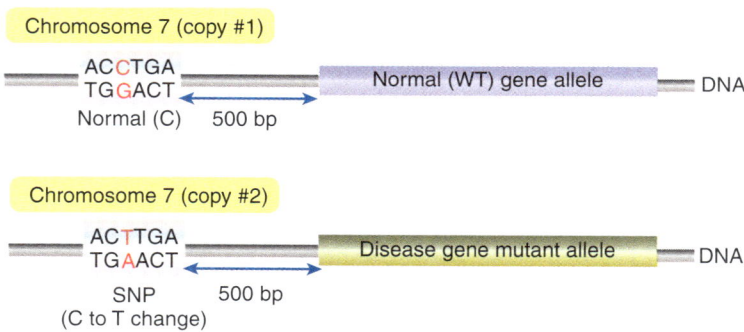

FIGURE 6-3 SNPs can be genetically linked to a disease gene This person inherited two copies of chromosome 7 that differ at several sites in the DNA sequence including the SNP site shown. It is important to note that this SNP sequence is located only about 500 bp from a mutant gene on one copy of chromosome 7 and is located only about 500 bp from the normal (wild type) copy of the same gene on the second copy of chromosome 7. In genetic terms this means that this specific SNP marker can be used to study the inheritance of the mutant gene that causes a particular human disease, just by monitoring that SNP marker in individual genomes.

UNIT 6 DNA Forensics and Epigenetic Reprogramming

> **KEY CONCEPT**
> The close proximity between an SNP marker and a mutant (disease) gene on the same DNA helix explains why the SNP and the disease gene allele are always inherited together. This is the way that certain SNP markers in the human genome can be used to identify the mutant genes that cause genetic diseases.

Detecting DNA Differences by DNA Fingerprint Analysis

The key to the DNA fingerprinting method is the use of highly specific DNA probes to base-pair with specific target sequences in the chromosome DNA. The base sequences of the DNA probes are designed to base-pair with selected chromosome sequences that are variable in sequence between human genomes (Figure 6-4). Each DNA probe is designed to detect the target DNA sequences at one chromosome location without interference from the millions of other DNA strands mixed together in the genome DNA sample. As a result, DNA fingerprinting can be used to distinguish between individual humans by detecting the specific sequence differences that vary between the genome DNAs.

Automated DNA fingerprinting and DNA chip technology (Unit 4), offer a rapid, inexpensive, and accurate method of generating DNA profiles for human identification. In the future genetic testing will take place using a DNA microarray chip containing thousands of genetic markers so that many genetic tests can be performed simultaneously (Figure 6-5).

Successful DNA fingerprint technology relies on the ability of specific DNA probes to base-pair to target regions of the chromosome DNA, which gives off a detectable signal to report the chromosome location of the base-paired

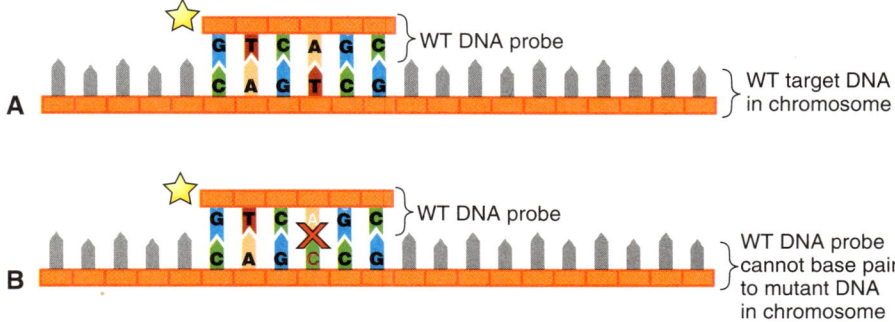

FIGURE 6-4 DNA probes can distinguish between different chromosomes DNA probes used in human identification are designed to base-pair with selected DNA sequences that vary between human genomes. These regions of variable DNA sequences can be used to distinguish among human genomes because each DNA probe is designed to detect the target DNA sequences at a single chromosome location. In this way, DNA fingerprinting can distinguish between the genomes of individual humans.

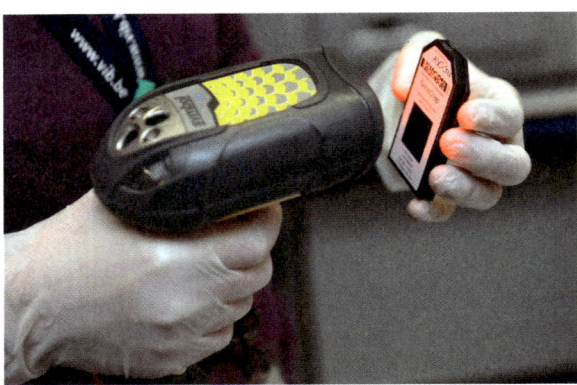

FIGURE 6-5 Genetic testing on a DNA microarray chip read with a barcode scanner Eventually it will be routine for genetic testing to take place on a DNA microarray chip cartridge with the results reported by a barcode scanner. The DNA microarray contains thousands of genetic markers so that many genetic tests can be performed simultaneously. The DNA microarray cartridge shown here is part of the Genome-Wide Human SNP Array 6.0 GeneChip made by Affymetrix.

DNA probes. The DNA fingerprint or DNA profile for each individual human genome is a graphic representation of a pattern of DNA fragments that are separated (usually on a gel) and visualized as dark bands (usually on an X-ray film) (Unit 2).

The pattern of alternating light and dark DNA bands on an X-ray film is one way to represent someone's unique DNA fingerprint. The alternating light and dark DNA bands in the fingerprint pattern resemble the black and white stripes on the familiar Universal Product Code (UPC) labels commonly found on products in the grocery store. The black stripes on the UPC label are scanned by a supermarket scanner, which reads the pattern and collects information about the loaf of bread, ingredients, price, vendor, last sell by date, etc.

The similarities between scanning the UPC label for product identification in the store and scanning DNA fingerprint patterns to identify humans, and many other biological species greatly intrigued Paul Hebert (University of Guelph in Ontario, Canada) who started the Barcode project in 2003. This is an international group of scientists working to establish standard DNA barcode sequences for all species so that eventually a DNA barcode database will be available to identify biological species.

> **KEY CONCEPT**
> In DNA fingerprinting, each type of DNA probe base-pairs to one region of variable DNA sequences in the chromosome. A series of about a dozen different DNA probes are used to generate a unique DNA profile that can be used to identify individual humans, beyond a reasonable doubt.

DNA Fingerprinting Rides a White Horse and Wears a White Hat

The DNA fingerprinting method was first used in a criminal investigation by its inventor, British scientist Sir Alec Jeffreys (University of Leicester, United Kingdom) in 1985 (Figure 6-6). Jeffreys had published a landmark paper on variable DNA sequences, which provided the foundation for developing DNA fingerprinting technology. Shortly thereafter a lawyer in London, Sheona York, contacted Jeffreys to find out if DNA fingerprinting might help her client. At the time York represented a family from Ghana who are UK citizens, but the British government was questioning the paternity of the children and claimed that the son was trying to enter Britain illegally.

Jeffreys conducted DNA testing to determine whether the boy in question was genetically related to the woman claiming to be his biological mother (the father was unavailable). The DNA profiles of the mother and her other sons were compared with the DNA profile of the son in question, and DNA fingerprinting clearly proved that the boy was the mother's biological child. This was the first time that DNA fingerprinting was presented in any court of law and it was used to establish a legal genetic connection among biological family members.

On November 21, 1983, 15-year-old Lynda Mann was raped and strangled in a small town in Britain. Police collected a semen sample belonging to a person with type A blood, but without additional evidence the case was not solved. Three years later in a nearby town, 15-year-old Dawn Ashworth

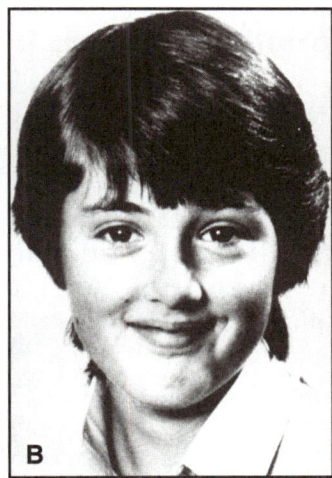

FIGURE 6-6 DNA fingerprinting catches a murderer for the first time (A) In 1984 Professor Alec Jeffreys developed DNA fingerprinting and pioneered the method for use in criminal forensics and child paternity. (B) Dawn Ashworth was a 15 year old British schoolgirl when she was murdered in 1986. The crime was solved by DNA fingerprinting. (C) DNA fingerprinting evidence helped to convict Colin Pitchfork of Dawn Ashworth's murder. He was sentenced to life imprisonment in 1988.

was raped and strangled, and once again police collected semen evidence that pointed to a perpetrator with type A blood (Figure 6-6). At the time the police had a prime suspect, 17-year-old Richard Buckland, who denied being involved in either murder. When the police asked Jeffreys for help, he generated DNA fingerprint patterns from the semen samples in evidence and compared them with Buckland's DNA profile. The answer surprised everyone. The DNA profiles from the semen evidence did not match Buckland's DNA. Just as surprising, comparison of the DNA fingerprinting patterns revealed that the DNA profiles of the two semen samples from the two murders matched, proving that the two girls were raped by the same man. In 1986, Richard Buckland became the first person to have his innocence established by DNA fingerprinting.

With the murders of Lynda Mann and Dawn Ashworth still not solved, the British police took the very unusual step of asking 5,000 local men from the surrounding towns to voluntarily provide samples for DNA testing. DNA fingerprinting was in its infancy and all the samples had to be processed by hand. It took over six months to complete the lab work and the DNA test results were highly anticipated by law enforcement and the public. Again the results were surprising. The DNA testing showed no matches between the DNA from the thousands of voluntary blood samples and the DNA from the semen evidence. Now people began to doubt the accuracy of the new DNA testing science. Then a woman called police with a tip, she had overheard a man bragging that he had been paid by the local baker, Colin Pitchfork, to substitute his blood for DNA testing. On September 19, 1987, when his DNA profile matched the DNA profile from both semen samples, Colin Pitchfork was arrested and charged with rape and murder (Figure 6-6). He was convicted and sentenced to life imprisonment with concurrent terms for rape and murder.

DNA Fingerprinting Depends on Highly Specific Base Pairing

The ability of DNA fingerprinting technology to act as a unique identifier of individual humans relies on the ability to detect DNA sequence differences between human genomes, including single DNA base-pair differences. At any one base pair position in the genome, the sequences from different individuals can contain any one of the four bases, A, T, G, or C. However, the comparison of individual human genome sequences shows that human genome sequences are much more similar to each other than they are different. So at any one position in the genome, most human chromosomes will have the same base while an occasional genome sequence will have a different base at that location, identified as an SNP. In some genomes the particular DNA base pair at a specific SNP site in an individual's genome is unique for that particular individual. Some individuals will have a C at that position in the genome, while other people will have a G, an A, or a T at that site. The DNA fingerprinting method reveals the specific base pair present at the polymorphic DNA sites in an individual's genome and reports this information as a DNA fingerprint or profile. The single base differences permit us to identify individual genomes.

The Role of Southern Blot Transfer in DNA Fingerprinting

The opportunity to review how DNA fingerprinting is performed in the lab often helps people to better appreciate why DNA testing is such a powerful tool with such a wide range of different applications. The descriptions of how the DNA is manipulated in the lab often provides insight into the methods that have revolutionized DNA research, including restriction enzyme digestion, gel electrophoresis, Southern and Northern blot transfers, and DNA detection.

The process of DNA fingerprinting in the lab:

(1) Target DNA preparation and separation by gel electrophoresis
(2) Southern blot transfer of target DNA from gel to membrane
(3) DNA probes hybridize to chromosome target DNA sequences
(4) Detection of the DNA fingerprint pattern (DNA profile)

(1) Target DNA Preparation and Separation by Gel Electrophoresis

The DNA fingerprinting method begins in the lab by cutting the chromosome DNA with a restriction enzyme to make shorter DNA fragments that are easily separated by gel electrophoresis. The DNA to be analyzed can come from blood, semen, saliva, hair follicles or any sample containing trace amounts of biological material. Small quantities of DNA collected from forensic evidence can be first amplified by PCR to obtain an additional source of identical DNA for analysis by DNA sequencing or Southern blot transfer.

(2) Southern Blot Transfers Target DNA from Gel to Membrane

The Southern blot transfer is a very powerful yet comparatively simple technique used to identify specific DNA sequences in the genome under study. This amazing application of gel electrophoresis and DNA hybridization allows scientists to detect and study a single DNA sequence hidden among billions of genome DNA sequences. The Southern blot technique was created by Edward M. Southern (Edinburgh University, Scotland) in the 1970s and quickly became an indispensable tool for studying DNA and genes.

To prepare chromosome DNA for Southern blot analysis, all the proteins are removed from the DNA, disrupting the chromosome structure and releasing the double-stranded DNA from the restraint of the packaging proteins. The "naked" DNA is cut with an appropriate restriction enzyme (depending on the design of the experiment), and the DNA products are separated by agarose gel electrophoresis (Unit 2). After electrophoresis the gel is stained with ethidium bromide, which reveals the smear of stain that represents millions of DNA fragments produced by cutting the chromosome DNA with a restriction enzyme (Figure 6-7). This DNA smear is evident in each lane that contains human genome DNA that was cut with a restriction enzyme and loaded onto the gel for electrophoresis (Figure 6-8).

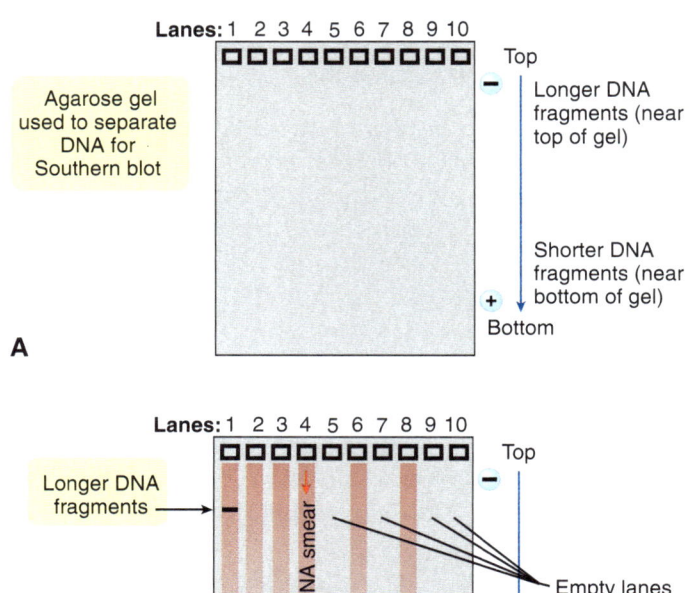

FIGURE 6-7 DNA fragments separated in preparation for the Southern blot transfer (A) The DNA is cut into millions of DNA fragments that are separated by length on an agarose gel in preparation for a Southern blot transfer. (B) The gel is stained with ethidium bromide (EthBr), causing the DNA fragments to fluoresce in UV light. The many DNA fragments produced by cutting the chromosome DNA appear as a smear of DNA extending down the gel. The blue arrow indicates the direction of the electric current.

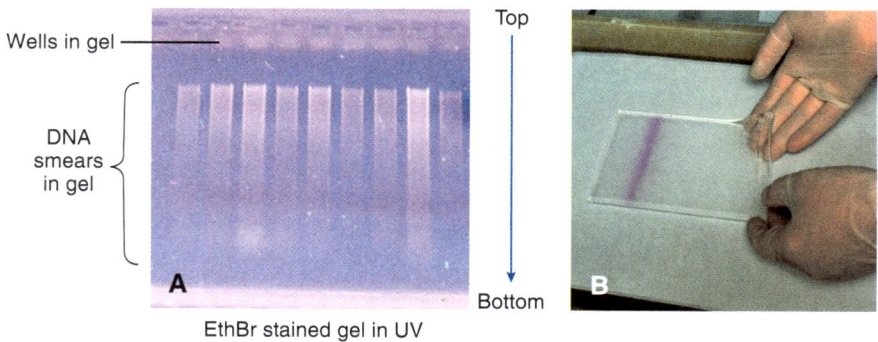

FIGURE 6-8 Agarose gel contains DNA for Southern blot (A) The DNA fragments fluoresce on the agarose gel illuminated with ultra-violet light. The DNA smears in several lanes represent large numbers of DNA fragments of different lengths that cannot be resolved on the stained gel but selected DNA fragments can be detected by Southern blot hybridization. (B) The agarose gel is fragile and must be handled carefully when assembling the Southern blot transfer. The purple-blue dye that was loaded on the gel at the same time as the DNA samples has migrated along with DNA fragments of a specific length and can be used to monitor the general progress of the gel electrophoresis experiment. The wells used to load the DNA on the gel are located at the opposite end of the gel from the dye.

The next step in the Southern blot method involves transferring the DNA strands out of the gel slab and onto a special membrane sheet. During this maneuver, it is essential that the DNA strands transfer to the same positions on the membrane as they were in the gel. To accomplish this trick, the fragile agarose gel is first soaked in a buffer that separates (denatures) the DNA strands *in situ* (in place) in the agarose gel (careful the base buffer makes the gel very slippery!) (Figure 6-8). The double-stranded DNA fragments denature into single strands inside the gel and do not change position in the gel.

Now the Southern blot transfer apparatus is assembled by placing the denatured gel on a low plastic support in a shallow dish containing transfer buffer (Figure 6-9). A sheet of nylon or nitrocellulose membrane is carefully positioned directly on top of the gel in one movement, and any air bubbles underneath the membrane are gently removed by rolling the surface of the membrane with a smooth glass rod. Once the membrane is laid on top of the gel it is very important not to reposition the membrane sheet on the gel to avoid multiple DNA transfer "prints" on the membrane. Mark the membrane to indicate orientation of the membrane relative to the wells in the gel. Now add 3 or 4 sheets of 3-mm filter paper on top of the membrane and add a stack of paper towels on top of the 3-mm paper, and finish by gently compressing the stack with a weight. The DNA strands will move out of the agarose gel as the transfer buffer moves through the membrane and into the paper towels. The DNA strands cannot pass through the membrane so the DNA becomes permanently attached to the membrane sheet in a pattern that is identical to the pattern of the DNA fragments before they transfer out of the gel (Figure 6-10).

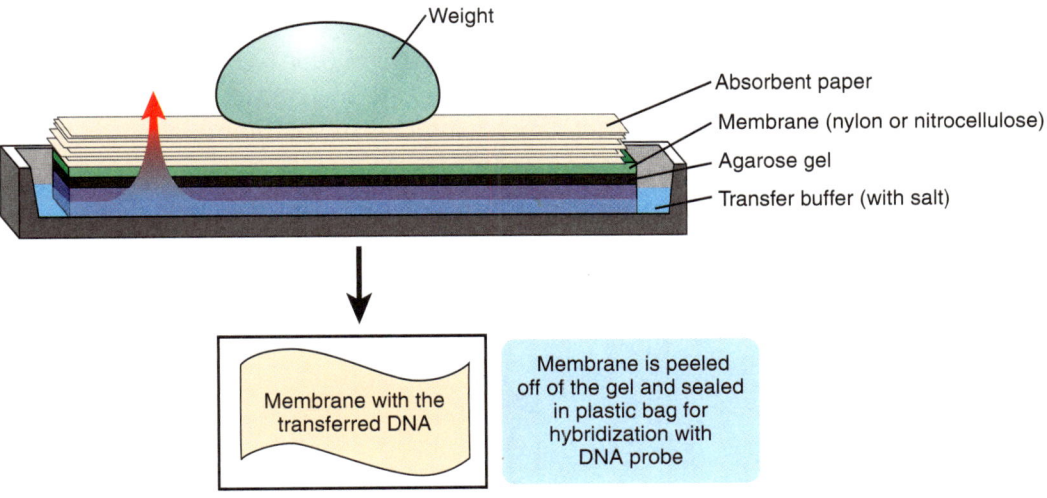

FIGURE 6-9 DNA transfers to the membrane Southern blot After electrophoresis the gel is treated with a basic solution to denature the DNA into single strands while in place in the gel (in situ). The denatured gel is placed on a plastic support in a shallow dish containing transfer buffer. A sheet of membrane is carefully positioned on top of the gel followed by 3 to 4 sheets of 3-mm filter paper, a stack of paper towels and a weight. The transfer buffer moves up through the gel, through the membrane, and into the stack of paper towels (red arrow). The DNA strands also move out of the gel but they stick to the sheet of membrane in a pattern that is identical to the pattern of the DNA fragments in the gel.

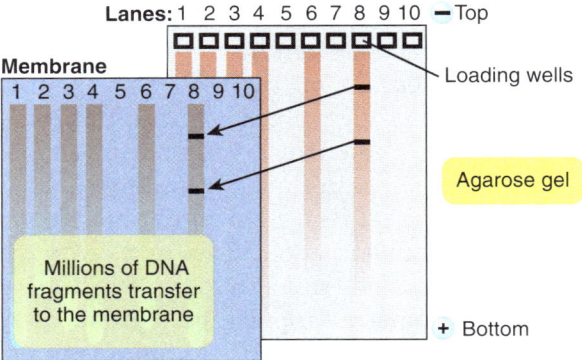

FIGURE 6-10 DNA pattern is the same on the gel and the membrane The DNA fragments transfer out of the gel and become permanently attached to the membrane at various positions that accurately represent the previous locations of the fragments in the gel. This means that the DNA pattern on the membrane is an exact copy of the DNA pattern on the gel.

(3) DNA Probes Hybridize to Chromosome DNA Sequences

In the next step the Southern blot transfer apparatus is disassembled and the membrane sheet is lifted carefully from the remains of the shrunken agarose gel. The membrane is then placed in a plastic bag containing a small amount of liquid hybridization buffer that just covers the sheet of membrane. Excess copies of a specific single-stranded DNA probe (radioactively labeled, or fluorescent) are added to the hybridization buffer, the bag is sealed and incubated at the hybridization (annealing) temperature with tilting for an appropriate length of time. The Southern blot hybridization reaction occurs on the membrane when the DNA probe base-pairs (anneals) to the complementary strand of target DNA attached to the membrane (Figure 6-11).

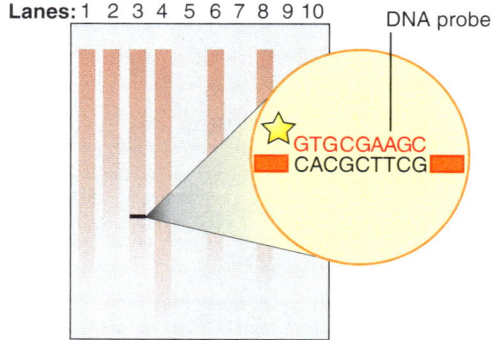

FIGURE 6-11 DNA probes base pair to target DNA sequences on the membrane After the DNA has transferred to the membrane, the sheet of membrane is placed in a plastic bag containing just enough liquid hybridization buffer to cover the membrane. The radioactively labeled (or fluorescently labeled) single-stranded DNA probes are added to the hybridization buffer and sealed in the bag with the membrane. The DNA probes base pair to the target complementary DNA strands attached to the membrane, while the membrane is incubated at the hybridization (annealing) temperature for the appropriate length of time.

(4) Detection of the DNA Fingerprint Pattern (DNA Profile)

When the membrane is removed from the bag the unbound DNA probes are rinsed off of the membrane. The DNA probes that have base-paired to specific DNA strands on the membrane do not wash off. In this way the Southern blot hybridization method is similar to the FISH method of labeling chromosomes except that the DNA probes used in FISH hybridize to chromosome DNA attached to glass slides, not a sheet of membrane. The Southern blot membrane is dried and exposed to X-ray film to detect the locations where the DNA probe had base-paired to the target DNA on the membrane (Figure 6-12). The X-ray film is a copy of the pattern of DNA fragments on the gel so the length of each DNA fragment can be determined by comparing the migration of an unknown DNA band with the migration of the known DNA markers. There is a one-to-one correlation between the DNA bands on the film and the positions of the corresponding DNA fragments on the membrane (and the DNA fragments previously separated in the gel).

The Southern blot transfer and hybridization method has been invaluable for studying the human genome and has been used in countless applications locating and studying human genes since the 1970s. The Southern blot transfer or a modification of this method is essential for many applications in DNA technology including DNA fingerprinting and genetic testing.

To generate the DNA fingerprinting profile of a given individual human, several different DNA probes must be used to test the genome sequences detected by several different SNPs. The current FBI standard indicates that if possible an individual's chromosome DNA or the DNA from semen or other evidence should be analyzed by a series of 13 different DNA probes that provide DNA profile information about 13 different SNP sites in the human genome. It is not always possible to obtain 13 loci and in some cases fewer loci are acceptable depending on other identification evidence. DNA fingerprint information from the first DNA probe is extended by repeating the hybridization procedure using the same sheet of membrane but hybridizing with a second DNA probe that has a different sequence. The new DNA probe will base-pair to a different target DNA strand and detect a different SNP locus on the membrane. These new DNA bands are detected directly by the X-ray film, in addition to the DNA bands previously detected by the first DNA probe (Figure 6-12). A third DNA probe with another DNA sequence can be hybridized to the same membrane to reveal another set of new DNA bands on the DNA fingerprint x-ray film.

> **KEY CONCEPT**
> Each time a different specific DNA probe is hybridized to the <u>same</u> membrane, additional DNA bands will potentially appear on the film, providing additional DNA fingerprint information. The additional DNA bands are revealed on the X-ray film with each new cycle using a new DNA probe.

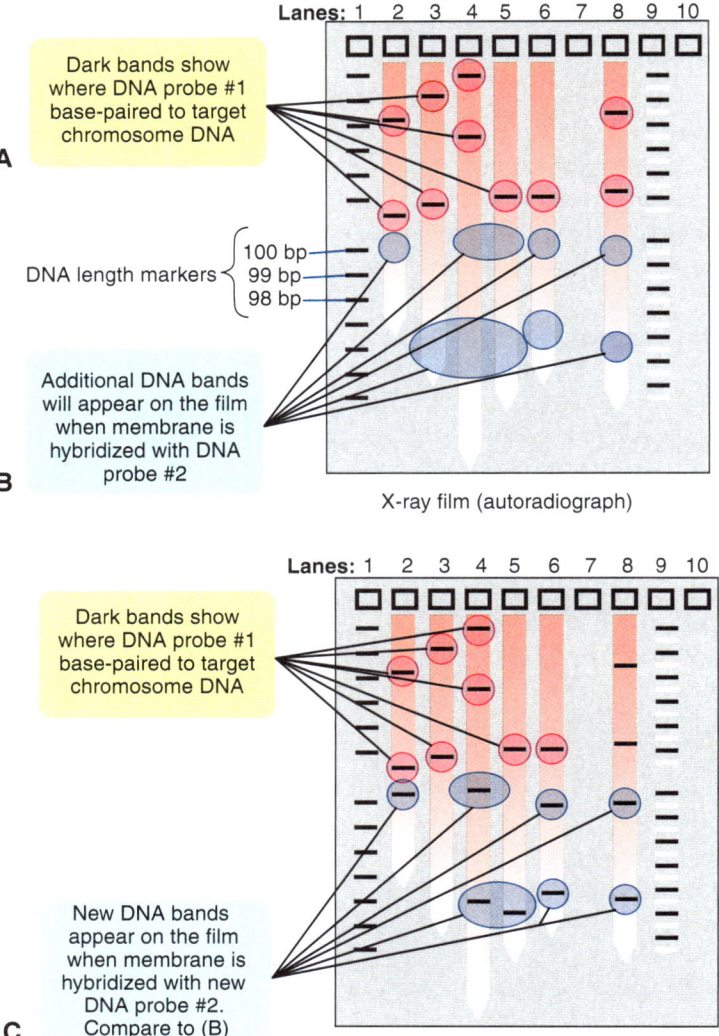

FIGURE 6-12 New DNA probes base pair to new target DNA sequences on the membrane
(A) The DNA bands revealed by the first DNA probe (#1) are shown on the x-ray film (pink circles). The migration of the DNA length markers are shown as black bands. (B) New DNA bands appear when the second DNA probe (#2) base pairs to different target DNA strands on the <u>same</u> membrane. (C) The new DNA bands detected by the second DNA probe (#2) appear as new, additional DNA bands on the x-ray film (blue circles).

The statistical probability that two unrelated people will have the same DNA fingerprint by coincidence becomes much less likely as each new DNA probe is used to screen the membrane and report on the additional SNP loci. The additional DNA bands appear on the film because the different DNA probes have hybridized to new target sites on the membrane. Each newl DNA probe adds more DNA bands to the fingerprint and increases the accuracy of a potential match.

DNA Profiles in the FBI COmbined DNA Index System (CODIS)

The FBI established a National DNA Databank, the COmbined DNA Index System (CODIS), under the DNA Identification Act of 1994. The CODIS database includes the Convicted Offender Index, which contains the DNA profiles of individuals convicted of violent crimes, and the Forensic Index, which contains DNA profiles developed from evidence collected at crime scenes. The CODIS computer software compares DNA profiles submitted from law enforcement agencies at the federal, state, and local levels. As of 2007 the CODIS DNA database contained over 5 million DNA profiles in its Convicted Offender Index, and about 188,000 DNA profiles in the Forensic Index had been generated from crime scene evidence.

The DNA profiles (DNA fingerprints) submitted to CODIS are generated from different sources including physical evidence from the crime scene and from people involved in the case, including the victim, suspects, and witnesses. The CODIS computer can compare the DNA profiles associated with the evidence and can compare the DNA profile of the suspect with the CODIS database of stored DNA profiles. If the computer finds a possible match between DNA profiles, the forensic technicians are notified, confirm the DNA match and obtain the identity of the potential perpetrator for investigators. DNA profiles obtained from the evidence are compared to the crime scene DNA profiles in the Forensic Index database, because a match between DNA profiles can potentially link two crimes to one suspect. A forensic DNA database match is sometimes sufficient to establish probable cause to obtain a court order authorizing the collection of a biological sample from a suspect and subsequent laboratory DNA analysis.

When the DNA profiles from two sources match, it strongly suggests that the two DNA samples come from the same source. However, it is possible that a match between two DNA profiles might occur by coincidence, so it is necessary to calculate the statistical probability that a match between two DNA profiles might occur by chance.

A standard FBI CODIS DNA profile involves testing 13 different alleles that are located on different human chromosomes and are independently inherited (genetically unlinked): CSF1PO, FGA, THO1, TPOX, VWA, D3S1358, D5S818, D7S820, D8S1179, D13S317, D16S539, D18S51, and D21S11 (Figure 6-13).

FBI forensic labs generate human DNA profiles by analyzing these human genome DNA repeats:

- PCR short tandem repeats (STRs)
- Y chromosome short tandem repeats (Y STR)
- Mitochondrial DNA (mtDNA) repeats

PCR Short Tandem Repeat Analysis

The polymerase chain reaction (PCR) produces millions of exact copies of a specific DNA strand by means of an enzymatic DNA amplification reaction in the lab (Unit 2). PCR basically increases the sensitivity of any

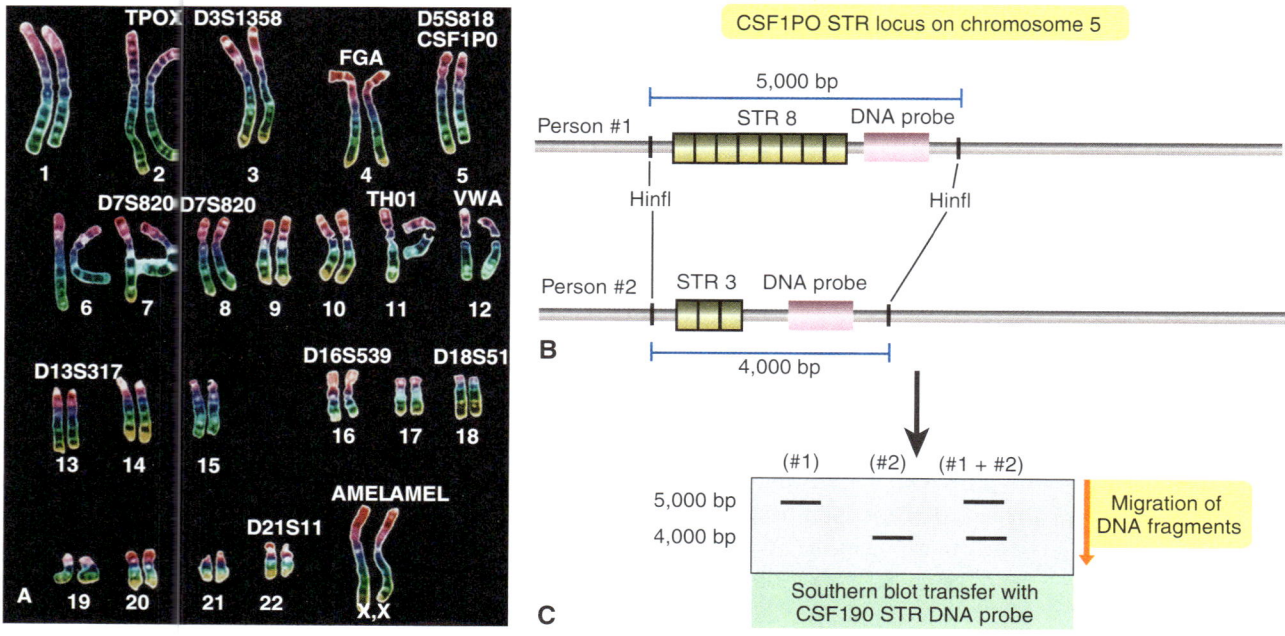

FIGURE 6-13 **FBI uses 13 DNA probes to identify people by DNA fingerprinting** (A) FBI DNA profiles are created by testing the STR alleles present on many human chromosomes. (B) This STR locus on chromosome 5 has different numbers of STR repeats in individual genomes as detected by DNA fingerprinting with a CSF1PO DNA probe. (C) Hinfl enzyme produces 5,000-bp DNA fragments from the both copies of chromosome 5 from person #1. The copies of chromosome 5 from person #2 produce 4,000 bp DNA fragments. The #1 and #2 DNAs cut with Hinfl were mixed and run in the far right lane on the gel.

type of DNA test where the amount of DNA subjected to the test is limited. In the lab, PCR has been shown to successfully amplify one molecule of DNA into millions of identical DNA copies visible on an ethidium-bromide stained gel. PCR has profoundly enhanced the capabilities of forensic science because PCR amplification allows scientists to perform DNA fingerprinting on identical DNA strands amplified from extremely tiny amounts of starting DNA in the evidence. This is a very important advantage because forensic analysis often involves testing small biological samples with just a few skin cells or a barely visible speck of semen or blood. PCR amplification provides an ample supply of identical DNA molecules copied from the tiny amount of DNA collected in evidence, which forensic scientists used to generate a DNA profile for CODIS analysis.

For purposes of human identification, the FBI Laboratory currently uses PCR-based STR technology, which combines the methods of PCR amplification and DNA fingerprinting with human genomics to generate human DNA profiles. The STR technology is used to detect the SNPs and other variable DNA sequences in the STR regions of the chromosome DNA (Figure 6-13). Different individual genomes can be identified by analyzing the different numbers of short tandem repeats (STRs) at each SNP site as identified by Southern blot transfer and hybridization with STR-specific DNA probes. The sequence of a DNA probes can be designed to base-pair exclusively with any target DNA sequence in the chromosome, including the DNA sequences at selected SNP sites.

The Southern blot experiment in Figure 6-13 shows how a region of chromosome 5 DNA from two different individuals can be tested to distinguish between the two people. The DNA samples from the individuals were cut with the HinfI restriction enzyme (in different tubes), and the DNA products were separated by gel electrophoresis and analyzed by Southern blot hybridization. In this experiment HinfI was used because the HinfI enzyme does not cut in between the ends of the STR DNA repeats; HinfI sites are located in DNA that flanks the STR DNA repeat (Figure 6-13). When HinfI cut the chromosome 5 DNA from person #1, it produced the 5,000-bp DNA fragments containing a locus with eight STRs as detected by the DNA probe. The DNA probe is complementary to the section of the DNA that is indicated in Figure 6-13. The chromosome 5 DNA copies from person #2 contain three instead of eight STR repeats so when the DNA from person #2 is cut with the HinfI enzyme, it produces the DNA fragment (4,000 bp) that is detected by base pairing with the DNA probe. Person #1 can be reliably distinguished from person #2 by STR DNA fingerprinting analysis using several different DNA probes. Assuming that the lengths of the DNA products in Figure 6-13 are accurate, can you determine the length in base pairs of a single STR unit?

FBI DNA Fingerprinting of Y Chromosome STRs

Human gender is determined by the X and Y sex chromosomes; mothers and daughters have two X chromosomes, while fathers and sons have an X and a Y chromosome. The DNA fingerprinting tests that analyze only the STR variations located on the Y chromosome DNA are limited for some purposes because they exclude the other chromosomes. However, the Y chromosome test is rapid, which is very important in missing person cases. The human Y chromosome is passed directly from father to biological sons but not to daughters. Examining the genetic alleles inherited on the Y chromosome is a very effective way to trace the biological relationships among human males. Information from tracing the Y chromosome often complements the genetic results from tracing the maternal transmission of mtDNA.

RFLP DNA Fingerprinting can Prove Paternity

Restriction fragment length polymorphism (RFLP) was one of the first methods used in to clearly distinguish between individual human chromosomes. RFLP continues to have applications in the research lab and is still used for paternity testing to determine the biological connections between parents and children. The more efficient and rapid DNA fingerprinting techniques have recently replaced the RFLP method of DNA profiling. Although these methods differ in the details, the SNP, STR and RFLP fingerprinting methods are really variations of the same basic approach to distinguish among people by DNA testing to detect differences. The RFLP method uses restriction enzyme sites to distinguish between individual chromosomes, the SNP method relies on single base pair differences between chromosomes and the STR approach detects variations in numbers of short DNA repeats.

RFLP (combined with a Southern blot) is an excellent way to analyze the DNA fragments produced when chromosome DNA is cut with a restriction enzyme. Since the restriction enzyme can only cut the chromosome DNA helix at a specific base-pair recognition sequence (Unit 1), the locations of these restriction enzyme cleavage sequences in the chromosome DNA will determine the number and lengths of the DNA fragments that are produced when the restriction enzyme cuts the chromosome DNA. If a restriction enzyme cleavage sequence is altered by a mutation, the enzyme will no longer recognize the DNA site and will not cut the DNA helix at that site.

The goal of RFLP fingerprinting is to identify specific restriction enzyme cut sites in the chromosome DNA where the DNA sequence is different in one person's chromosome DNA compared to the same site in the same chromosome DNA of a different person. Of course, the human genome contains thousands of cut sites for each restriction enzyme, and cutting the human chromosome DNA with even just one restriction enzyme will generate millions of DNA fragments of different lengths. However, the Southern blot transfer method uses DNA probes to specifically visualize only certain DNA fragments that are complementary to a given DNA sequence of the DNA probe. The occasional differences in the DNA sequences of individual human genomes sometimes alter a restriction enzyme cut site in the DNA, creating a polymorphism that is detected by RFLP.

A combination of RFLP and Southern blot analysis was used to prove the paternity of a child from four different candidates (Figure 6-14). The DNA samples from the people in question were cut separately with the same restriction enzyme and the DNA products of the reactions were loaded into separate wells and run in adjacent lanes on the gel. After gel electrophoresis the DNA fragments separated in the gel were transferred to a membrane by the Southern blot method, the membrane was hybridized with a DNA probe and then the membrane was used to expose a sheet of X-ray film. Comparing the DNA bands on the RFLP x-ray film reveals the actual biological relationships between the mother, father and children in question (Figure 6-14). The father's DNA profile contains HinfI DNA bands that match DNA bands in the sample from child C2.

A biological son inherits half of his chromosome DNA from Mom and the other half of his chromosome DNA from Dad. This also means that a biological son inherits half of his DNA fingerprint bands from Mom and inherits the other half of his DNA fingerprint bands from Dad. Each DNA fingerprint analysis such as the paternity test shown in Figure 6-14, would not be acceptable without additional testing. Also remember that the DNA visualized on this Southern blot represents only those DNA regions of the genome that are complementary to the very short DNA probe, hardly a widespread sample of the genome. Additional tests with additional DNA probes are not shown.

RFLP can Distinguish Between Chromosome Copies

RFLP analysis allows researchers to compare chromosomes from different individuals without having to determine the DNA sequences of entire genomes. In this example the RFLP method was used to examine a region of

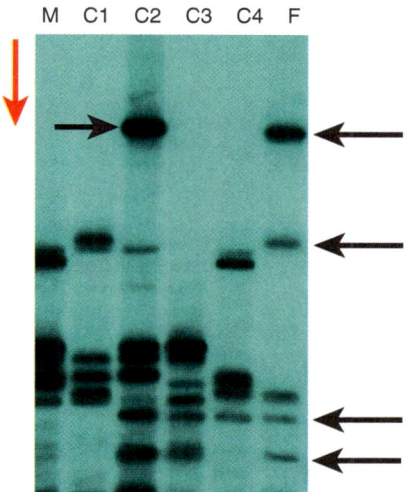

FINGER 6-14 RFLP DNA testing for paternity This x-ray film shows part of an RFLP DNA test used to prove child paternity. The DNA was cut, separated by gel electrophoresis, transferred by Southern blot and annealed to a probe that revealed a pattern of DNA bands. In cases where the DNA profiles belong to biologically related people, some of the DNA bands will be present in both RFLP patterns. The lanes containing the DNA bands from the mother (M), child (C1, C2, C3, C4) and father (F) are marked on the film. A biological child will share half the DNA bands with Mom and half the DNA bands with Dad. The black arrows indicate the father's DNA bands that match bands in the DNA from child C2. The red arrow indicates the direction of DNA migration in the gel.

chromosome 3 DNA from two different individuals, person #1 and person #2 (Figure 6-15). This region of different copies of human chromosome 3 contains either two or three BamHI sites, indicated on the map and labeled BamHI-1, BamHI-2 and BamHI-3 (Figure 6-15). The DNA samples obtained from these two individuals were cut with the BamHI restriction enzyme and the DNA products of the cleavage reactions were separated by gel electrophoresis.

The next step in RFLP analysis involves cutting the DNA with BamHI and separating the DNA products on a gel. All of the DNA fragments resolved on the gel were visualized by staining with ethidium bromide. When illuminated with ultra-violet light the smear of fluorescent stain in the gel represents the millions of DNA fragments separated by length and distributed down the gel (see the DNA smear in Figure 6-15). Hidden under the smear of millions of DNA fragments on the stained gel are the specific DNA fragments of interest from chromosome 3, which were generated by cutting the chromosome 3 DNA with BamHI enzyme.

A single base-pair change altered the BamHI-2 restriction enzyme cut site in the chromosome DNA, creating an RFLP site that can easily be detected (Figure 6-15). In this case the RFLP is also an SNP, because the BamHI-2 cleavage site was altered by a single base pair change. Of course, most SNPs are not also RFLPs because most single base pair changes do not also coincidentally alter a restriction enzyme cleavage sequence in the genome DNA. Results from testing a single RFLP site in the genome cannot be used

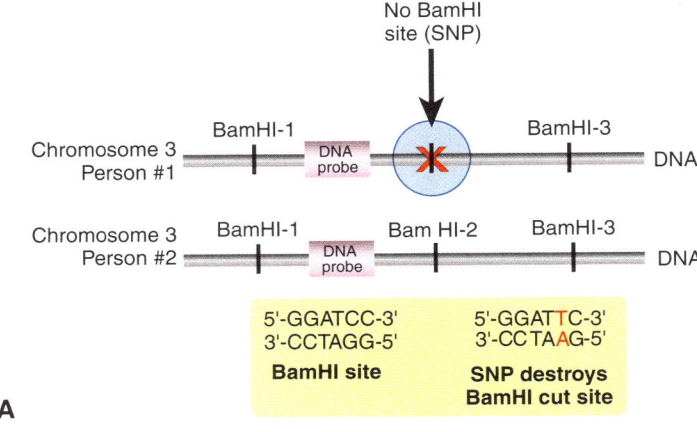
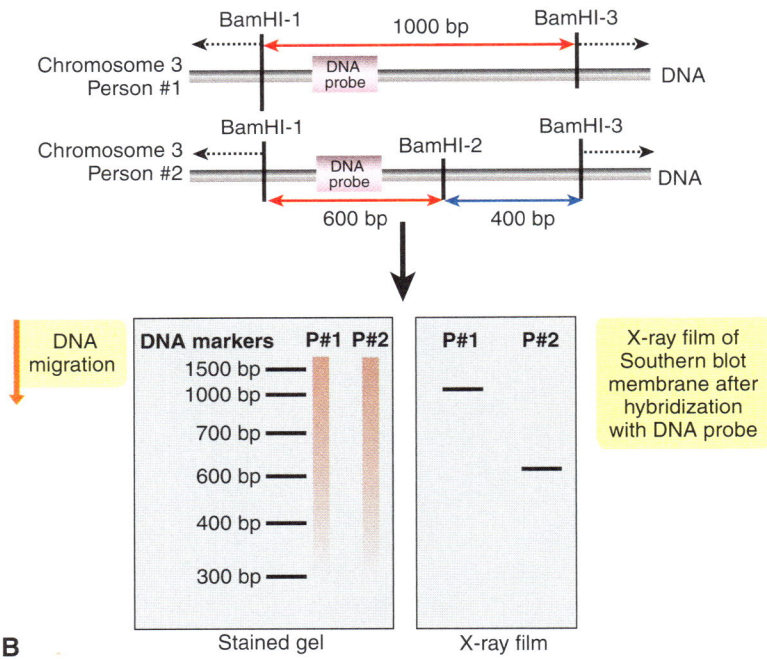

FIGURE 6-15 SNPs and RFLPs distinguish between individual human chromosomes
(A) RFLPs and SNPs were used to examine a specific region of chromosome 3 DNA in person #1 and person #2. The BamHI cleavage sites in this region are labeled as BamHI-1, BamHI-2 and BamHI-3. (B) DNA samples obtained from these two individuals were cut with BamHI, the DNA products were separated by gel electrophoresis and the DNA was visualized by staining the gel with ethidium bromide and Southern blot analysis. The DNA fragments are represented by a smear of fluorescent stain representing millions of DNA fragments separated by length in the gel. Hidden beneath the smear of millions of DNA fragments on the gel are the few specific DNA fragments of interest produced by cutting chromosome 3 DNA, and visualized on the x-ray film after hybridization.

to identify a specific individual human genome among many possibilities but a standard collection of RFLPs with changes distributed across the entire human genome are used for purposes of human identification.

In the RFLP experiment, the Southern blot is used to identify the specific DNA fragments of interest hidden among the countless DNA fragments produced by cutting the entire human genome with BamHI (Figure 6-15). The use of Southern blot hybridization permits the targeted region to be detected by a chromosome 3-specific DNA probe (Figure 6-15). On the X-ray film the scientists visualize *only* the target DNA fragments that overlap with the specific DNA probe used in the hybridization. The target DNA fragments are easily visualized among the millions of other DNA fragments produced by cutting the entire human genome with BamHI (Figure 6-15). The film shows that the chromosome-3-specific DNA probe detected a 1,000-bp DNA fragment (person #1) and a 600-bp DNA fragment in the DNA from person #2. Both of these DNA sequences overlap with the DNA probe, except for the 400-bp DNA fragment from person #2 (Figure 6-15). Do you know why the 400 bp DNA fragment is not detected on the X-ray film in Figure 6-15?

Mitochondrial DNA is a Genetic Link to Mom

Mitochondrial DNA (mtDNA) is a very useful tool that offers some advantages for human DNA identification and forensic investigation. With each generation the mtDNA and the mitochondrial genes are passed from the mother to both sons and daughters by maternal transmission, with no contribution from the father's mitochondrial DNA, or any other part of the father's cytoplasm. The biological father contributes only nuclear DNA, which can be traced by following the genetic inheritance of the Y sex chromosome.

Human mitochondrial DNA is a 16.5-kb (16,500 bp) double-stranded circular molecule that encodes a small number of mitochondrial genes required for mitochondrial function. Proteins encoded by the mtDNA chromosome are expressed and translated in the mitochondrion. Some mitochondrial genes code for enzymes involved in oxidative phosphorylation, the biochemical reactions in mitochondria that converts oxygen and sugar into adenosine triphosphate (ATP). ATP is an important source of energy in eukaryotic cells. Other proteins required for the functions of mitochondria are coded by genes carried on chromosomes and are expressed in the cell's nucleus like other nuclear genes. Proteins destined to function in the mitochondria but are translated in the cytoplasm contain special protein domains that make sure the new protein is imported into the mitochondrion.

Although mtDNA comprises <1% of the total DNA in a cell, compared to the nuclear DNA the mtDNA and the mitochondrial genes are present in high copy numbers. Also the circular mtDNA molecules are much more resistant to degradation and physical breakage than the long, linear double-stranded chromosomes found in the nucleus. This is an advantage when working with DNA evidence obtained from degraded biological samples. Human hair (and pet hair) is often collected as evidence and submitted to the hair and fiber scientists for forensic analysis. This evaluation might include mtDNA testing but the hairs under analysis often have little or no root cells attached and the cells in the hair shaft are dead and lack nuclei. The mtDNA from the bone marrow cells obtained from human long bones and the dental pulp from the inside of the teeth, will often produce degraded nuclear DNA, but will successfully generate a mtDNA profile, possibly because of the large amounts of mtDNA in the mitochondria.

MtDNA testing is an especially valuable forensic tool for investigating unsolved "cold" cases, which often involve evidence that is many years old, because mtDNA is resistant to degradation by the environment. The mtDNA test results are also valuable to make maternal connections to children when investigating missing persons. Human remains can often be identified by comparing the mtDNA profiles from two people, a potential maternal relative and the mtDNA profile of the deceased person.

Testing mitochondrial DNA is also appropriate when trying to identify a person who has drowned or continually submerged in water for a period of time because mtDNA can survive such wet conditions without degrading. But mitochondrial DNA testing cannot be used to identify individual humans and exclude others the way that nuclear DNA fingerprinting can. However, the maternal inheritance of mtDNA does provide genetic links between a mother and her children, sons and daughters. Tracing this maternal lineage also requires testing the biological samples from the other relatives involved.

Children inherit mitochondria and mtDNA entirely from their mother by maternal genetic inheritance, which is the result of events that occur when the mother's egg cell is fertilized by the father's sperm (Figure 6-16). The sperm cell nucleus and the 23 chromosomes are carried in the sperm head but the mitochondria are located in the neck and tail regions of the sperm. The mitochondria generate the energy to power the tail of the sperm as it swims toward the egg, propelled by microtubules in the cytoskeleton of the tail (Figure 6-16). Most of the time only one sperm wins the "race" and successfully fertilizes the egg. Immediately the cell wall surrounding the egg called the zona pellucida becomes very thick and prevents the penetration of the competing sperm cells that are also trying to fertilize the egg. The sperm cell head contains a nucleus with 23 chromosomes that enters an egg cell, and fuses with the egg cell nucleus for a total of 46 chromosomes in the zygote nucleus. However, the sperm tail and the contents of the paternal cytoplasm including the mitochondria remain on the outside of the egg cell and are discarded (Figure 6-16).

As the new zygote begins to divide and develop into a human embryo, the embryo cells read and interpret nuclear genome instructions that were inherited from both parents. The mitochondria in the developing embryo cells are directly descended from the mitochondria in the mother's egg cell. Inheriting mtDNA from a maternal source does not affect the gender of the embryo, although it sometimes explains inheriting a mitochondrial genetic disease.

> **KEY CONCEPT**
> Mitochondrial DNA (mtDNA) is a direct genetic link from mother to sons and daughters. Mothers have the same mitochondrial DNA as their biological offspring, but fathers do not contribute mitochondria to their offspring.

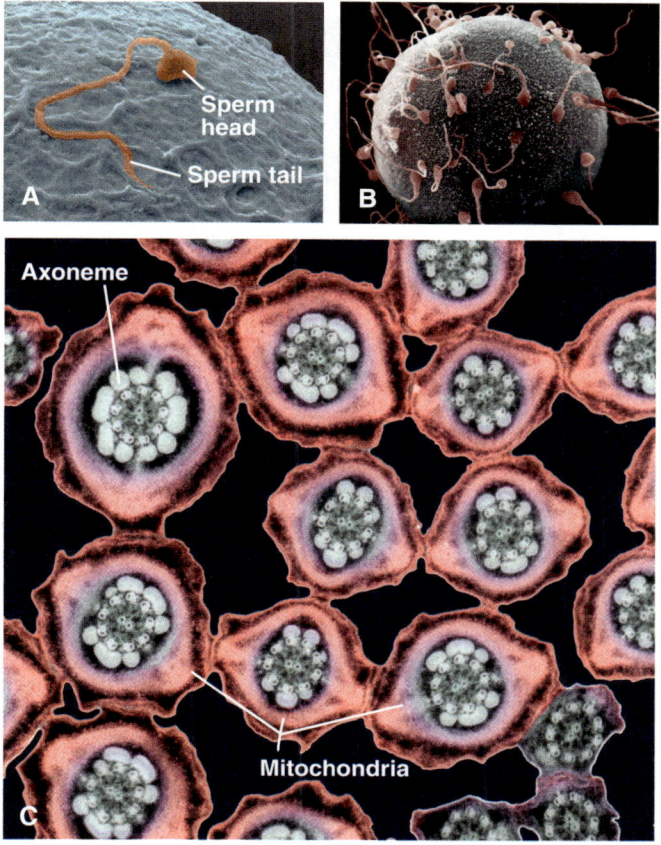

FIGURE 6-16 Maternal inheritance excludes the father's mitochondria (A) The head of a sperm cell (brown) penetrates the surface of the egg cell (ovum). After fertilization, the outer layer of the egg cell prevents other sperm cells from fertilizing the same egg. (B) The competing sperm continue unsuccessfully to try to penetrate the egg. (C) Cross section through sperm tails show the cytoskeleton composed of a central axoneme with two single microtubules in the center and nine double microtubules around the outside. The mitochondria (pink) produce energy for the sperm to move. The sperm tail assembly is wrapped in an outer sheath (red).

DNA Testing Holds Up in U.S. Courts

The nuclear DNA testing evidence that seems almost routine to people living in 2011, was eyed with great suspicion when it was first used in a courtroom in the United States in 1987. It was many more years before DNA fingerprinting techniques were finally accepted by the legal community. The vital importance of DNA testing evidence in criminal trials became center stage in 1995 when football star O.J. Simpson was charged in a high-profile double murder. This was the first time a murder trial was broadcast on live television and it occupied the attention of many Americans for months. The defense attorneys also put DNA evidence on trial, arguing about the potential limitations and the dangers of testing contaminated or degraded DNA evidence.

Whether or not the public and the professionals agree with the verdict in the O.J. criminal trial (which was different from the verdict in the civil trial), the O.J. murder case played an important role in emphasizing the urgent need to properly train the lab personnel who work with DNA and to establish standardized rules for the collection of biological evidence and the storage of DNA samples (Figure 6-17). Genome DNA fingerprinting and mitochondrial DNA test results are now commonly accepted as evidence in legal proceedings worldwide. In criminal cases, DNA testing will reveal if the accused person's DNA does not match the DNA evidence found at the crime scene and the person can be excluded from the list of possible suspects. DNA testing can also show when two DNA profiles match, revealing that an individual could be guilty beyond a reasonable doubt, but remember that a DNA match is not an indication of automatic guilt. All the evidence must be fairly considered by the jury.

PCR amplification of DNA is an extremely powerful investigative tool that has had an enormous impact on research science and has dramatically changed many aspects of law enforcement, forensic science and the U.S. legal system. However, the strengths of the PCR method can also be a huge drawback if people are not extremely careful to avoid contaminating PCR samples with external DNA (Figure 6-18). The polymerase enzyme that performs the PCR will amplify any DNA molecule in the reaction, regardless of the source of the DNA. This caution is even more important as PCR technology continues to develop increasingly sensitive DNA fingerprinting methods every year (Unit 2).

> **KEY CONCEPT**
> When the appropriate precautions are taken while collecting, storing and manipulating the DNA evidence it is quite clear that the results of DNA testing are reliable, reproducible, and accurate and have the potential to change people's lives in the most profound ways.

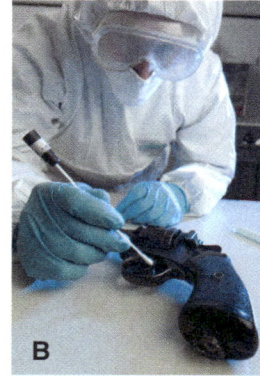
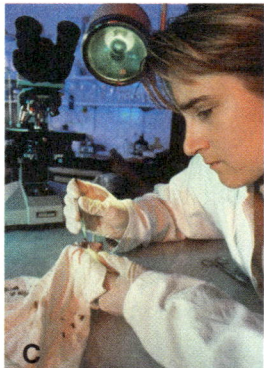

FIGURE 6-17 Forensic scientists collect samples for evidence (A) Forensic scientists collect samples from blood-stained clothes for DNA testing. (B) The evidence collected from the trigger of a handgun might include skin cells from a potential shooter that can provide a DNA fingerprint. (C) Tiny bits of material from a crime scene called trace evidence are collected and processed for forensic information.

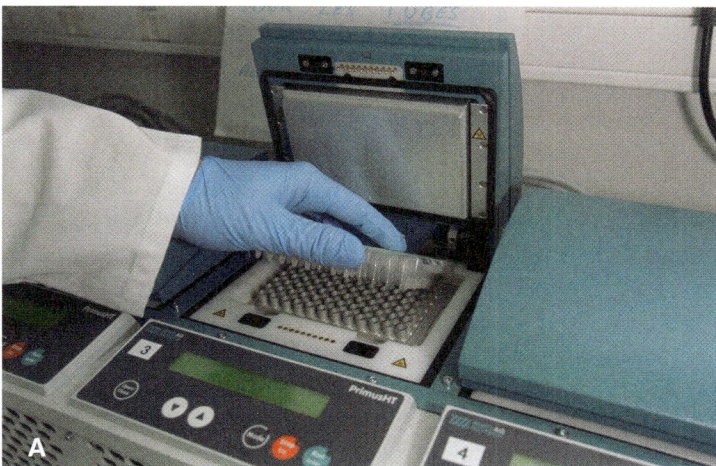

FIGURE 6-18 PCR used in DNA fingerprinting (A) This scientist is loading DNA samples into a thermocycler machine that is programmed to automatically perform a series of PCR cycles that amplify the DNA molecules in the tiny tubes. The thermocycler rapidly increases and decreases the temperature of the samples according to the cycles directed by a computer. PCR is often used to amplify the amount of template DNA available for the subsequent DNA sequencing reactions. (B) The DNA fingerprinting image on an x-ray film is enhanced by computer and the information is stored digitally to search for DNA fingerprint matches.

The Innocence Project Lawyers and DNA Testing Save Lives

DNA fingerprint analysis and other types of DNA testing technologies can do more than just help to convict people who have committed crimes. The Innocence Projects are amazing organizations that use the power of DNA testing to exonerate people who were wrongly convicted and are incarcerated in prison. Not only have the lawyers at the Innocence Projects righted many wrongs, they have also uncovered some serious problems with the U.S. legal system. As a result, the Innocence Project lawyers have worked to reform the standards and protocols for how forensic DNA samples are handled and processed, and they have had an important impact on efforts to reform the process of eyewitness identification in the legal system.

The first Innocence Project was founded by attorneys Barry Scheck and Peter Neufeld at the Benjamin N. Cardozo School of Law, Yeshiva University. Barry Scheck was one of the lawyers on the O.J. Simpson 'dream team' criminal defense, but his interest in DNA and legal justice has led him to establish a way to use DNA testing to help exonerate innocent convicted prisoners. The lawyers work with the prisoners to win the right to have old trial evidence subjected to DNA testing, or have the evidence retested with modern PCR DNA testing methods that are much more sensitive than earlier methods of DNA testing. This can be a long and tedious process with frequent disappointment. DNA evidence that is lost can't be tested and DNA evidence degrades when stored incorrectly, which is often the case after a conviction.

In the 18 years since it began, the Innocence Project lawyers armed with DNA testing have exonerated 258 people who served an average of 13 years in U.S. prisons before they were released. Seventeen of the 258 people who were exonerated had served time on Death Row. There are now Innocence Projects in every state in the United States except for Oregon and Tennessee, and similar programs exist in Australia, New Zealand, Canada, and the United Kingdom. The Innocence Projects are staffed by attorneys and law students who select appropriate cases for review, and they also provide the necessary legal representation. The efforts of the people behind the Innocence Projects have made a huge difference in the lives of the many people who were freed and their families. The research performed by Innocence Project lawyers has also had an important impact on efforts to reform legal standards, especially concerning issues involving eyewitness identification.

> **KEY CONCEPT**
> DNA testing has exonerated people on death row, proving their innocence and instantly raising more questions about the use of the death sentence in the United States.

"But I Saw it with My Own Eyes"

When the Innocence Project lawyers had exonerated 239 convicted people, the lawyers released a research report in which they analyzed the factors contributing to these wrongful convictions (Figure 6-19). The Innocence Project lawyers discovered that the 239 mistaken convictions were not just random error, most were caused by systematic problems in the way that the eyewitnesses to a crime are processed by the criminal justice system.

The research by the lawyers showed that while a range of factors influenced the incorrect convictions, by far the most frequent errors were mistaken eyewitness identifications (77%) (Figure 6-19). This result is supported by years of research performed by social scientists long before human DNA testing became possible. Social scientists have shown that eyewitness identification is often very unreliable and is easily influenced by inaccurate recall. Neuroscientists have only begun to understand the intricate molecular mechanisms at work in the human brain, but research shows that human memories, especially memories of traumatic events, are very fragile and are easily contaminated by interfering memories.

There are many factors that can impact the accuracy of eyewitness memories and identification, including the physical characteristics of the crime scene, such as the physical distance between the witness and the suspect, the lighting at the scene, and the time of day when the interaction took place. The studies also show that more subtle factors can influence the accuracy of eyewitness identifications. For instance, people tend to make far more mistakes when identifying someone of a different race. Also, mistakes in memory occur more frequently when a witness becomes

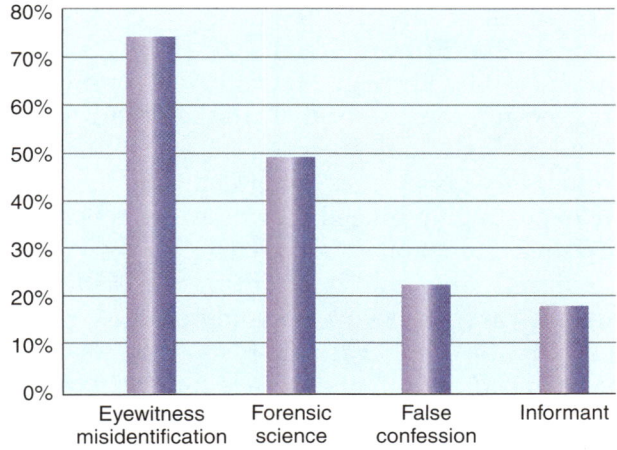

FIGURE 6-19 Eyewitness identification is often wrong! Research by the Innocence Project found that the primary reason that innocent people are wrongfully convicted and go to prison is faulty eyewitness identification. The second most prevalent reason for wrongful convictions is fraud in forensic science testing, followed by false confessions and problems with informants.

distracted, which impacts negatively on both short-term memory retention and long-term recall.

The Innocence Project is working to reform the methods used by law enforcement to retrieve information from a witness's memory, and they have recommended ways to reform the current witness identification processes, including physical lineups and photo arrays of the suspect. The lawyers at the Innocence Project are also trying to set consistent nationwide standards for witness identification, designed to obtain the most highly accurate eyewitness memories possible.

> **KEY CONCEPT**
> Eyewitness identification is very unreliable. Studies show that many eyewitnesses are mistaken and that the memories of eyewitnesses are easily contaminated by outside influences.

Fraud and Misconduct in Forensic Science

The Innocence Project research indicated that the second most common mistake leading to a wrongful conviction was fraud and misconduct in forensic science investigation and testing (52%) (Figure 6-19). Dishonesty in forensic science has an especially devastating affect on the public's ability to trust the legal system and science in general. It corrupts the legal system because fraudulent DNA science can mislead a jury trying to decide between life and death. In cases of forensic scientific fraud, the Innocence Project lawyers traced some wrongful convictions to scientists who falsified the DNA test results submitted by forensic laboratories, committed perjury when testifying in court, and misrepresented statistical data. In the courtroom the forensic DNA evidence is usually presented by scientific experts whose testimony is appropriately given more weight by the jurors. But in cases of fraud, the forensic experts give biased testimony or miss-represent the facts in court. Forensic misconduct means that the jurors are misled by the very people who are directly involved in handling the evidence in the case.

Safeguards are necessary to avoid human errors made during evidence collection at a crime scene. For this reason the forensic technicians are required to record a specific chain of evidence for every piece of forensic evidence, from the time it is collected until it is logged into the police evidence room. Specific laws that vary from state to state actually impose the local standards for handling forensic evidence. The Innocence Project lawyers have made recommendations for improvements to be adopted by forensic labs, law enforcement, and the courts to help prevent misconduct in forensic science. The identification of factors that contribute to the large numbers of wrongful convictions in the U.S. criminal justice system is an important step toward promoting the changes needed to prevent such injustices in the future.

Novel Applications of DNA Fingerprinting and DNA Testing

In addition to using DNA fingerprinting to identify criminal suspects for forensic investigations, DNA testing has become an important part of the medical tests needed to diagnose and treat many diseases. Organ and tissue transplants routinely require testing for genetic matches between the donors and recipients. However, the power of DNA testing extends well beyond diagnosing diseases and has been used in many novel and unique applications.

Super Bowl Footballs and Souvenirs from the Summer Olympics

For the past decade, DNA technology has been used by the National Football League (NFL) as a tag to identify the authentic footballs used in Super Bowl XXXIV. The footballs and other items were tagged with an invisible strand of unique synthetic DNA that can be verified in the future using a special laser. In this way these footballs are forever identified and authenticated, which is an important protection against fraud. In a similar approach, all of the official products and souvenirs of the 2000 Summer Olympic Games were marked with ink mixed with a sequence of human genome DNA taken from several unnamed athletes. This type of DNA technology has also been used to mark artwork and other unique and expensive items.

DNA Testing to Identify United States Armed Forces Personnel

It has long been a priority of the United States Armed Forces to accurately identify the men and women who die while serving in the U.S. military. In 1991 the Armed Forces DNA Identification Laboratory (AFDIL) was established by the U.S. Department of Defense and has worked on identifying remains from Vietnam, Korea, and World War II, as well as victims from natural disasters. The Armed Forces use the latest advances in DNA fingerprint analysis and dental identification, including PCR, but even such sophisticated technology cannot identify all remains.

DNA samples and other biological information are now routinely collected from all personnel in the U.S. Armed Forces, with the goal of never using the Tomb of the Unknowns in the future (Figure 6-20). Individual stories help to bring home the importance of using DNA science to help people understand the past. Michael Joseph Blassie received his officer's commission in the United States Air Force in June 1970 at 22 years old. About two years later, First Lieutenant Blassie's aircraft was shot down in Vietnam. Months later a South Vietnamese Army patrol found human remains, part of a flight suit and parachute and sent the items to the U.S. Army Central Identification Laboratory in Hawaii. Although the remains were tentatively identified as those of Lt. Blassie, they were mistakenly classified as Unknown number X-26. In 1998, the Blassie family petitioned the Department of Defense to use the new PCR-based DNA testing methods to try to determine the identity of the human remains designated X-26.

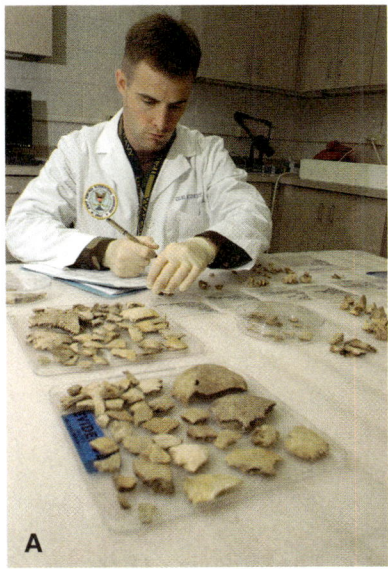

FIGURE 6-20 DNA testing to identify United States Armed Forces personnel (A) This forensic odontologist (dentist) is examining the teeth from someone who might have been in the U.S. military. Often pieces of skull and bone are also examined for potential sources of DNA for DNA fingerprinting. (B) A soldier saluting at the Tomb of the Unknown Soldier.

Military forensic anthropologists extracted mitochondrial DNA from the X-26 bone marrow and compared the mtDNA profile with the mtDNA profiles submitted by First Lieutenant Blassie's mother and sister, and the mtDNA profiles clearly matched. On July 11, 1998, First Lieutenant Michael Blassie was buried with full military honors in Jefferson National Cemetery, Missouri, near his hometown, and next to his father, another example of how DNA testing can be used to bring a different kind of justice.

Identification of People Killed in the World Trade Center and Pentagon Terrorist Attacks

The identification of people killed in the World Trade Center and Pentagon terrorist attacks on September 11, 2001, was a particularly difficult challenge for forensic investigators. The identities of many victims who were visiting the buildings were unknown, and the intense heat generated by the jet fuel explosions and the widespread fires left behind only fragments of burned bone and tissue for identification.

On September 11, 2001, the United States had no system in place to identify the victims of such a large disaster. In response to these attacks, the U.S. National Institutes of Justice assembled a panel of experts that was given the job of developing procedures to identify victims of future national disasters. The panel began with the DNA samples collected at Ground Zero at the WTC. These experts established the necessary protocols and distributed the evidence kits needed to collect the samples and to store reference DNA. They established a database of DNA profiles containing over 20,000 samples of human remains from the WTC site. The scientists also developed new information technologies to transfer DNA data between the state police and medical examiner's offices. Despite these efforts, by 2005 only 1,585 of the 2,792 people who died that day were identified. In 2007, the development of PCR and other DNA testing technologies finally allowed some additional victims to be identified.

Identifying Missing and Murdered Children in Argentina

Between 1976 and 1983 the military junta government in Argentina kidnapped and murdered almost 20,000 men, women, and children that the people in Argentina call the "disappeared" (*los desaparecidos*). When the new civilian government finally came to power in Argentina, it asked the international scientific community for help identifying missing people, and the American Association for the Advancement of Science (AAAS) sent scientists who are experts in the use of DNA testing to identify human remains. It turned out that some of the children kidnapped between 1976 and 1983 were still alive and had been raised by their kidnappers. The DNA testing helped the grandparents to find children they had tried for decades to locate. The discovery of mass graves in Argentina led to The Argentine Forensic Anthropology Team of forensic scientists that was established to investigate cases of human rights abuse in Argentina. The forensic investigators used a combination of archaeological techniques, forensic methods and DNA fingerprinting to identify human remains in the mass graves (Figure 6-21).

Liliana Pereyra was 21-years-old and five months pregnant when she disappeared in Argentina on October 6, 1976. Just before she was kidnapped Liliana had a tooth extracted, which was an important clue for those investigators who were trying to identify her remains. In 1985, a forensic odontologist discovered a skull and jaw with a tooth missing recovered alongside a partial skeleton that was eventually identified as Liliana. Although no fetal bones were found, the pelvis contained a groove revealing that Liliana had given birth to a live baby before her death, but the child was never seen again. The scientists in Argentina continue to use DNA fingerprinting to establish the biological family relationships involving the "disappeared" children (Figure 6-21). Blood samples are collected from all the potential relatives of each missing child, which provide the reference DNA profiles needed for comparison with the unknown child's DNA fingerprint. Justice was finally served for Liliana Pereyra and thousands of other victims when the former leaders of the Argentinean military junta went to trial and were convicted in 1985.

FIGURE 6-21 DNA testing to identify Argentina's disappeared children The 'disappeared' children in Argentina were sometimes born in captivity, and their kidnapped parents were killed by the former military regime. A collection of blood samples from all the potential relatives of the missing child are used to provide the DNA profiles needed to compare with the child's DNA fingerprint. These DNA testing protocols involve processing large numbers of DNA samples at the same time using a multi-channel pipette.

Identifying the Son of Louis XVI and Marie Antoinette

In 2000 DNA testing solved a great historical mystery and proved that the son of French King Louis XVI and Marie-Antoinette had in fact died as a child in prison. For over 200 years the fate of Louis-Charles de France was hotly debated among historians who proposed theories about whether the boy had actually died in a Paris prison in 1795 or had somehow escaped during the French Revolution. Scientists were given the presumed heart of Louis-Charles, preserved for over two centuries, and they compared the DNA fingerprint profile made from the cardiac tissue with the DNA fingerprints from biological samples collected from the current members of the royal family. The DNA tests were also performed on a lock of Marie-Antoinette's actual hair, which confirmed the maternal genetic link to the boy's heart.

Another historical mystery solved by DNA testing involved another royal family, this time in Russia. In 1917 after the Russian Revolution the entire Romanov family, Tsar Nicholas II, his wife Tsarina Alexandra Feodorovna and their five children, were executed by the Bolsheviks. The bodies of the royal family were buried in a secret location in Russia until 1992, when the bodies were discovered and positively identified by DNA testing and facial reconstructions (Figure 6-22). The remains of the royal family were reburied in St Petersburg.

Scientists Use Mummy mtDNA to Trace very Old Family Trees

DNA testing is applied to the science of anthropology to study the remains of people who lived thousands of years ago. In 1995 climbers on Mt. Ampato in the Andes discovered the frozen mummy of a teenage girl, thought to be a sacrifice to the Incan gods by Incan priests about 500 years ago. The "Ice Maiden" mummy is especially well preserved, offering scientists the rare

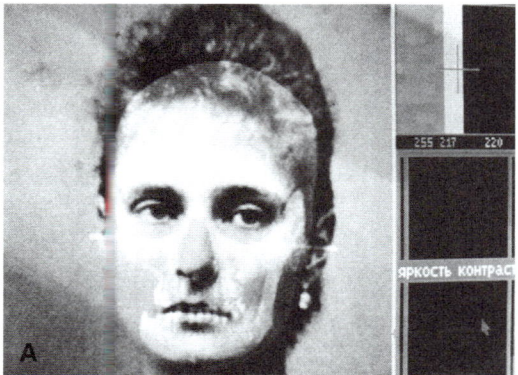
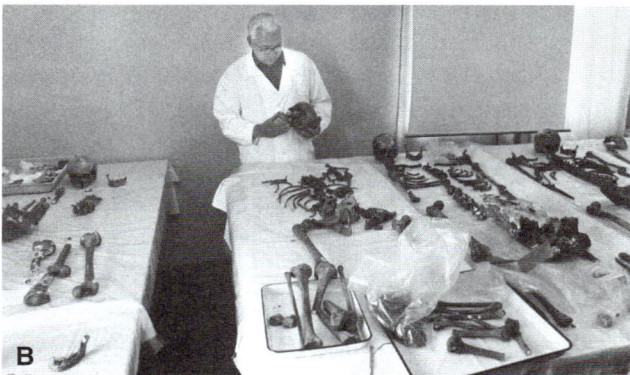

FIGURE 6-22 Tsar Nicholas II and his wife Tsarina Alexandra (A) In 1917 Tsar Nicholas II Romanov, and his entire family, were executed by the Bolsheviks and the bodies were buried in secret locations until 1992 when their bodies were discovered. Forensic technologies including facial reconstructions were used to confirm the identity the remains. (B) Forensic examination of the Romanov skeletons.

opportunity to test her DNA to learn more about her genetic ancestors and the origins of the Incan people. For mtDNA testing, the scientists extracted mitochondrial DNA from the heart cells because all muscle cells, including cardiac muscle cells, contain additional mitochondria to produce the energy needed to for physical movement.

The maternal inheritance of mitochondrial DNA allowed scientists to trace Native Americans to four original populations of Paleo-Indians that migrated from Asia into the Americas. MtDNA testing on the Ice Maiden shows that she has genetic ties to the Ngobe Bugle, one of the seven indigenous tribes of Panama, and she has connections to Taiwan and Korea, supporting the idea that the original Paleo-Indians actually came from Asia. Since the Ice Maiden was found, additional frozen mummies have been discovered in the Andes, and scientists predict that additional research will provide the information needed to trace the genetic origins of the Ice Maiden's family.

The remains of a mummy recovered from Peru were studied by researchers from the University of Rome Tor Vergata in Italy (Figure 6-23). The scientists wanted to analyze the mummy DNA to generate a mummy DNA fingerprint to help researchers to better understand the genetics of the Peruvian pre-Hispanic population. The hot, dry desert climate of Peru is ideal for the mummification process, which preserves the remains by desiccation, and also helps to preserve the DNA molecules. The scientists extracted the DNA from the pulp of a tooth from the mummy and used gel electrophoresis to analyze the DNA (Figure 6-23).

DNA Testing, Dog Genes and Man's Best Friend

The domesticated dog is the genetic product of over 100,000 years of human domestication, including many recent decades of selective breeding. Modern dog breeds represent a huge diversity and range of animals that includes the 6-pound Chihuahua and the 120-pound Great Dane, the high-energy, spirited Jack Russell Terriers and the calm, well-mannered basset hounds.

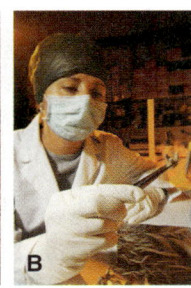

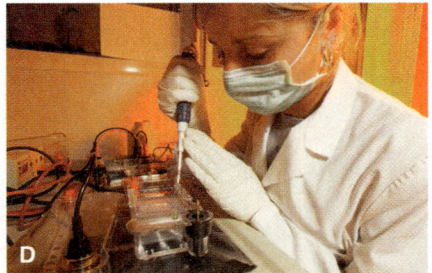

FIGURE 6-23 Peruvian mummy donated a tooth for DNA testing (A) Several intact mummys have been recovered from Peru because of Peru's ideal desert climate for mummification by desiccation. (B) Researcher wanted to extract DNA from the pulp region of one of the mummy's teeth. (C) The tooth extracted from the mummy. (D) Researchers analyze the DNA from the mummy's tooth using gel electrophoresis to separate any DNA fragments. The researchers must dress to avoid contaminating the mummy DNA, which is particularly essential for PCR amplification of the mummy DNA.

Individual dogs of different breeds have exhibited amazing abilities from the precise herding skills of Shetland sheepdogs to the mixed breed mutts that can smell danger in the forms of explosives in suitcases, cancer cells in the body, and even low blood sugar levels in diabetics. Humans have long bred dogs for desirable physical and behavioral traits. However, while these breeding practices do preserve the desired traits from one generation to the next, they also predispose many pure bred dogs to serious genetic disorders, including heart disease, cancer, blindness, cataracts, epilepsy, hip dysplasia and deafness. Health problems often result because pure bred dogs inherit two copies of the recessive mutant genes that cause these genetic problems. Dog breeders now routinely have purebred dogs screened by a commercial testing service to avoid breeding dogs that are carriers of these gene variants and to prevent offspring that are predisposed to genetic problems.

In 2005 a boxer called Tasha was the first dog to have her genome DNA sequence published. The boxer genome was adopted by scientists as a reference sequence for dog genome studies because the boxer breed represents the average purebred dog. Scientists compared the boxer genome DNA with the genome sequences from 10 other dog breeds, as well as genomes from the gray wolf and the coyote, and found many single nucleotide differences among the genomes. These canine SNPs are used as genetic markers to locate dog genes encoding proteins that help scientists to learn about similar genes and proteins that cause human diseases.

The dog genome is not just important for understanding the genetics of pedigree dogs, it is also important for research on human diseases that are mimicked and studied in dogs. Narcolepsy is a form of sleeping sickness that occurs in humans and also occurs naturally in some animals, including a few breeds of dogs. Stanford University researchers studied the genetics of the narcoleptic dogs and found two different mutations in the same gene that cause narcolepsy in two different breeds of dogs. These canine studies were instrumental in identifying a similar gene in humans that is also mutated in people with narcolepsy. The gene altered in narcolepsy in both dogs and people codes for the hypocretin receptor 2 protein (HRP2), which functions to transmit sleep signals to the brain. Studies on dogs have helped scientists to learn about many other genes that affect both man and man's best friend.

Janet Foad's problem with her dogs was not a medical emergency, but she felt fortunate that it could be easily solved by DNA testing. Foad's business is breeding pedigree Siberian husky dogs and she was quite surprised when her Siberian female, Kynda became pregnant (Figure 6-24). In addition to DNA testing for genetic diseases, DNA testing can also be used to determine paternity. Foad needed to be certain which male dog was the father of the puppies before she could complete the paperwork to register the proof of pedigree for the litter. Cellmark Diagnostics lab performed the DNA fingerprinting work and identified Ali as the happy, proud father of the pedigree litter of puppies (Figure 6-24).

DNA Testing to Help Endangered Animals

The increase in endangered animals around the world, combined with the threat that climate warming will further endanger animal habitats, has prompted people to start different projects to help endangered animals

FIGURE 6-24 DNA fingerprinting for dog pedigree Janet Foad breeds pedigree Siberian husky dogs. She was surprised when her Siberian female, Kynda (upper right) became unexpectedly pregnant. In order to register the proof of pedigree for the new litter, Foad needed to know which of her male dogs was the father. Foad hired Cellmark Diagnostics lab to perform DNA fingerprinting, which showed that the dog, Ali (upper left) is the father of Kynda's puppies.

survive now and in the future. Unusual types of biological banks have been established to preserve samples of sperm, ova, embryos, tissue samples, and serum as well as genome DNA samples collected from living endangered animals. Most biological materials can be kept frozen in liquid nitrogen indefinitely, which inspired zoos and conservation groups around the world to collect and preserve genetic material from endangered animals. The largest DNA bank for endangered animals is called Frozen Zoo and is based at the San Diego Zoo. Since 1975 the Frozen Zoo has added samples from more than 7,000 endangered species, including more than 13,000 samples of semen, oocytes (eggs), and embryos. The Cincinnati Zoo and Botanical Garden and the Audubon Society also support a DNA bank that contains about 95 different endangered animal species and 150 plant species. In 2004, three British institutions began the Frozen Ark project to provide a genetic repository for all endangered species on the World Conservation Union's (IUCN) Red List, which contains more than 7,200 species.

Many animals are endangered because their habitat is threatened by the encroaching human populations. Grizzly bears were recently granted threatened species status because of concerns about the declining grizzly bear numbers in Canada and the United States (Figure 6-25). The difficulty of counting individual grizzly bears and other kinds of bears in the wild has affected efforts to preserve the species and has an impact on the number of annual hunting permits made available to hunt grizzly bears. The researchers studying bears in the western United States have devised a simple way to collect information on the bears in the wild without direct encounters with the dangerous animals. The scientists hang short strings of barbed wire alongside the trails commonly used by the bears. When the bears pass by they snag bits of fur on the wire, which is left behind for scientists to collect

FIGURE 6-25 Helping endangered animals survive (A) Grizzly bears recently became endangered because of recent declines in the grizzly bear populations in the U.S. and Canada. (B) The rarest big cat in the world is the Amur leopard in the mountains of northern Russia where fewer than 50 animals were counted in 2006. (C) The giant panda must eat for most of its waking life because it exists on bamboo shoots and leaves. Bamboo is abundant in the panda's native home in the mountains of southern central China.

and analyze. The DNA fingerprints developed from the fur left behind has provided a wealth of information about the bear population and can be used to track the movements of individual bears.

The endangered Amur leopard, which numbered just 50 animals in 2006, could be a candidate for programs that clone endangered animals (Figure 6-25). The giant panda is also an endangered species, but its habitat is protected by the Chinese, who have also sponsored Panda breeding programs and have a successful program to loan pandas to zoos around the world and to reintroduce giant pandas into the wild in China.

Unfortunately, unlike the Amur leopard and the giant panda, many animals are endangered because they are hunted by poachers who profit from the animals they kill and sell for the illicit trade in animal products. The white rhinoceros are still killed for their horns, which are thought to be an aphrodisiac by Asians who grind the horns into a powder that is sold illegally for use in traditional medicines (Figure 6-26). The rhino horns are actually made of keratin just like human fingernails, and the horns also grow continuously like human fingernails. The park wardens cut off the rhino's horns to try to protect the animals from poachers, but the horns grow back.

Poaching reduced the white rhino to near extinction by the end of the nineteenth century and African elephants have long been poached because of the value of their ivory tusks (Figure 6-26 and Figure 6-27). The game wardens display large collections of elephant and rhino skulls from animals killed for their tusks and horns. The numbers of these animals have dramatically decreased due to poaching, and currently both African elephants and the white and black rhinos have been given endangered species status. Poaching still continues despite the national parks and reserves that were established in Zambia to try to protect the wildlife.

UNIT 6 DNA Forensics and Epigenetic Reprogramming

FIGURE 6-26 Endangered animals threatened by poachers (A) The Galapagos Giant Tortoise is endangered even though their lifespan is 150 to 200 years. When allowed to live, these tortoises are the longest living vertebrates known. (B) The horns of the white rhinoceros are made of solid keratin like human fingernails. Poachers kill the rhinos for their horns. Sometimes the game keepers cut off the rhinos horns like the rhino in this picture to avoid poaching, but the horns grow back. (C) African elephants have long been the targets of poaching because of their long ivory tusks.

DNA Barcode Technology

In addition to using DNA fingerprinting to identify criminal suspects for forensic investigations, DNA testing has become an important part of the medical tests needed to diagnose and treat many diseases. Organ and tissue transplants routinely require testing for genetic matches between the donors and recipients. However, the power of DNA testing extends well beyond diagnosing diseases and has been used in many novel and unique applications. DNA barcoding allows all biological organisms to be identified using the different

FIGURE 6-27 DNA testing helps track illegal trade in endangered animal products
(A) Elephants and rhinos have been killed by poachers for their tusks and horns. (B) Piles of African elephant tusks. (C) These rows of the jawbones and skulls represent the African elephants and rhinos killed illegally. (D) Forensic lab director inspects wildlife products.

DNA sequences of a short region of DNA. The organism's entire genome does not need to be sequenced in order to use the DNA barcode system.

The Barcode project was started by Paul Hebert at the University of Guelph in Ontario, Canada in 2003 who heads an international group of scientists with the goal of establishing a standard DNA barcode sequence to identify each species. This molecular approach to species identification differs from the more traditional approach, which often relies primarily on the examination of the morphological characteristics of the sample including size, shape and color.

The scientists chose a 648 bp region of mitochondrial DNA as the standard DNA barcode for almost all animal groups because this region of mtDNA has variable sequences in different species as needed to assign a unique barcode sequence to each species. Also, mtDNA is much more abundant than nuclear (chromosome) DNA in human cells. The CO1 barcode DNA is located in the mitochondrial cytochrome c oxidase 1 gene encoded in the mtDNA genome. The COI DNA barcode can be used to identify birds, butterflies, fish, flies and many other animal groups, but cannot be used to identify different plant species. Two regions of chloroplast DNA were identified as standard DNA barcode regions for use identifying plant species.

DNA barcoding has the advantage that species can be identified without having to determine the DNA sequence of the entire genome. In addition, hand held scanners are under development to detect the short mtDNA sequence in the field. To make the DNA barcoding system appropriate for field applications, the barcode scientists have developed a straightforward hands-on method to isolate mtDNA from different types of biological samples. Even non-scientists use this method to generate an accurate DNA barcode from a small amount of tissue sample. DNA barcoding is now an essential addition to the tools traditionally used by taxonomists as well as experts in other fields to determine species identification.

> **KEY CONCEPT**
> DNA fingerprinting and DNA barcoding are different methods with overlapping goals and applications. DNA fingerprinting provides a DNA profile with a series of DNA bands that represent genetic markers throughout the entire genome and can be used to identify individual people and rule out others. DNA barcoding is a method of categorizing species according to the sequences of a 648 bp region of mtDNA, which is a little bit different in each species.

Epigenetic Modifications can Alter Gene Expression for Many Generations

DNA barcoding and DNA fingerprinting technologies both involve determining differences in the DNA sequences of human genomes. These DNA sequence differences can alter the functions of genes, proteins and cells and can be transmitted to biological offspring. Scientists now know quite a lot about how eukaryotic cells control the expression of the genes in order to develop into new and different types of cells with different shapes and functions in the body.

Even though very sophisticated genetic systems control the transmission of genetic diseases and the inheritance of mutant genes, it is the epigenetic mechanisms that can have a far-reaching impact on gene and genome control for generations to come. Imagine if the environmental events experienced by someone's great-great grandparents made permanent changes in their genomes (but no changes in DNA sequences), which were transmitted to their children and grandchildren by epigenetic inheritance. It is an intriguing possibility that the current childhood obesity and diabetes epidemics might in part be caused by alterations in the sperm DNA of ancestral fathers who lived a sedentary lifestyle and ate a high-fat diet. How could this happen without altering the DNA sequences of the genes involved?

Epigenetic gene regulation involves two types of chemical modifications that involve the DNA helix and the histone proteins. These epigenetic mechanisms alter the functions of the main protein components of the nucleosome and regulate gene expression by chemical modification, but they do not change the DNA base-pair sequence (Unit 3). A system of enzymes that recognizes specific regions of chromosome DNA, including the promoter sequences, will add and remove small methyl groups, modifying the genome and changing the expression of many genes. The DNA methyltransferase enzyme (DNMT) adds a methyl group ($-CH_3$) to the C base in CpG base sequences in promoter regions of the DNA helix. The methyl groups are added to the external backbone strands of the DNA helix and do not change the base sequence of the DNA helix (Figure 6-28). However, the addition of small chemical methyl groups to the promoter DNA region of a gene will typically repress the expression of the methylated gene (Figure 6-28).

The second important epigenetic mechanism involves chemical modification of the histone proteins in the nucleosome (Figure 6-29). The central regions of eight core histone proteins, H3, H4, H2A and H2B fold together to make the octamer core of each nucleosome, with the amino (N-terminal)

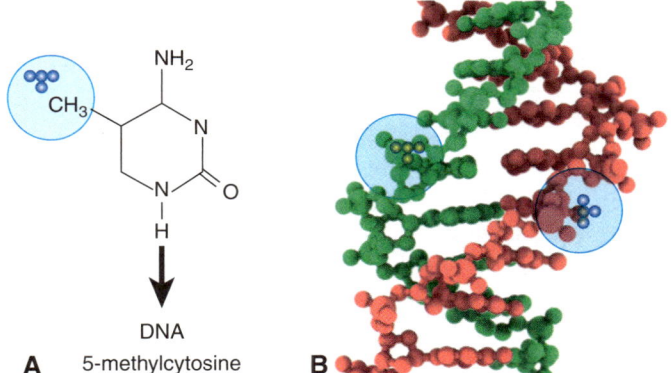

FIGURE 6-28 Methylation does not change the DNA sequence (A) Changes in DNA methylation occur when special enzymes move methyl groups (-CH3) from one site to another site on the chromosome DNA. (B) The DNA sequence remains unchanged by the addition or removal of methyl groups from the DNA helix.

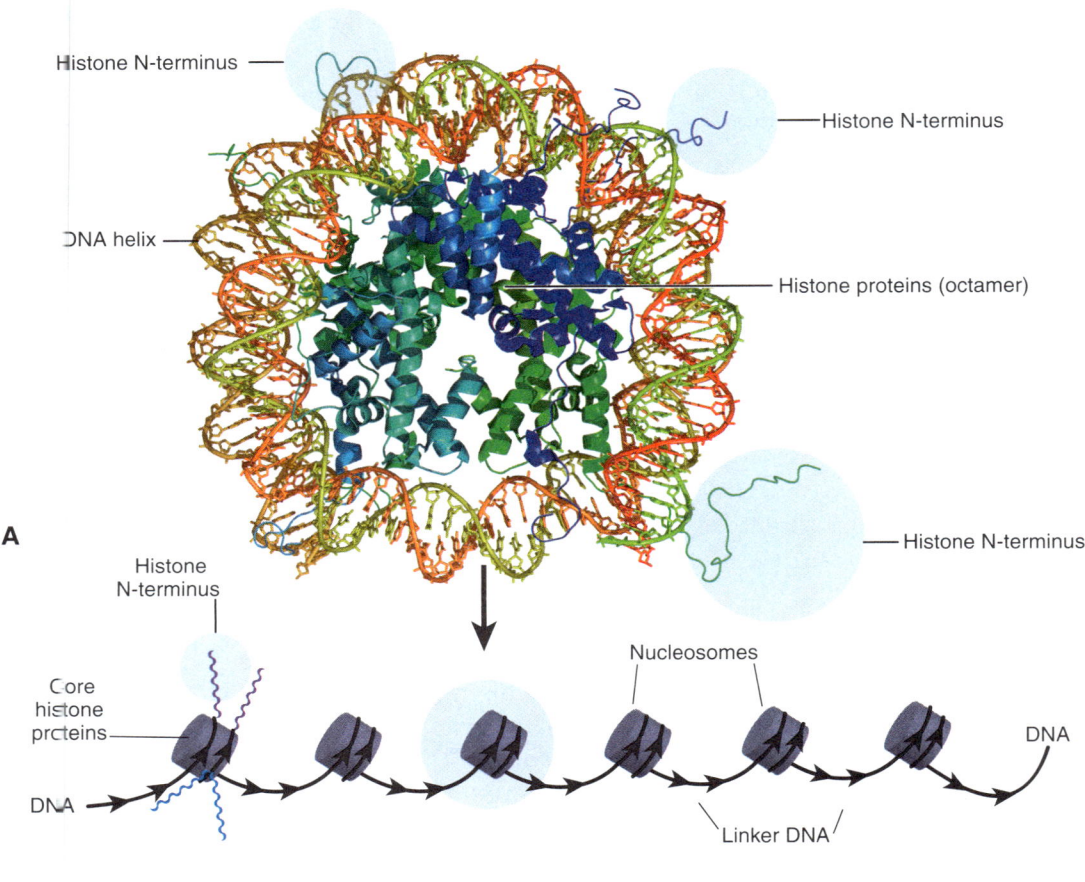

FIGURE 6-29 Histones are DNA binding proteins that package chromosomes (A) Nucleosomes are DNA-protein complexes in eukaryotic chromosomes. Each nucleosome contains a core of eight histone proteins (H3, H4, H2A, H2B) with the histone globular regions assembled into the core, and the amino (N-) terminal ends of the histone proteins extending away from the nucleosome core. A completely assembled nucleosome has about 146 bp of DNA helix wrapped twice around the outside of an octamer core. The amino (N-) termini of the core histone proteins extend away from the nucleosome core. (B) Chromatin fibers contain nucleosomes arranged like beads on a string with short regions of linker DNA between the nucleosomes

ends of the histone H3 proteins extending outside the nucleosome. The amino terminal ends of the histone core proteins interact with the N-termini of the histone proteins in neighboring nucleosomes. Special enzymes add small methyl and acetyl chemical groups to specific amino acids in the amino (N-terminal) ends of the histone proteins in the nucleosome core, altering the interactions between proteins.

Adding and removing these chemical modifications alters the charges on the ends of the histone proteins. Changes in charge alters the interactions between the histone proteins in adjacent nucleosomes. The histone N-termini with the same charge tend to repel each other, whereas the histone N-termini with opposite charges will attract each other. These interactions between histones in adjacent nucleosomes in beads on a string can

UNIT 6 DNA Forensics and Epigenetic Reprogramming

effectively cause the nucleosomes to push apart and extend the chromatin fiber, while attraction between neighboring nucleosomes will tend to contract and condense the chromatin fiber containing the DNA helix. Histone protein modifications that alter the charges on the ends of the histone proteins can also change the expression of associated genes (Figure 6-29).

The presence or absence of methyl groups at the many DNA methylation sites (CpG) in the human genome DNA are controlled in part by the DNA methyltransferase enzymes (DNMT). These enzymes move the methyl groups from one region of the chromosome DNA, such as a promoter, to another region in nearby DNA, where there is a region to store methyl groups on the DNA (Figure 6-30). For a gene to be expressed (transcribed) the methyl groups must be removed from the promoter DNA by the DNMT enzymes, because the transcription factor proteins and the RNA polymerase enzyme cannot bind to the methylated promoter DNA. The chromatin structure in

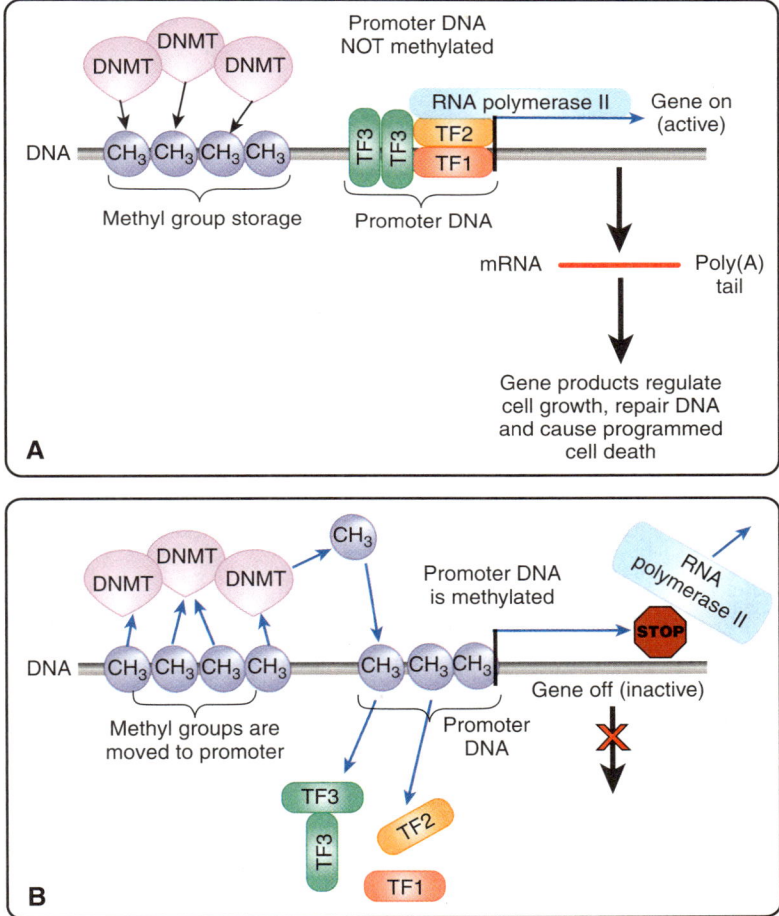

FIGURE 6-30 Enzymes change DNA methylation on the gene and alter gene expression (A) Methylation enzymes add and remove methyl groups (-CH_3) from the DNA helix, but do not change the DNA sequence. The addition of the methyl groups to the promoter DNA blocks access to RNA polymerase II and represses gene expression. (B) The same methylation enzymes can remove the methyl groups from the promoter DNA, which then allows the gene to be expressed.

the region of a gene influences the activity of the DNMT enzymes because the DNMT enzymes can methylate only DNA that has been packaged into accessible chromatin structures.

The DNA methylation and replication mechanisms in eukaryotic cells ensure that the methylation patterns in the genome DNA are maintained through generations of cell divisions. After the chromosome DNA has replicated, each strand of the original DNA helix molecule is base-paired with the newly synthesized complementary DNA strands. These DNA helices are hemimethylated because they each contain one methylated DNA strand and one unmethylated DNA strand (Unit 2) (Figure 6-31). The DNA methyltransferase 1 enzyme (DNMT1) specifically recognizes the hemimethylated CpG dinucleotides and transfers a methyl group onto the unmethylated cytosines to convert the hemimethylated DNA to fully methylated replicated DNA.

> **KEY CONCEPT**
> DNA methylation is not random in the human genome. The pattern of DNA methylation in the genome is altered by the action of specific protein enzymes, which also changes gene expression.

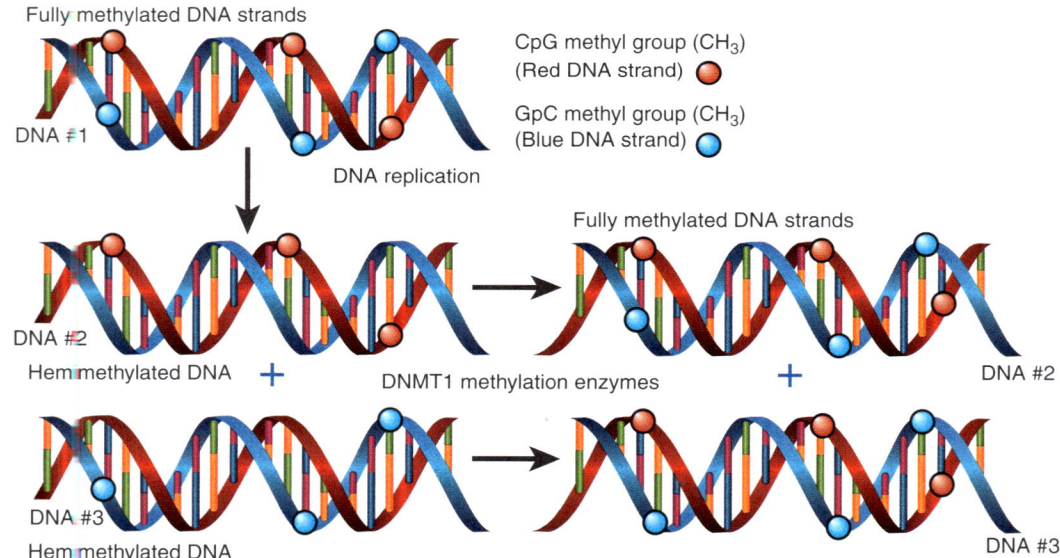

FIGURE 6-31 Patterns of DNA methylation are actively maintained on the DNA Following DNA replication in eukaryotic cells, the regions of hemimethylated DNA are repaired. The hemimethylated DNA is a DNA heteroduplex that contains one methylated DNA strand base paired to one unmethylated DNA strand. The DNA methyltransferase 1 enzyme (DNMT1) specifically recognizes the hemimethylated CpG dinucleotides in the DNA and transfers a methyl group onto the unmethylated cytosines (C), converting the hemimethylated DNA into fully methylated replicated DNA.

Insulin gene expression is regulated by TFs, H3 methylation and chromatin structure. A good example of the impact of chromosome structure on gene regulation involves the insulin gene, which is packaged into compact, silent chromatin in all types of cells in the human body, except for the islet cells in the pancreas, the organ that normally makes insulin in healthy adults. The mechanism that permits the insulin gene in the pancreatic islet cells to escape chromatin mediated suppression of transcription is not yet completely understood. Insulin gene expression in the pancreatic islet cells is epigenetically regulated by modification of the local histone H3 proteins during early embryogenesis. At that time transcription of the insulin gene is activated by the addition of methyl groups to the N-termini of the histone H3 proteins (H3K4).

The pancreatic islet cells express the PDX-1 protein, which binds to the insulin promoter DNA. PDX-1 recruits the Set7/9 methyltransferase enzyme to the insulin gene promoter, where the methyltransferase enzymes add methyl groups to the amino-termini of H3 histone proteins in nearby chromatin. The addition of methyl groups causes the chromatin fiber containing the insulin gene to extend, allowing the RNA polymerase II enzyme to have access to the promoter to transcribe the insulin gene.

The pattern of histone H3 methylation of the insulin genes occurs during early embryogenesis and has established the chromatin structures necessary to ensure that active insulin promoters are functioning when the progenitor stem cells differentiate into specialized cells. After the differentiated state is established by the cells, in this case pancreatic islet cells, the specialized functions must be maintained through many cell divisions, including stages when many tissue-specific expressed genes are temporarily repressed.

The status of a transcriptionally active gene like the insulin gene is maintained during mitotic cell division by epigenetic memory, which relies on the epigenetic modifications of both the histone proteins and the genome DNA to duplicate the local chromatin structures. Further research is needed to explore the relationships between the chromatin modification events that control gene expression in pancreatic islet cells and in other cell types.

The timing, type and locations of different epigenetic modifications on genome DNA and histone proteins are affected by many factors, including age, diet, conditions *in utero*, nutrition during childhood, and exposure to environmental radiation, chemicals, and drugs. Scientists have discovered from experiments with lab mice that the methylation level of chromosome DNA can be altered by changing the methyl group donors present in food sources. In turn, the methyl groups can activate or repress certain genes. Modification of the histone proteins changes interactions in the chromatin that activate associated genes, which can have a direct impact on an individual's health, potentially causing cancer, autoimmune diseases, mental disorders, or diabetes.

Transgenerational epigenetic inheritance

Dr. Michael Skinner and his team (Washington State University) performed experiments on lab rats and discovered that certain epigenetic changes in the rat's sperm DNA could be transmitted to future generations. In this research the team began by exposing the male rats to a fungicide, vinclozolin, to determine whether the fungicide could induce epigenetic changes in the sperm DNA. The researchers found that vinclozolin altered the methylation patterns in the rat's genome DNA, and they confirmed that the epigenetic changes caused by the fungicide were transmitted to the rat's offspring by transgenerational epigenetic inheritance. Amazingly, the scientists also found that transgenerational epigenetic changes in the DNA can be caused by early life experiences even experiences in the womb, which permanently change the epigenetic program of the sperm genome.

The molecular mechanisms responsible for transgenerational epigenetic inheritance differ from the more familiar mutations that change the DNA base sequence of a gene. Exposing the rats to the fungicide altered the epigenetic on-off master switches controlling gene expression in the sperm DNA. These master switches are regulated by the presence or absence of methyl groups on the control regions of the DNA. The addition of a methyl group will turn the DNA switch off and prevent transcription, silencing the gene. However, removing methyl groups from a DNA switch will turn the gene on, and the gene will be actively transcribed once again.

Skinner carefully monitored the changes in the methylation patterns of the sperm DNA from the male rat pups that were exposed to the toxin while *in utero*. The researchers analyzed the sperm DNA from the male pups exposed to the fungicide *in utero*, and they found that 16 of the CpG methylation sites in the rat sperm DNA were altered (switched ON) when the normal position was OFF, or the CpG sites were (switched OFF) when the normal position was ON.

The male pups exposed *in utero* were allowed to grow and breed and the 16 different DNA sites were analyzed in the sperm DNA of the male offspring. Surprisingly, the same changes were found in the sperm DNA of the males exposed *in utero* and in the sperm DNA of their sons and in the sperm DNA of their sons' sons as well. At fertilization, the sperm and egg genomes come together and create an epigenome, the new genome programmed for embryo development. Until recently, scientists thought that the methyl groups added to the genes before fertilization would be 'reset' or 'erased' in the epigenome, blocking or removing any impact of the environmental events on the next generation. However, recent research reveals that while some methyl groups are removed from the genome, other methyl groups remain covalently bonded to the DNA backbone as 'permanent' genome modifications. In this experiment the rat offspring exposed to the fungicide and their grandsons exhibited the effects of exposure to the fungicide, including many changes in gene expression, such as the activation of silent genes and repression of active genes. Some male descendants also experienced abnormal development of the testes, prostate, and kidneys.

In human terms, this animal research suggested that environmental events experienced by grandparents and even great-grandparents could potentially transmit permanent genome changes to their children and grandchildren by epigenetic inheritance. However, this research does not prove that exposure *in utero* to the fungicide would cause the same problems in humans. Scientists have found that exposure to low levels of some environmental chemicals had a measurable impact on four generations of human male descendents. These "common human life experiences" include smoking cigarettes, childhood starvation, and obesity in adults, experiences that can potentially cause specific changes in DNA methylation in the human sperm genome that can be transmitted to subsequent generations. It is an intriguing possibility that the current epidemic in childhood obesity and type II diabetes in children could be caused by alterations in the sperm DNA of ancestral fathers who were sedentary and ate a high-fat diet. The 73% increase in obese infants since 1980 cannot be completely explained by the couch-potato lifestyle and fattening diets of newborn infants!

Transgenerational epigenetic inheritance can also have positive effects. For example, when 15-day-old female mice were allowed to frolic for two weeks in a stimulating environment with exercise wheels, novel objects, and other mice for socialization, the results show that the experience greatly strengthened the mechanism in the brain that builds memory. Subsequent experiments showed that the animals raised in the enriched environment also exhibited an increased memory for running mazes. Amazingly, research by Larry Feig (Rush University Medical Center, Chicago) demonstrated that this improved memory for running mazes is transferred to the biological offspring, even if the pups had never lived in an enriched environment!

Household Toxins can Make Epigenetic Changes in Our Children's Children's DNA

Scientists wanted to find out if the offspring of a pregnant mouse exposed to a household toxin would show signs of epigenetic changes similar to the methylation changes in the sperm DNA of the rats exposed to a fungicide *in utero*. Transgenerational epigenetic inheritance involves traits that are inherited through several generations without an associated change in the sequence of the genome DNA. Transgenerational epigenetic inheritance suggests the possibility that exposure of an embryo or fetus to a toxin, even *in utero,* can not only potentially alter the health of the individual but also potentially impact the health of that individual's descendants.

Bis-phenol-A (BPA) was chosen for this experiment because this household chemical mimics the sex hormone estrogen and can cause cancer. BPA is used in many plastic consumer products, including drinking bottles and baby bottles. In 2008 the governments of several countries issued reports questioning the safety of BPA, and BPA is now banned in Europe and Canada. BPA is not banned in the United States, although the U.S. Food and Drug Administration (FDA) raised concerns in 2010 about exposing the public to BPA in common household items.

Scientists wanted to explore the possible affect of exposure to BPA on mouse fetuses developing *in utero*. Scientists decided to use the range of coat colors available on Agouti mice as a way of using the coat color gene to monitor DNA methylation on the coat color genes. Using this idea, scientists designed an experiment to test whether maternal exposure to the common chemical BPA would affect gene methylation and expression patterns during fetal development *in utero*.

To picture how the Agouti coat color gene can be used to monitor gene methylation, consider two genetically identical twin mice that have identical DNA genomes, yet the twin mice have very different physical appearances.

One of the twin mice is slender and has a brown coat, while the other twin mouse is obese and has a yellow coat.

Although these twin mice have genomes with identical DNA sequences, they clearly have different gene expression patterns. In the slender twin mouse with the brown coat the methylated coat color genes are repressed, preventing or reducing the expression of certain genes. The coat color genes in the obese twin with the yellow coat are not methylated and therefore are expressed, so this twin mouse is born with a yellow coat. The gene expression controlling the development of the different body types of the twin mice, slender and obese, is also affected by the changes in DNA methylation.

BPA Exposure Alters Gene Expression in Lab Mice

In the experiment to test BPA, the control group included slender, brown pregnant female mice that were not exposed to BPA. As expected these slender brown female mice gave birth to litters of mouse pups that were predominantly slender and had brown coats. The experimental group of slender brown female mice that were exposed to BPA while pregnant gave birth to predominantly obese offspring with mostly yellow coats. These results indicate that exposure to BPA *in utero* changed the expression of the coat color genes presumably because of the addition and removal of methyl groups on the DNA of the affected genes.

The scientists also sequenced the DNA of the coat color genes in the offspring to confirm that the DNA sequences of these genes were not mutated when the genomes of the mice were exposed to BPA *in utero*. Additional research confirmed that the coat color genes did have changes in DNA methylation compared with the genes from mice that were not exposed to BPA. These studies concluded that the changes in gene expression in the baby mice were a direct result of the changes in the DNA methylation of the fetal DNA caused by exposure to BPA *in utero*.

The brown and yellow twin mice have littermates with mottled yellow and brown coats, which were actually made up of a mixture of coat color skin cells. About half of the cells carried repressed methylated genes that were silent while the other half of the cells expressed unmethylated genes that were expressed. Gene control by DNA methylation does not always have an all-or-nothing impact on gene expression. In a population of cells, some variation in DNA methylation exists, and it is possible that some copies of a particular gene will be fully methylated while other copies of the same gene might not have as many methyl groups. For any given gene, the methylation pattern might be retained when the gene is replicated and transmitted to offspring, but sometimes the inherited DNA methylation patterns are altered. These observations show that the epigenetic changes involving DNA methylation do not alter the base sequence of the genome DNA.

In the next BPA study using mice the scientists wanted to find out if they could counteract the effect of BPA on gene expression *in utero* by providing the pregnant female mice with an excess of methyl donors in their diet. A group of the slender, brown pregnant mice was fed a methyl-rich diet supplemented with methyl donors to provide the developing fetuses with an abundant supply of methyl groups. The slender, brown pregnant mice were divided into two groups: one group was exposed to BPA, and the other group was not exposed to BPA. Both groups continued to eat a methyl-rich diet. A control group of pregnant mice was exposed to BPA, but these females were not fed a methyl-rich diet to duplicate previous results obtained for BPA.

The scientists were surprised to find that the group of slender, brown female mice that was exposed to BPA and fed a high-methyl donor diet gave birth to litters of slender mice with brown coats! In contrast, the offspring born to females that were exposed to BPA but without the protection of a high-methyl diet, were born predominantly obese with yellow coats, clearly reflecting the unmethylated state of the coat color genes in these offspring.

The success of the methyl donor experiments to protect pregnant mice exposed to BPA raised the idea that a similar mechanism might operate in humans. In the future, OB/GYN doctors might routinely recommend that pregnant women supplement their diets with nutrients known to counteract the potential genetic risk of exposure to dangerous environmental toxins like BPA.

> **KEY CONCEPT**
> Women who are planning to become pregnant even in the future should heed the warnings of doctors to take specific vitamins and other supplements. Do not use 'natural' or 'herbal' replacements or supplements without discussing it with your OBGYN doctor.

MyDNA Book has Methylated Bookmarks!

The DNA sequence of an individual's genome remains unchanged from embryo to adulthood over many years, as the cells grow and develop into tissues and organs according to the preprogrammed gene expression plans recorded in the human genome. Sometimes body cells acquire a chromosome DNA mutation as a result of exposure to a mutagen, a mistake in DNA replication or the failure of the cell to repair damaged chromosome DNA. But with these exceptions the amazing human genome DNA functions equally well in a developing 4-cell embryo and an 80-year-old grandmother, without any significant changes taking place in the DNA base sequence of the individual genome. However, epigenetic changes that occur during the life span of a human can alter gene expression for generations to come.

There are important differences in the epigenetic genomes from an embryo cell and an adult cell. The methylated DNA sites in the embryonic and adult genomes all function as bookmarks or placeholders in the genome (Figure 6-31). However, the new genome DNA in the developing embryo cells contains very few methylated bookmarks compared with the adult version of the same genome DNA, which contains many more methyl group bookmarks (Figure 6-31). Is this why Dolly the cloned sheep died young (Unit 5)?

When a nucleus undergoes somatic cell nuclear transfer (SCNT), it ends up inserted into the empty cytoplasm of an empty egg cell. In this confusing situation the adult somatic cell nucleus contains a genome with a full complement of methylated bookmarks represented by specific methylated sites in the DNA genome. This highly marked adult genome must change dramatically in order to convey the instructions needed to convert the adult nucleus into an embryonic nucleus, in a process called nuclear or genome reprogramming. A similar situation is encountered by the adult skin cells that express the external DNA genes to convert adult skin cells into induced pluripotent stem (iPS) cells (Unit 5).

Like the book markers in a printed, bound book, the methylated markers in the DNA genome are used to help the cells to navigate the 46 chromosomes, 3.2 billion base pairs, and 18,000-20,000 genes in the human genome (Figure 6-32). The cells in the human body all have a copy of the same genome DNA book, but different cell types need to use different pages from the genome book. The methyl group bookmarks are guideposts to help control cell and tissue development and to make the correct proteins, depending on the type and functions of the cell. The chromosome DNA genomes in fully differentiated cells have many methyl group bookmarks with important roles in controlling gene expression (Figure 6-32).

Researchers at Johns Hopkins University analyzed the changes in the genome methylation patterns of DNA samples obtained from people living in Iceland over a period of 11 years. The team analyzed 4.5 million human DNA methylation sites and determined which of the DNA sites contained methyl groups and which sites were not methylated in each genome. They identified 227 sites in the variably methylated regions (VMRs) of the human genome and found that the methylated state of 119 of the 227 methylation

A   B

FIGURE 6-32 Methyl groups are MyDNA book markers The methylated DNA markers in the human genome book help the cells to express the correct genes at the correct times and in the correct amounts. (A) All the cells in the human body have a copy of the same genome DNA book, but different cell types need to use different pages. (B) The methyl group bookmarks in the genome and in the MyDNA book are guideposts to help the cell to control gene expression and make the correct proteins to perform cell functions.

sites had not changed during the 11-year period, while the remaining VMRs did change, possibly as a result of diet, age, and environmental exposure.

The scientists also identified four DNA sites in the genome where the presence of methylated DNA at that site was correlated with genes involved in obesity and large body mass. These methylation sites are located adjacent to genes that are normally involved in weight control and type II diabetes. Further studies are needed to determine whether these epigenetic marks actively cause obesity and/or diabetes or whether they are a side effect of another factor in the study, such as body mass index.

We tend to think of chromosomes in terms of genes and proteins made that are needed to run the cell. But the vast majority of the DNA in human chromosomes is non-coding DNA, and much of the DNA sequences are highly repeated and do not contain genes. Some DNA sequences provide functions required for chromosome functions. For example, the telomere DNA sequences form structures that are necessary to replicate the ends of eukaryotic chromosomes. These telomere sequences vary somewhat in different eukaryotic species, but otherwise the telomere sequences provide the same functions in all species. The highly repeated satellite sequences flanking the centromere regions of chromosomes are packaged into very highly condensed heterochromatin that does not decondense much during the cell cycle.

Eukaryotic chromosomes are actually dynamic structures that contain specialized chromosome DNA sequences designed to assist with the duplication and maintenance of each human genome. Human chromosomes carry and express our genes, controlled not only by conserved DNA sequences on the genes but also by changes in DNA and protein methylation. Epigenetic control is exerted over future generations without altering the DNA sequences of the genes.

Unit 6 Questions

1. Mitochondrial DNA (mtDNA) is a direct genetic link from mother to daughters but not from mother to sons because:
   a. a mother has the same mitochondrial DNA as biological daughters and sons
   b. a father contribute mitochondria to his son only
   c. a mother has the same mitochondrial DNA as her biological daughter only
   d. a mother has the same mitochondrial DNA as her biological son only

2. Which of the following statements about SNPs are true?
   a. SNPs can be linked to human disease genes, but SNPs do not usually cause the disease.
   b. SNPs are used for human identification by forensic DNA testing.
   c. SNPs are sometimes located in repeated regions of the human chromosome DNA.
   d. All of the above.
   e. None of the above.

3. The DNA fingerprinting analysis used to identify individual people depends on detecting which DNA sequences?
   a. DNA sequences that are identical among all the individual genomes in the human population
   b. DNA sequences that differ (are variable) among the individual genomes in the human population
   c. DNA sequences that are made by the cells at the end of a human finger (in the area of the skin fingerprint)
   d. DNA sequences from the mtDNA genomes in maternally inherited mitochondria

4. The very first time that human DNA fingerprinting was used as evidence in a courtroom, it was used to:
   a. Prove that a particular suspect committed a murder
   b. Prove that a person could be exonerated early in an investigation
   c. Prove that the contested biological family relationships were authentic
   d. Prove that a convicted person is innocent of the charges used to send him to prison

5. The Southern blot method allows scientists to:
   a. Separate DNA on a gel and transfer the DNA from the gel to a membrane
   b. Separate RNA on a gel and transfer the RNA from the gel to a membrane
   c. Transfer the DNA from a membrane to the gel for further analysis
   d. Transfer the RNA on a Northern blot membrane to a new gel

6. DNA probes are very specific for target DNA sequences because the DNA probes base-pair to:
   a. Highly repeated DNA sequences in the target DNA
   b. Target DNA sequences that are exactly the same as the DNA probe sequence (i.e., probe: GCTAATGCCA; target: GCTAATGCCA)
   c. Complementary sequences in the target DNA
   d. Complementary proteins in the chromosomes

7. What does the phrase 'beads on a string' refer to in the context of eukaryotic chromosome structure?
   a. The way that chromosomes are arranged in preparation for cell division, they de-condense and become fibers draped around the nucleus.
   b. It refers to the chains of amino acids that are made on the ribosomes.
   c. The beads are the nucleosomes along the DNA 'string', the fundamental structure of eukaryotic chromatin.
   d. The spindle apparatus assembles before cell division and connects each chromosome on to the same spindle fiber like beads on a string.

8. Epigenetic gene regulation involves two types of chemical additions that modify these nucleosome components:
   a. The DNA helix and the histone proteins
   b. The replication fork DNA and the polymerase enzyme
   c. The splicing proteins and the small RNAs
   d. The telomere DNA and the telomere binding proteins

9. DNA barcoding is a DNA testing method that:
   a. measures the amount of DNA contained in mature RBCs in the blood
   b. analyzes the DNA sequence of a short region of mitochondrial DNA in cells
   c. counts the number of times that DNA winds around nucleosomes near an expressed gene
   d. alters the chromatin structure of the gene being tested

10. In a Southern blot transfer the purpose of the membrane is to:
    a. Trap the DNA strands and stop them from passing through the membrane
    b. The membrane stops the mRNAs by binding to the poly(A) tails on the mRNAs
    c. Collect all the longest DNA strands at the center of the membrane sheet
    d. Allow the DNA strands to pass through but the membrane stops any tRNA and rRNA molecules in the sample

Credits

Unit 1

Page 1: DEV CARR, CULTURA /SCIENCE PHOTO LIBRARY; Athanasia Nomikou / Shutterstock; Page 2: Pasieka / Photo Researchers, Inc.; Carlyn Iverson / Photo Researchers, Inc.; Page 5: lculig / Shutterstock; ynse / Shutterstock; martan / Shutterstock; lculig / Shutterstock; Page 6: Roberto Sanchez / Shutterstock; Page 8: theromb / Shutterstock; Roger J. Bick & Brian J. Poindexter / UT-Houston Medical School / Photo Researchers, Inc.; R. BICK, B. POINDEXTER, UT MEDICAL SCHOOL/ Photo Researchers, Inc; Roger J. Bick & Brian J. Poindexter / UT-Houston Medical School / Photo Researchers, Inc.; Roger J. Bick & Brian J. Poindexter / UT-Houston Medical School / Photo Researchers, Inc.; alxhar / Shutterstock; Page 9: DR ELENA KISELEVA/ Photo Researchers, Inc; Page 10: Blamb / Shutterstock; Photo Researchers, Inc; Martin Oeggerli / Photo Researchers, Inc.; David Goodsell / Photo Researchers, Inc.; Page 13: Courtesy of Dr. Brad J. Marsh, Division of Molecular Cell Biology, Institute for Molecular Bioscience, The University of Queensland, Brisbane, Australia.; Page 14: Meder Lorant / Shutterstock; Pichugin Dmitry / Shutterstock; Eric Isselee / Shutterstock; Subbotina Anna / Shutterstock; Tischenko Irina / Shutterstock; Alhovik / Shutterstock; Eric Isselee / Shutterstock; Eric Isselee / Shutterstock, Eric Isselee / Shutterstock; Page 16: A. Barrington Brown / Photo Researchers, Inc.; Science Source; Omikron / Photo Researchers, Inc.; Page 17: Gustoimages / Photo Researchers, Inc.; Page Roger Harris / Photo Researchers, Inc.; lculig / Shutterstock; Laguna Design / Photo Researchers, Inc.; Page 20: lculig / Shutterstock; Page 21: Kenneth Eward / Photo Researchers, Inc.; lculig / Shutterstock; Page 23: Phantatomix / Photo Researchers, Inc.; Page 27: Carlyn Iverson / Photo Researchers, Inc.; Page 31: From *Forensic Science for High School*, Second Edition by Barbara Ball-Deslich and John Funkhouser. Copyright ©2009 by Kendall Hunt Publishing Company. Reprinted by permission. Page 32: Robyn Wilson / Shutterstock; lculig / Shutterstock; Page 33: Nicemonkey / Shutterstock; Page 35: Juergen Berger / Photo Researchers, Inc.; Omikron / Photo Researchers, Inc.

Unit 2

Page 45: Biophoto Associates / Photo Researchers, Inc.; Photo Researchers, Inc; Page 47: Phantatomix / Photo Researchers, Inc.; Page 48: Kenneth Eward / Photo Researchers, Inc.; Page 51: Laguna Design / Photo Researchers, Inc.; Laguna Design / Photo Researchers, Inc.; Page 62: Peter Menzel / Photo Researchers, Inc.; Dr. Tim Evans / Photo Researchers, Inc.; Page 79: Robert Moyzis, University of California, Irvine, CA; U.S. Department of Energy Human Genome Program, http://genomicscience.energy.gov.

Unit 3

Page 91: RAMON ANDRADE 3DCIENCIA/ Photo Researchers, Inc; LAGUNA DESIGN/ Photo Researchers, Inc; Page 93: LAWRENCE BERKELEY NATIONAL LABORATORY/ Photo Researchers, Inc; Page 103: Laguna Design / Photo Researchers, Inc.; Page 104: Kenneth Eward / Photo Researchers, Inc.; Page 109: Laguna Design / Photo Researchers, Inc.; Laguna Design / Photo Researchers, Inc.; Laguna Design / Photo Researchers, Inc.; PASIEKA/ Photo Researchers, Inc; Dr. Tim Evans / Photo Researchers, Inc.; Mark Lorch / Shutterstock; Kenneth Eward / Photo Researchers, Inc.; PASIEKA/SCIENCE PHOTO LIBRARY; Page 113: SPL / Photo Researchers, Inc.; Steve Gschmeissner / Photo Researchers, Inc.; Page 115: Don W. Fawcett / Photo Researchers, Inc. Colorization by: Jessica Wilson; Page 117: Dr. Stephen Cusack, EMBL / Photo Researchers, Inc.; Inger Andersson / Laboratory of Molecular Biophysics, Oxford / Photo Researchers, Inc.; Page 118: Laguna Design / Photo Researchers, Inc.; Laguna Design / Photo Researchers, Inc.; Page 119: Roger J. Bick & Brian J. Poindexter / UT-Houston Medical School / Photo Researchers, Inc.; ROBERT MCNEIL, BAYLOR COLLEGE OF MEDICINE / Photo Researchers, Inc; Dr. Torsten Wittmann / Photo Researchers, Inc; Roger J. Bick & Brian J. Poindexter / UT-Houston Medical School / Photo Researchers, Inc.; Page 120: Thomas Deerinck, NCMIR / Photo Researchers, Inc.; Dr. Gopal Murti / Photo Researchers, Inc; Page 121: DR PAUL ANDREWS, UNIVERSITY OF DUNDEE/ Photo Researchers, Inc; DR PAUL ANDREWS, UNIVERSITY OF DUNDEE/ Photo Researchers, Inc; Laguna Design / Photo Researchers, Inc.; Page 123: RAMON ANDRADE 3DCIENCIA/ Photo Researchers, Inc; SPL / Photo Researchers, Inc; Dr. Tim Evans / Photo Researchers, Inc.; Page 125: Kenneth Eward / Photo Researchers, Inc.; Page 126: Henning Dalhoff/Bonnier Publications/Photo Researchers, Inc.; Kenneth Eward/Photo Researchers, Inc.; Don. W. Fawcett / Photo Researchers, Inc.; Laguna Design / Photo Researchers, Inc.

Unit 4

Page 133: Pasieka / Photo Researchers, Inc.; Page 137: U.S. National Library of Medicine; U.S. National Library of Medicine; Page 140: U.S. National Library of Medicine; U.S. National Library of Medicine; U.S. National

Credits

Library of Medicine; Page 145: L. Willatt, East Anglian Regional Genetics Service / Photo Researchers, Inc; Page 147: L. WILLATT EAST ANGLIAN REGIONAL GENETICS SERVICE/ Photo Researchers,Inc; Page 148: Addenbrookes Hospital / Photo Researchers, Inc; SPL / Photo Researchers, Inc; Addenbrookes Hospital / Photo Researchers, Inc; Page 151: Addenbrookes Hospital / Photo Researchers, Inc; Page 152: Stills / Photo Researchers, Inc; Page 153: Ralph C. Eagle, Jr. / Photo Researchers, Inc.; Page 154: Dan Gerber/ shutterstock.com; Dr. Najeeb Layyous / Photo Researchers, Inc.; SCI-COMM STUDIOS/ Photo Researchers, Inc; MedicalRF / Photo Researchers, Inc.; Page 156: Volker Steger / Photo Researchers, Inc.; Page 161: Bill Longcore / Photo Researchers, Inc.; Page 162: Roberto Sanchez / Shutterstock; Page 164: Susumu Nishinaga / Photo Researchers, Inc.; Omikron / Photo Researchers, Inc.; Page 166: ST BATHOLOMEW'S HOSPITAL/ Photo Researchers, Inc; Page 168: theromb / Shutterstock; theromb / Shutterstock; Page 169: Medical Body Scans / Photo Researchers, Inc; Alfred Pasieka / Photo Researchers, Inc; Page 170: Thomas Deerinck, NCMIR / Photo Researchers, Inc.; Leonard Lessin / Photo Researchers, Inc; Leonard Lessin / Photo Researchers, Inc; Page 172: DR MARK J. WINTER/ Photo Researchers, Inc; Page 174: Alfred Pasieka / Photo Researchers, Inc.

Unit 5

Page 180: SPL / Photo Researchers, Inc.; Juergen Berger, Max-planck Institute / Photo Researchers, Inc.; Page 182: MedicalRF / Photo Researchers, Inc; Steve Gschmeissner / Photo Researchers, Inc.; Dan Gerber / Shuttertock.com; S.Borisov / Shutterstock.com; Page 185: Innerspace Imaging / Photo Researchers, Inc.; Stephanie Schuller / Photo Researchers, Inc.; Thomas Deerinck, NCMIR / Photo Researchers, Inc.; Page 186: Pan Xunbin / Shutterstock.com; Roger J. Bick & Brian J. Poindexter / UT-Houston Medical School / Photo Researchers, Inc.; PROF. P. MOTTA/DEPT. OF ANATOMY/ UNIVERSITY "LA SAPIENZA", ROME/ Photo Researchers, Inc; Quest / Photo Researchers, Inc.; Pasieka / Photo Researchers, Inc.; Steve Gschmeissner / Photo Researchers, Inc.; Professor Pietro M. Motta / Photo Researchers, Inc; Photo Researchers, Inc; Quest / Photo Researchers, Inc.; ZEPHYR/ Photo Researchers, Inc; Page 188: Omikron / Photo Researchers, Inc; Dr. David Furness, Keele University / Photo Researchers, Inc.; David Becker / Photo Researchers, Inc.; Thomas Deerinck, NCMIR / Photo Researchers, Inc; Page 190: Sven Hoppe / Shutterstock.com; SPL / Photo Researchers, Inc.; Simon Fraser / MRC Unit, Newcastle General Hospital / Photo Researchers, Inc.; Page 195: Steve Gschmeissner / Photo Researchers, Inc.; Tek Image / Photo Researchers, Inc.; Klaus Guldbrandsen / Photo Researchers, Inc.; Page 196: vetpathologist / Shutterstock.com; M. I. Walker / Photo Researchers, Inc; M. I. Walker / Photo Researchers, Inc.; Li Wa / Shutterstock.com; vetpathologist / Shutterstock.com; Page 197: theromb / Shutterstock.com; Page 201: ZEPHYR/ Photo Researchers, Inc; Lysia Forno / Photo Researchers, Inc.; John Bavosi / Photo Researchers, Inc.; Thomas Deerinck, NCMIR / Photo Researchers, Inc.; Page 203: Thomas Deerinck, NCMIR / Photo Researchers, Inc; Steve Gschmeissner / Photo Researchers, Inc.; Page 206: Oguz Aral / Shutterstock.com; Omikron / Photo Researchers, Inc.; Page 207: James

King-Holmes / Photo Researchers, Inc.; Philippe Psaila / Photo Researchers, Inc.; Page 209: theromb / Shutterstock.com; Page 211: Gusto / Photo Researchers, Inc.; Geoff Tompkinson / Photo Researchers, Inc.; Novosti / Photo Researchers, Inc.; Page 213: Philippe Psaila / Photo Researchers, Inc.; Mauro Fermariello / Photo Researchers, Inc.; Mauro Fermariello / Photo Researchers, Inc.; Page 215: Simon Fraser / Photo Researchers, Inc; Damien Lovegrove / Photo Researchers, Inc.; Dr. Jeremy Burgess / Photo Researchers, Inc.

Unit 6

Page 221: CARLOS GOLDIN / Photo Researchers, Inc; Kateryna Larina / Shutterstock.com; Dmitriy Shironosov / Shutterstock.com; Page 223: MedicalRF / Photo Researchers, Inc; Dan Gerber / Shutterstock.com; Page 226: AN VAN DE VEL / Photo Researchers, Inc; Page 227: David Parker / Photo Researchers, Inc; Neville Chadwick / Photo Researchers, Inc; Neville Chadwick / Photo Researchers, Inc; Page 230: Pascal Goetgheluck / Photo Researchers, Inc; James King-Holmes / Photo Researchers, Inc; Page 226: L. Willatt / Photo Researchers, Inc.; Page 239: David Parker / Photo Researchers, Inc.; Page 243: Thierry Berrod, Mona Lisa Production / Photo Researchers, Inc.; Thierry Berrod, Mona Lisa Production / Photo Researchers, Inc.; Steve Gschmeissner / Photo Researchers, Inc.; Page 244: Peter Menzel / Photo Researchers, Inc; JIM VARNEY / Photo Researchers, Inc; Dr. Jurgen Scriba / Photo Researchers, Inc; Page 345: Philippe Psaila / Photo Researchers, Inc.; PETER MENZEL/Photo Researchers, Inc; Page 249: US DEPARTMENT OF DEFENSE / Photo Researchers, Inc; zulufoto / Shutterstock.com; Page 251: CARLOS GOLDIN / Photo Researchers, Inc; Page 252: RIA Novosti / Photo Researchers, Inc.; RIA Novosti / Photo Researchers, Inc.; Page 253: David Nunuk / Photo Researchers, Inc.; Pasquale Sorrentino / Photo Researchers, Inc.; Pasquale Sorrentino / Photo Researchers, Inc.; Pasquale Sorrentino / Photo Researchers, Inc.; Page 255: DAVID PARKER/ Photo Researchers, Inc; Page 256: CLFProductions / Shutterstock.com; Stuart Berman / Shutterstock.com; Eric Isselee / Shutterstock.com; Page 257: Ryan M. Bolton / Shutterstock.com; Photo Researchers, Inc; Carleton Chinner / Shutterstock.com; William Ervin / Photo Researchers, Inc.; John Reader / Photo Researchers, Inc.; John Reader / Photo Researchers, Inc.; David Weintraub / Photo Researchers, Inc.; Page 259: lculig / shutterstock.com ; Page 260: LAGUNA DESIGN / Photo Researchers, Inc.; Page 262: Booka / Shutterstock.com; Page 268: Quang Ho / Shutterstock.com

Index

A

A-T base pairs, 22–23
ABCA4 gene, 205
Abl gene, 150
Abl-bcr gene, 150
Adenomatous polyposis coli (APC) gene, 169
Adult progenitor stem cells (ASCs)
    characterization of, 186
    life span of, 187
    potential of, 186–187
    research on, 187–188
    types of, 186
Advanced Cell Technology (ACT), 205
Aequorea victoria, 110
Agarose gel, 230
Agarose gel slab, 30–31
Agouti mice, 265–266
Alleles. *See also* Mutations
    CODIS, 235
    defined, 40
    formation, 71
    mutations and, 135
    STR, 236
    wild types, 137–138
    Y chromosome, 237
Alpha tropomyosin gene, 101–102
Alpha-fetoprotein (AFP), 142
Alpha-globin proteins, 6
Alu I DNA repeats
    characterization of, 82–83
    distribution of, 81–82
    elements of, 82–83
    origins of, 83
Alzheimer's disease
    cause of, 167
    mechanisms, 168
    prions and, 174
    progression of, 167–169
    treatments for, 168
American Association for the Advancement of Science, 250
Amino acids
    CFTR sequence, 166
    hydrophilic, 112
    hydrophobic, 110
    mutations and, 136
    protein bonds, 107
    side chains, 111–112
    transfer of, 104
Amniocentesis, 150–152
Amniotic fluid, 194
Ampicillin, 41
AmpS cells, 41
Amur leopard, 256
Ancestry.com, 159
Aneuploidy, 139
Animal cloning
    blastocyst divergence in, 209
    cats, 212
    cow's milk, 213
    defined, 207
    dogs, 212–213
    endangered species, 214–215
    ethics, 215–216
    genetic modification and, 212
    inefficiency of, 208
    long-term health problems, 216
    mule, 212
    pigs, 213–214
    processes of, 208–209
    risks of, 215
    sheep, 210–212
    therapeutic cloning *vs.,* 208–209
Animal models
    BPA exposure, 265–267
    gene sequencing with, 74–75
    human disease studies, 204, 211–212
    transgenerational inheritance, 264
Annealing reaction, 23, 47
APOE gene, 168
Argentine Forensic Anthropology Team, 250–251
Aricept, 168
Armed Forces DNA Identification Laboratory (AFDIL), 249
ASD. *See* Autism spectrum disorder (ASD)
Ashworth, Dawn, 227–228
Atoms
    bonding of, 19–20
    types of, 18–19
ATP (adenosine triphosphate), 116, 241
Audubon Society, 255
AUG codon, 105–106
Autism spectrum disorder (ASD), 72–73
Automated DNA sequencing
    data generation, 59, 62
    PCR, 61–63
    RT-PCR, 63–65
    typical approaches to, 59
Autosomal chromosomes mutations, 135–137
Autosomal dominant disorders, 139
Autosomal recessive disorders, 139
Axon nerve fibers, 203

277

B

Bacteriophage, 34
BamHI, 35–36, 239–241
Barcode project, 226
Base pairs
　CFTR, 166
　changes, detection of, 239–240
　characterization of, 21–22
　cutting, 34–35
　epigenetic modifications, 259
　mutations, 132, 134, 143–144
　number of, 67
　order of, 53
Bcr gene, 150
Beta-globin proteins, 6
Bidirectional DNA replication forks, 51
Binding proteins, 28–28
Bis-phenol-A (BPA), 265–267
Blassie, Michael Joseph, 249–250
Blastocyst embryo, 183, 209
Bloom's syndrome, 153
Bolsheviks, 252
Bone marrow stem cells, 195–197
BRCA1 gene, 171–172
BRCA2 gene, 171–172
Brown, Louise Joy, 195–196
Buckland, Richard, 228
Bush, George W., 193

C

Caenorhabditis elegans, 66, 74
CAG codon, 170–171
Calcium, 7
Cancer, 139
Carboxy-terminal domain (CTD), 91–92
Cardiac disease
　genetic causes, 172–173
　hESC research, 192–193
Cardiac muscle cells, 188
Catabolite activator protein (CAP), 122–124
Cell cycle, 13
Cell division. *See* Mitosis
Cell membranes, 114–115
Cell replacement therapy, 209
Cellmark Diagnostics, 254
Cells
　biochemical processes, 119
　compartments of, 7–8
　major types of, 6–7
　secondary structure formation, 26–29
　specialized, 7
Cellular metabolism, 116
Central dogma, 11
Centromeres, 78–79, 120
Chemical bonds, 19
Chorionic villus sampling (CVS), 150
Chromatin, 13, 263
Chromosome 7, 224
Chromosomes
　autosomal, 135–136
　centromere in, 78–79
　copies, distinguishing, 238–239, 241
　DNA fingerprinting of, 225–226
　DNA helix in, 2, 46–47
　genes location on, 67–68
　jumping genes on, 80
　kinetochore in, 79
　length of, 46
　minisatellite repeats on, 79–80
　mutation types, 132–134
　number, effects of, 74
　painting, 147
　rearrangements, 148–150
　satellite repeats on, 78–79
　segregation of, 120
　structures, 269
　telomeres on, 79
　translocations, 148–150
　whole, genetic testing with, 144–146
Cincinnati Zoo and Botanical Garden, 255
Cloning. *See* Recombinant DNA cloning
Codominant inheritance, 141
Collagen proteins, 109–110
Collins, Francis, 65
Colon cancer, 144, 169
COmbined DNA Index System (CODIS), 235
Comparative genomics, 73–74
Complementary DNA (cDNA)
　analysis of, 96
　genetic testing and, 143
　human cloning and, 207
Complex multigenic disorders
　inheritance of, 141
　types of, 139
Complex RNA, 98–99
Computer animation, 18
Control regions, 134–135
Copy Cat, 212
Copy number variation (CNV), 71–73
Covalent bonds, 19–20
Cre enzyme, 199
Creutzfeldt-Jakob disease (CJD), 170
Crick, Francis, 16, 18
Cyclic AMP (cAMP), 122
Cysteine, 112
Cystic fibrosis
　cause of, 166
　characterization of, 165–166
　inheritance of, 168
　prevalence of, 164
　tests for, 158
Cystic fibrosis transmembrane conductance regulator (CFTR)
　carriers of, 158
　function of, 166–167
　location of, 166
　mutations in, 114–115
Cytoskeleton
　functions of, 118–119
　major components of, 118
　proteins, 119, 170

D

Deep brain stimulation, 201
Dengue hemorrhagic fever, 110
Diabetes
　characterization of, 12
　epigenetic modification, 265
　heart disease and, 173
Diploid cells, 67
Disappeared. *See Los Desaparecidos*
DNA
　ancient, tracing of, 159–160
　chemical structure of, 26
　elements, 80
　gel electrophoresis, 30–31
　location of, 2
　non-coding, 269
　structure of, 15–16
DNA banks, 255
DNA binding proteins
　function of, 122–123
　sequence-specific, 123
　types of, 122
DNA chips
　advantages of, 155
　commercial use of, 157
　development of, 154–155
　fingerprinting with, 225–226
DNA expression. *See* Gene expression
DNA fingerprinting
　automated analysis methods, 225–226
　barcode technology from, 257–258
　endangered species, 256
　genome differences and, 224

Index

legal system use of, 38–39, 227–228
PCR-based STR analysis, 236–237
RFLP-based, 237–241
Southern blot transfer
    fragment to membrane transfer, 231
    hybridization reaction, 232
    pattern detection, 233–234
    preparation, 229–230
Y chromosome STRs, 237
DNA helix
    annealing reaction, 23
    backbone strands, 20–21
    base pairs, 21–22
    chromosomes and, 46–47
    denature of, 23
    discovery of, 16–18
    epigenetic modifications, 259
    overlapping, 60
    shape of, 19
DNA language, 25–26
DNA ligase, 39–40
DNA loop, 122–123
DNA methyltransferase (DNMT), 259
DNA polymerase, 49, 50–52
DNA probes
    base-paired
        design of, 225
        DNA fingerprinting with, 225–226, 232
        genetic testing with, 143–144
        specificity of, 228
    STR-specific, 236–237
DNA profiles, 233–234
DNA replication
    accuracy of, 49–50
    fingerprints and, 223
    mechanisms, 262
    process of, 47–49, 47–50
DNA replication forks, 50–53
DNA sequences
    assemble of, 60
    mutation effects on, 131–135
    repeated, 77–78
    structures, 269
    ultra conserved, 77
    vectors from, 199
DNA sequencing, 3–4. *See also* Human Genome Project
    animal models, 74–75
    automated, 57, 59–67
    earliest strategies for, 57–59
    genotyping *vs.*, 157
    process of, 47–50
    products, 57
    protocol, 53
    reading, 58
    Sanger chain termination method, 54–59
    steps of, 54–57
DNA testing. *See also* DNA fingerprinting
    anthropological use, 252–253
    Argentine "disappeared," 250–251
    armed forces identification, 249–250
    barcode technology from, 257–258
    court acceptance of, 243–244
    dog genes, 253–254
    endangered species protection, 254–256
    Ice Maiden mummy, 252–253
    Innocence Project use of, 245–248
    Louis-Charles de France remains, 252
    mtDNA, 241–243
    NFL use of, 248
    Olympic Games use of, 248
    PCR-based, 235–237
    Romanov family remains, 252
    terrorist victims, 250
DNA transposons, 80–81
DNA-protein interactions, 92
DNMT. *See* DNA methyltransferase (DNMT)
Dog genomes, 253–254
Dolly, cloning of, 210–212
Down syndrome, 120, 139
Drosophila melanogaster, 66, 74, 94
Dynein motors, 119
Dystrophin, 69–70

E

Ear hair cells, 189
EcoRI, 39
Elephants, 256
Embryogenesis
    fruit fly, 94
    process of, 179, 182
    stages, 73
Embryonic stem cells (ESCs). *See also* Adult progenitor stem cells (ASCs)
    alternatives to, 193
    amniotic fluid, 194
    bone marrow, 195–197
    characterization of, 180
    development role of, 180–181
    division of, 180–181
    iPS approach, 197–199
    origins of, 181–183
    overview of, 179
    renewal of, 183–185
    research (*See* Human embryonic stem cells (hESC) research)
    specialized cells generated by, 184–185
    umbilical cord, 194
Embryos
    early development, 142
    extra, 195
    genetic tests for, 153–154
    mtDNA, 141
Endangered species
    biological banks for, 255
    habitat loss and, 255–256
    illegal trade in, 257
    increase in, 254–255
    poaching of, 256
Enhancer elements, 92–93
Enzymes. *See also specific enzymes*
    characterization of, 116
    function of, 116–117
Epigenetic genes
    differences in, 267
    regulation of, 138
    reprogramming of, 216–217
Epigenetic modifications. *See* Transgenerational epigenetic inheritance
Escherichia coli, 9, 122
ESCs. *See* Embryonic stem cells (ESCs)
EthBr (ethidium bromide), 33
Eukaryotic cells
    characterization of, 6–7
    division of, 13
    DNA genomes in, 8–9
    function of, 6–7
    mechanism hierarchy, 88
    nuclei, 8
    viewing, 12
Exons, 69–70, 135
Eye color, 15
Eyewitness identifications, 246–247

F

Familial adenomatous polyposis (FAP), 144, 169
Fat cells, 7
Feig, Larry, 265
Feodorovna, Alexandra, 252
Fertilization. *See also In vitro* fertilization
    epigenome formation, 164
    gene copying during, 138
    inheritance, 243

natural, 181–182, 216
stages of, 182
zygote from, 132
Fetal blood, 194
Fingerprints. *See* DNA fingerprinting; Skin fingerprints
Fluorescence in situ hybridization (FISH), 146–147
Foad, Janet, 254
Folded proteins, 5
Forensics. *See also* DNA fingerprinting; DNA testing
fraud in, 247–248
military, 249–250
mtDNA use, 242–243
Fox, Michael J., 193
Franklin, Rosalind, 16–17
Frozen Ark project, 255
Frozen Zoo, 25

G

G-bands, 145
G-C base pairs, 22–23, 145
GAG codon, 163
Gel electrophoresis
band detection, 33
DNA fingerprinting with, 229
interpretation, 32–33
process, 30–31
Gender determination, 15
Gene expression
binding proteins in, 122–127
BPA exposure and, 266–267
cellular control of, 88
control of, 92–94
histone proteins in, 125–127
molecular changes from, 15
motor protein transport in, 119–121
mutation effects on, 136–137
occurrence of, 25
pathway, 12
primary role of, 115–117
protein production and, 10–11
regulation of, 7
RNA ribosomes in, 117–118
transcription process, 24–25
typical, 88–89
Gene inheritance. *See also* Transgenerational epigenetic inheritance
complex multigenic disorders, 141
copy number variation in, 71–73
cystic fibrosis, 168
diseases, patterns for, 139–141
parental contribution, 70–71

Genes. *See also specific genes*
codominant inheritance, 141
comparisons of, 73–74
defined, 2
family similarity, 63
human, quantity of, 68
number, effects of, 74
reprogramming, 199
sequencing of, 3–4
silent, 28
transgenerational inheritance, 264
Genetic counseling, 161–162
Genetic disorders, 4
Alzheimer's disease, 167–170
breast cancer as, 171–172
carriers of, 137–138
causes of, 132
chromosome abnormalities, 139
colon cancer as, 169
complex multigenic diseases, 139
cystic fibrosis as, 164–167
heart disease as, 172–173
Huntington's disease as, 170–171
inheritance patterns for, 139–141
rearrangements and, 148–150
risk factors for, 161–162
sickle cell anemia as, 162–164
single gene, 138–139
translocations and, 148–150
Genetic information
flow of, 87–90
storage of, 88
Genetic Information Nondiscrimination Act (GINA), 157
Genetic testing
adult populations, 152–153
applications of, 142
approaches to, 142
cautions, 156–158
data from, 158
DNA chips for, 154–156
drug store kits, 160–161
FISH, 146–147
genealogical use, 159–160
importance of, 158
karyotype displays, 144–146
maternal, 160
molecular DNA probes for, 143–144
newborns, 150–151
prenatal, 150–151
Genotyping
defined, 155
DNA sequencing *vs.*, 157
limits of, 157
process of, 155–157
Geron Corporation, 205
Glial cells, 188, 203

Glutamate-aspartate transporter (GLAST) enzymes, 189
Glyceraldehyde-3-dehydrogenase (GAPDH), 116
Glycine, 112
Glycolysis, 116
Glycoproteins, 109
GM2 gangliosidosis. *See* Tay-Sachs disease
Golgi membranes, 113
Green fluorescent protein (GFP), 109–110
Grizzly bears, 255–256
GROEL chaperone, 109–110
Guthrie, Woody, 170

H

H3 proteins, 263
Hair cells, 189
Hammerhead ribosome, 118
HD gene, 170–171
Heart muscle cells, 7
Heat shock proteins, 109
Helix-turn-helix (HTH), 122–123
Hematopoietic stem cells (HSCs), 187
Hemoglobin
complex, 162–163
function of, 4–5
molecules, 66
subunits, 6
Hereditary nonpolyposis colon cancer (HNPCC), 169
Hexosaminidase A deficiency. *See* Tay-Sachs disease
High-density lipoproteins (HDLs), 108–109
HindIII, 39
Histones
characterization of, 126
epigenetic gene regulation and, 259–260
function of, 125–126
nucleosome assembly, 126–127
HIV/AIDS, 74
Host cells, 207
Human characteristics, 10–11
Human cloning
goals of, 206–207
illegality of, 208
types of, 206
Human development, 12
Human embryonic stem cells (hESC) research
beginning of, 189
clinical trials, 204–205
federal funding for, 193

Index

growth of access, 189–190
Parkinson's disease treatment, 199–202
promise of, 190–193
spinal cord paralysis treatment with, 202–204
therapeutic cloning with, 208
Human genome
characterization of, 8–9
differences in, 224
DNA repeats on, 81–83
flexibility of, 99–102
identical twins, 221–222
LINEs in, 83–84
methylated markers, 268
mitochondria in, 141
non-coding DNA discovery, 75–76
retrotransposons on, 81
sperm in, 182
Human Genome Project (HGP)
accomplishments of, 16, 47
controversy over, 65–66
first species used, 66
goals of, 65
government role, 65
lessons from, 67–69
private sector role, 66–67
Huntington's disease, 170–171
Hybridization buffer, 232
Hydrogen bonds
breaking of, 22–23
characterization of, 19
formation of, 22
specificity of, 143
Hydrophilic amino acids, 112
Hydrophobic amino acids
distribution of, 113
function of, 112
proteins containing, function of, 114
Hydroxyl group (OH), 21, 53
Hypocretin receptor 2 (HRP2) gene, 254

I

Ice Maiden mummy, 252–253
Identical twins, 222–223
In vitro fertilization
cloned animals, 211
extra embryos from, 195
mutants, 153–154
process, 181–182
Inactive pre-hormone protein, 12
Induced pluripotent stem cells (iPS)
development of, 197
drawbacks, 199
function of, 197–198

Parkinson's disease and, 202
skin cells, 197
specialized cell generation by, 198–199
Inner cell mass (ICM), 183
Innocence Project
forensic fraud research, 247–248
founding of, 245
reform goals of, 247
tests conducted by, 246
Insulin, 12
Insulin genes, 263
Integrin proteins, 110
Introns, 69–70
Ionic bonds, 19
iPS. *See* Induced pluripotent stem cells (iPS)
Iron-storage disease, 144

J

Jeffreys, Alec, 227
Jumping genes, 80

K

Karyotype displays, 144–146
Keirstead, Hans, 203–204
Kinesin motors
characterization of, 119–120
function of, 120–121
Kinetic polymerase chain reaction (KPCR), 65
Kinetochores, 79
Knock-out pigs, 213–214
KPCR. *See* Kinetic polymerase chain reaction (KPCR)

L

lac repressor protein, 122
Leopards, 256
Leptin hormone, 108–110
Leucine zipper (LA), 124
Lewy body, 201
Ligation reaction, 39–40
LINEs. *See* Long interspersed repeated nuclear elements (LINEs)
Long interspersed repeated nuclear elements (LINEs)
characterization of, 81
function of, 83–84
Los Desaparecidos, 250–251
Louis XVI, 252
Louis-Charles de France, 252
loxP sites, 199

Lynch syndrome. *See* Hereditary nonpolyposis colon cancer (HNPCC)

M

Major grooves, 21
Malaria, 164
MALDI-TOF (matrix-assisted laser desorption/ionization time of flight), 18
Mammalian features, 14–15
Mann, Lynda, 227–228
Marie Antoinette, 252
Maternal lineage tests, 160–161
MC1R gene, 15
Meiosis, 181
Melanin, 15
Membranes
double, 8
layers of, 7
proteins, functions of, 112–115
messenger RNA (mRNA). *See also* Pre-mRNA
analysis of, 96
AUG codon, 105–106
codons, 103–107
determination of, 98
function of, 87
isolation of, 95–96, 96
mutations and, 135
precursors, 69–70, 95
protein translation mechanism, 102–105
protein-coding genes expression of, 64
transport, 13–14
tRNA codon link, 05
Methylation
enzymes role, 261–262
mechanism, 262
mutations during, 138
patterns, 268
sequencing and, 259
sites, 267, 269
vinclozolin and, 264
Microtubule-associated proteins (MAPs), 118
Minisatellite, 79–80
Minor grooves, 21
Miss-sense mutation, 136
Mitochondrial DNA (mtDNA), 26
advantages of, 241–242
characterization of, 241
forensic use of, 242–243
Ice Maiden mummy, 253
inheritance of, 141, 160–161, 242

Mitosis
  characterization of, 181
  ESCs, 183
  function of, 26
  mistakes during, 120
  preparation for, 119
Molecular DNA probes, 143–144
Molecular genealogy, 159–160
Molecules
  fat-like, 7
  organic atoms in, 18–19
Monogenic diseases, 138–139
Monomers chromosome DNA, 2–3
Motor proteins
  function of, 121
  location of, 118
  types of, 119–121
Mt. Ampato mummy, 252–253
Mullis, Kary, 61–62
Multifactorial genetic disorders, 141–142
Multipotent progenitor stem cell, 183–185
Muscle contraction, 121
Mutations
  affects, evaluation of, 136–137
  alleles and, 135
  amino acids and, 136
  APC gene, 169
  BRCA1/BRCA2 genes, 169
  canine, 254
  control regions alteration by, 134–135
  environmental factors, 132–134
  GAG/GTG codons, 162–163
  gene sequence changes and, 131–134
  HD gene, 170–171
  inherited, 144
  preexisting, 132
  protein coding alteration by, 134–135
  testing for, 142–143
  types of, 132–134
Myelin sheath, 203
Myelogenous leukemia, 150
Myosin V, 120–121

N

NADH (reduced nicotinamide adenine dinucleotide), 116
Namenda, 168
Narcolepsy, 254
National Football League (NFL), 248
Nerve cells
  destruction of, 199
  generation of, 204–205
  Parkinson's disease and, 200–201
  spinal cord paralysis and, 230
  types of, 188
Neufeld, Peter, 245
Neurotransmitters, 201
NFL. See National Football League (NFL)
Nicholas II, 252
Niemann-Pick disease, 153
Northern blot transfer, 98
Nuclear magnetic resonance (NMR), 18
Nuclear pores, 8
Nuclear reprogramming, 216–217
Nucleotide bases, 4

O

Obama, Barak, 193
Oct4, 194
Oligodendrocyte cells, 203
Onco genes, 151

P

Palindromes, 35
Pancreas cells, 12–13
Pancreatic islet cells, 263
Pandas, 256
Parkinson's disease
  iPS cells, 202
  neural damage from, 201
  protein aggregates of, 170
  symptoms of, 200
  treatments for, 200–201
Paternal cytoplasm, 242
Paternity test, 238–239
Pentagon terrorist attack victims, 250
Peptide bonds, 107
Pereyra, Liliana, 251
Personalized medicine, 16
Philadelphia (Ph1) chromosome, 148, 150–151
Phosphodiester bonds, 21
Phospholipid (PL) molecules, 113–114
Phosphorylation, 113
Photoreceptor cells, 206
Pitchfork, Colin, 227–228
Plant cloning, 215
Plasma membranes, 113–114
Poaching, 256
Poly(A) tail, 95–96
Polygenic diseases, 139
Polymerase chain reaction (PCR)
  advantages of, 62–63
  development of, 61
  evidence, court acceptance of, 61–62244
  KPCR, 65
  performance of, 61–62
  reverse transcriptase, 63–65
  routine use of, 41
  STR-based technology, 236–237
  strand amplification by, 235–236
Positive electrode, 32
Post-translational modification, 113
Pre-mRNA
  contents of, 64
  copy mechanism of, 25
  gene expression by, 69
  poly(A) tail, 96
  splicing, 98–101
  transcription, 95
Preimplantation genetic diagnosis (PGD)
  development of, 196
  process of, 153–154
  stem cell research and, 196–197
Prenatal genetic diagnosis (PGC), 150
Prions
  characterization of, 173–174
  function of, 174
  harmfulness of, 174–175
Progeny cells, 120
Programmed cell death, 118–119
Prokaryotic cells
  characterization of, 6–7
  DNA genomes in, 8–9
Protein aggregates, 170
Protein complexes
  characterization of, 3
  function of, 3–4
  subunits, 4
Protein folding
  factors influencing, 110–115
  functional shape, 110
  post-translational modification, 113
  prion effects on, 173–175
  region specificity, 108
  regulation of, 108–109
  ribosome synthesis and, 107–108
  shapes formed by, 100
Protein synthesis
  AUG codon role, 105–106
  initiation of, 105
  location of, 102
Protein-coding genes
  contents of, 69–70
  distribution of, 68–69
  expression, RNA polymerase II role, 88–89

Index

regions, mutations, 134–135
splicing mechanism of, 98–99
structure of, 89
transcription of, 98
Proteins
  amino acid bonds, 107
  folded, 5
  gene family, 73
  genetic instruction role, 11–12
  primary structure of, 108
  production of, 10–11
  RNA export regulating, 8
  translation of, 102–105
PrPSc prion proteins, 174
Punnett squares, 167
Purkinje nerve cells, 188
Pyrococcus furiosus, 62

R

R-bands, 145
Randall, John, 16
Random mutations, 133
Reagan, Nancy, 193
Reagan, Ronald, 193
Recombinant DNA cloning
  defined, 206
  goals of, 296–207
  process of, 39–41
Red blood cells, 163–164, 187
Repair enzymes, 134
Repeated DNA sequences
  LINEs, 81
  location of, 77–78
  minisatellite, 79–80
  most populous, 81–83
  number of, 77
Reproductive cloning. *See* Animal cloning
Restriction enzymes
  digestion, 36
  DNA analysis with, 36–39
  function of, 34
  research role of, 34–35
  sticky end production, 35
Restriction fragment length polymorphism (RFLP)
  chromosome comparison with, 238–239, 241
  goals of, 238
  mechanism of, 237–238
  paternity test, 239
Retinal pigment epithelium (RPE), 205
Retinal stem cells, 206
Reverse transcriptase PCR, 63–65
RFLP. *See* Restriction fragment length polymorphism (RFLP)

Rhinoceros, 256
Rhodopsin, 206
Ribonuclear-protein (RNP), 76, 99
ribosomal RNA (rRNA)
  characterization of, 76
  function of, 28–29
  gene expression and, 95–96
  protein coding gene expression and, 89
  sequence characterization, 29
Ribosomes
  amino acids transfer to, 104
  function of, 117–118
  protein folding, 107–108
  protein synthesis and, 105–106
Ribozymes. *See* Ribosomes
RNA
  binding proteins, 28–29
  chemical name for, 26
  export regulation of, 8
  RT-PCR use of, 64
  single-stranded, 26
  strand folding, 29
  structures, 28
RNA polymerase, 24, 122
RNA polymerase I, 89, 95
RNA polymerase II
  characterization of, 90
  DNA controlled regions, 89
  function of, 92
  LINEs in, 83
  major types of, 76
  methylation and, 263
  mutations, 134
  pre-mRNA transcription of, 95
  processing, 94–95
  promoter regions, 90, 92
  protein coding genes expression and, 88–89
  splicing, 94–95
  transcription factors, 91
  transcription process and, 95
RNA polymerase III, 82, 95
RNA splicing
  accuracy of, 100
  alternative, 99–102
  mechanisms, 98–99
  mutations and, 135
  process of, 70
Rough endoplasmic reticulum (RER), 113
Russian Revolution, 252

S

Saccharomyces cerevisiae, 9, 66
San Diego Zoo, 255
Sanger chain termination
  advantage of, 58–59
  key to, 57
  steps, 54–55
Sanger chain termination method, 58–59
Satellite DNA repeats, 78–79
Scheck, Barry, 245
Schwann cells, 203
SCNT. *See* Somatic cell nuclear transfer (SCNT)
Secondary structure formation, 26–29
Selfish DNA, 80
Seryl tRNA synthetase, 116–117
SH group, 112
Short interspersed repeated elements (SINEs), 81–82
Short tandem repeats (STRs), 236–237
Sickle cell anemia disease, 162–163
Side chain groups, 111–112
Silencer DNA elements, 92
Silent genes, 28
Simpson, O. J., 244
SINE retrotransposition, 81–82
SINEs. *See* Short interspersed repeated elements (SINEs)
Single nucleotide polymorphisms (SNPS)
  canine, 245
  changes in, 76–77
  chromosome comparison with, 240
  genealogical use of, 159–160
  genetic testing with, 155, 159
  stability of, 224
Skin cells, 187–188, 197
Skin fingerprints
  DNA replication and, 223
  in identical twins, 222–223
  limitations of, 221–222
  universal uniqueness of, 222
  use of, 221
Skinner, Michael, 263–264
Small nuclear RNA (snRNA), 76, 98–99
SNPs. *See* Single nucleotide polymorphisms (SNPS)
Somatic cell nuclear transfer (SCNT), 268
  live animal cloning with, 211
  process of, 207–208
  use of, 208–209
Southern blot transfer
  development of, 229
  DNA fingerprinting with fragment to membrane transfer, 231

hybridization reaction, 232
pattern detection, 233–234
preparation, 229–230
mRNA analysis, 96, 98
RFLP with, 238
Southern, Edward M., 229
Specialized cells
ESC generation of, 184–185
functions of, 188
iPS generation of, 198–199
types of, 188–189
Sperm tail, 242
Sphingolipidosis. *See* Tay-Sachs disease
Spinal cord paralysis, 203–206
Spinal cord-brain relationship, 202–203
Spindle fibers, 120
Spontaneous abortion, 183
Stargardt's macular dystrophy (SMD), 205
Sticky ends, 35
STRs. *See* Short tandem repeats (STRs)
Substantia nigra, 199
Sugar-phosphate repeating units, 20
Summer Olympic Games, 248
Syndecan-4 protein, 110

T

TaqL polymerase, 62
Tasha (the dog), 254
Taste buds, 188–189
TATA box, 89, 91
Tau proteins, 118
Tay-Sachs disease, 152–153
Telomeres
identification of, 26
location of, 79
sequence, 269
Template strand, 25
TFIIE proteins, 91
TFIIF proteins, 91
TFIIH proteins, 91
Therapeutic cloning
blastocyst divergence, 209
goals of, 210

hESCs in, 208
SCNT processing in, 208
Thermus aquaticus, 62
Thomson, James, 189
Tissue-specific gene transcription, 93
Traits, 14–15
Transcription
barriers, 69
control of, 92, 94
initiation of, 76
quantifying, 65
read-through, 69
reduction of, 134
regulation of, 88
stop signal, 207
Transcription factor proteins, 125
Transfer RNA (tRNA)
amino acids transfer by, 104
anticodons, 111
characterization of, 76
discovery of, 103
function of, 76
gene expression and, 95–96
mRNA codon link, 105
seryl tRNA synthetase, 116–117
translation process, 103–105
Transgenerational epigenetic inheritance
environmental factors, 264–265
examples of, 263
mechanisms of, 264
positive effects of, 265
process of, 259–263
research on, 263–264
timing of, 263
toxins effects, 265–266
transgenerational inheritance, 263–265
types of, 259–260
Trisomy 21. *See* Down syndrome
Tubulin proteins, 120

U

UAG codon, 135–136
Umbilical cords, 194–195
UPC label scanning, 226

UTR coding region, 94–95
UUC codon, 135–136
UUG codon, 135–136
UUU codon, 135–136

V

Vaccines, 72
Venter, J. Craig, 66
Vinclozolin, 264

W

Watson, James, 16, 18, 65, 67
Wild-type alleles, 137–138
Wilkins, Maurice, 16, 18
World Conservation Union, 255
World Trade Center victims, 250

X

X-linked dominant disorders, 141
X-linked recessive disorders, 139–140
X-ray diffraction, 16–18

Y

Y chromosome, 237
Y-DNA test, 159
York, Sheona, 227

Z

Zhang, Su-Chun, 198
Zygotes
cloning of, 210
development, 179, 209
division of, 132, 179
embryogenesis of, 182
gene expression, 216–217